Lecture Notes in Computer Science   8942

Commenced Publication in 1973
Founding and Former Series Editors:
Gerhard Goos, Juris Hartmanis, and Jan van Leeuwen

More information about this series at http://www.springer.com/series/7407

Jaime Gutierrez · Josef Schicho
Martin Weimann (Eds.)

Computer Algebra and Polynomials

Applications of Algebra and Number Theory

 Springer

Editors
Jaime Gutierrez
University of Cantabria
Santander
Spain

Martin Weimann
University of Caen
Caen
France

Josef Schicho
Ricam Linz
Linz
Austria

ISSN 0302-9743   ISSN 1611-3349   (electronic)
Lecture Notes in Computer Science
ISBN 978-3-319-15080-2   ISBN 978-3-319-15081-9   (eBook)
DOI 10.1007/978-3-319-15081-9

Library of Congress Control Number: 2014960202

LNCS Sublibrary: SL1 – Theoretical Computer Science and General Issues

Springer Cham Heidelberg New York Dordrecht London

Printed on acid-free paper

Springer International Publishing AG Switzerland is part of Springer Science+Business Media
(www.springer.com)

Preface

This textbook regroups selected papers of the Workshop on Computer Algebra and Polynomials, which was held in Linz at the Johann Radon Institute for Computational and Applied Mathematics (RICAM) during November 25–29, 2013, on the occasion of the Special Semester on Applications of Algebra and Number Theory. The workshop included invited talks and contributed talks. Authors of selected contributed talks were invited to submit a paper to these proceedings.

This workshop focuses on the theory and algorithms for polynomials over various coefficient domains that may be (but is not restricted to) a commutative algebra, such as a finite field or ring. The operations on polynomials in the focus are factorization, composition and decomposition, basis computation for modules, etc. Algorithms for such operations on polynomials have always been of central interest in computer algebra, as it combines formal (the variables) and algebraic or numeric (the coefficients) aspects.

The plan was to bring together a mix of experts for the various coefficient domains in order to explore similarities as well as differences. Also experts for applications of manipulation of polynomials were invited, such as polynomial system solving or the analysis of algebraic varieties.

The workshop contributions were selected through a rigorous reviewing process based on anonymous reviews made by various expert reviewers. There were usually two reviewers for one submission. The process was simple blind as authors did not know the names of the reviewers evaluating their papers. We have chosen 12 articles from the many excellent submissions we received. We hope that the reader will find an interesting perspective of this rich and active area. Let us mention here a few words about each of the selected papers.

The expository paper by Felix Breuer gives an introduction to Ehrhart theory and takes a tour through its applications in enumerative combinatorics. The paper by Carlos D'Andrea presents several methods and open questions for dealing in a more efficient way with the implicitization of rational parameterization. The survey paper by Joachim von zur Gathen and Konstantin Ziegler presents several counting results for indecomposable/decomposable polynomials over finite fields. Willem A. de Graaf describes methods for dealing with the problem of deciding whether a given element of the vector space lies in the closure of the orbit of another given element. The paper by Georg Grasegger and Franz Winkler presents a new and rather general method for solving algebraic ordinary differential equations. The paper by Manuel Kauers, Maximilian Jaroschek, and Fredrik Johansson shows a Sage implementation of Ore algebras. The paper by Zoltán Kovács and Bernard Parisse presents several changes for solving equation system of the GeoGebra software. Ragni Piene's paper studies several concepts of the classical polar varieties. The paper by Cristian-Silviu Radu solves an open problem about modular polynomials of levels 3 and 5. The survey paper by Carsten Schneider presents algorithms and their efficiency for some parameterized telescoping problems. The paper by J. Rafael Sendra, David Sevilla, and Carlos Villarino provides sufficient conditions for a parameterization to be surjective and computing a set of the points not covered by the parameterization. Finally, the paper by Maria-Laura Torrente presents an overview of the problem of the representation of rational surface as set theoretic complete intersection and also an original proof that the rational normal quartic is set-theoretically complete intersection of quadrics.

All accepted papers, except one, were presented at the workshop during talks of 25 or 45 min. There were also some talks with no contribution to the proceedings, see the website http://www.ricam.oeaw.ac.at/specsem/specsem2013/workshop3/ of the workshop for the list of all speakers and abstracts.

We would like to thank all the speakers for their contributions to the program, and all the authors who have submitted their precious manuscripts to this book. We would also like to ask for their understanding for our possible mistakes. The workshop and this volume would not have been possible without the contributions of numerous individuals and organizations, and we sincerely thank them for their support.

November 2014

Jaime Gutierrez
Josef Schicho
Martin Weimann

Organization

Organizers and Scientific Committee

Jaime Gutierrez   University of Cantabria, Spain
Josef Schicho   Johann Radon Institute, Austria
Martin Weimann   University of Caen, France

Contents

An Invitation to Ehrhart Theory: Polyhedral Geometry and its Applications in Enumerative Combinatorics

Felix Breuer[(✉)]

Research Institute for Symbolic Computation, Johannes Kepler University,
Altenberger Str. 69, 4040 Linz, Austria
felix@felixbreuer.net
http://www.felixbreuer.net

Abstract. In this expository article we give an introduction to Ehrhart theory, i.e., the theory of integer points in polyhedra, and take a tour through its applications in enumerative combinatorics. Topics include geometric modeling in combinatorics, Ehrhart's method for proving that a counting function is a polynomial, the connection between polyhedral cones, rational functions and quasisymmetric functions, methods for bounding coefficients, combinatorial reciprocity theorems, algorithms for counting integer points in polyhedra and computing rational function representations, as well as visualizations of the greatest common divisor and the Euclidean algorithm.

Keywords: Polynomial · Quasipolynomial · Rational function · Quasisymmetric function · Partial polytopal complex · Simplicial cone · Fundamental parallelepiped · Combinatorial reciprocity theorem · Barvinok's algorithm · Euclidean algorithm · Greatest common divisor · Generating function · Formal power series · Integer linear programming

1 Introduction

Polyhedral geometry is a powerful tool for making the structure underlying many combinatorial problems visible – often literally! In this expository article we give an introduction to Ehrhart theory and more generally the theory of integer points in polyhedra and take a tour through some of its many applications, especially in enumerative combinatorics.

In Sect. 2, we start with two classic examples of geometric modeling in combinatorics and then introduce Ehrhart's method for showing that a counting function is a (quasi-)polynomial in Sect. 3. We present combinatorial reciprocity theorems as a first application in Sect. 4, before we talk about cones as the basic building block of Ehrhart theory in Sect. 5. The connection of cones to rational

Felix Breuer was supported by Austrian Science Fund (FWF) special research group *Algorithmic and Enumerative Combinatorics* SFB F50-06.

© Springer International Publishing Switzerland 2015
J. Gutierrez et al. (Eds.): Computer Algebra and Polynomials, LNCS 8942, pp. 1–29, 2015.
DOI: 10.1007/978-3-319-15081-9_1

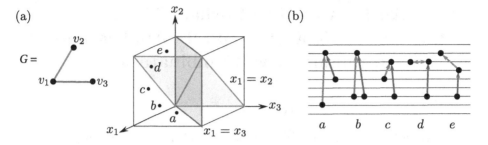

Fig. 1. (a) The graphic arrangement of the graph G. An edge between vertices v_i and v_j corresponds to a hyperplane $x_i = x_j$. (b) Points in the cube correspond to colorings. They can be visualized by drawing the graph G in a coordinate system such that the height of a vertex v_i is given by its color $x_i \in \{1, \ldots, k\}$. This induces an orientation of the edges from the vertex with smaller color to the vertex with the larger color. Moving from coloring a through coloring b to coloring c we pass the hyperplane $x_2 = x_3$ which is not part of the graphic arrangement; in b vertices v_2 and v_3 are at the same height. Moving on through d to e we pass the hyperplane $x_1 = x_2$ which is in the graphic arrangement; in d two adjacent vertices are at the same height, so d is not proper. Moving from c to e thus reverses the orientation of the edge between vertices v_1 and v_2 (Color figure online).

functions is the topic of Sect. 6, followed by methods for proving bounds on the coefficients of Ehrhart polynomials in Sect. 7. Section 8 discusses a surprising connection to quasisymmetric functions. Section 9 is about algorithms for counting integer points in polyhedra and computing rational function representations, in particular Barvinok's theorem on short rational functions. Finally, Sect. 10 closes with a playful look at the connection between the Euclidean algorithm and the geometry of $\mathbb{Z}^2$.

2   Geometric Modeling in Combinatorics

Many objects in combinatorics can be conveniently modeled as integer vectors that satisfy a set of linear equations and inequalities. In applied mathematics, this paradigm has proven tremendously successful: the combinatorial optimization industry rests to a large part on mixed integer programming. However, also in pure mathematics this approach can help to prove theorems. We illustrate this approach of constructing geometric models of combinatorial objects and problems on two of the most classic counting functions in all of combinatorics: The chromatic polynomial of a graph and the restricted partition function.

The chromatic polynomial $\chi_G(k)$ of a given graph G counts the number of proper k-colorings of G. Let V be the vertex set of G and $\sim$ its adjacency relation. A k-coloring is a vector $x \in [k]^V$ that assigns to each vertex $v \in V$ a color $x_v \in [k] := \{1, \ldots, k\}$. Such a k-coloring x is proper if for any two adjacent vertices $v \sim w$ the assigned colors are different, i.e., $x_v \neq x_w$. This way of describing a coloring as a vector rather than a function already suggests

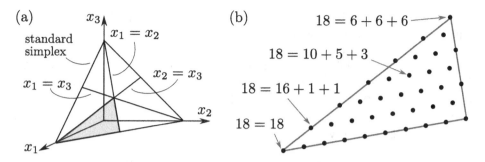

Fig. 2. (a) The partition polytope is cut out from the standard simplex $\{x \mid x_i \geq 0, \sum x_i = 1\}$ by the braid arrangement of all hyperplanes $x_i = x_j$. (b) The integer points in the partition polytope for $k = 18$ and $m = 3$ correspond to the partitions of 18 in at most 3 parts.

a geometric point of view (Fig. 1). Define the *graphic arrangement* of G as the set of all hyperplanes $x_v = x_w$ for adjacent vertices $v \sim w$. Then the chromatic polynomial counts integer points $x \in \mathbb{Z}^V$ that are contained in the half-open cube $(0, k]^V$ but do not lie on any of the hyperplanes in the graphic arrangement of G, i.e.,

$$\chi_G(k) = \#\mathbb{Z}^V \cap \{x \in \mathbb{R}^V \mid 0 < x_v \leq k \text{ and } x_v \neq x_w \text{ if } v \sim w\}. \qquad (1)$$

The restricted partition function $p(k, m)$ counts the number of partitions of k into at most m parts.[1] This can be modeled simply by defining a partition of k into at most m parts as a non-negative vector $x \in \mathbb{Z}^m$ whose entries sum to k and are weakly decreasing. For example, in the case $m = 5$ and $k = 14$ the partition $14 = 7 + 5 + 2$ would correspond to the vector $(7, 5, 2, 0, 0)$. In short,

$$p(k, m) = \#\mathbb{Z}^m \cap \{x \in \mathbb{R}^m \mid x_1 \geq x_2 \geq \ldots \geq x_m \geq 0 \text{ and } \sum_{i=1}^{m} x_i = k\}. \qquad (2)$$

Geometrically speaking, the restricted partition function thus counts integer points in an $(m - 1)$-dimensional simplex in m-dimensional space. This is visualized in Fig. 2. Note that the constraints that all variables are non-negative and that their sum is equal to k already defines an $(m - 1)$-dimensional simplex, bounded by the coordinate hyperplanes. The *braid arrangement*, i.e., the set of all hyperplanes $x_i = x_j$, subdivides this simplex into $m!$ equivalent pieces; the definition of the restricted partition function then selects the one piece in which the coordinates are in weakly decreasing order.

It is interesting to observe that both constructions work with the braid arrangement. Indeed, there are a host of combinatorial models that fit into this setting. A great example are *scheduling problems* [21]: Given a number k of time-slots,

---

[1] It is easy to adapt the following construction to the case of counting partitions with exactly m parts by making one inequality strict.

how many ways are there to schedule d jobs such that they satisfy a boolean formula ψ over the atomic expressions "job i runs before job j", i.e., $x_i < x_j$? E.g., if $\psi = (x_1 < x_2) \to (x_3 < x_2)$ then we would count all ways to place 3 jobs in k time-slots such that if job 1 runs before job 2, then job 3 also has to run before 2. We will return to scheduling problems in Sect. 8. However, the methods presented in this article are not restricted to this setup as we will see.

3   Ehrhart Theory

For any set $X \subset \mathbb{R}^d$ the *Ehrhart function* $\mathrm{ehr}_X(k)$ of X counts the number of integer points in the k-th dilate of X for each $1 \le k \in \mathbb{Z}$, i.e.,

$$\mathrm{ehr}_X(k) = \#\mathbb{Z}^d \cap (k \cdot X).$$

Both our constructions from the previous section are of this form, since (1) and (2) are, respectively, equivalent to

$$\chi_G(k) = \#\mathbb{Z}^V \cap k \cdot \{x \in \mathbb{R}^n \mid 0 < x_v \le 1 \text{ and } x_v \ne x_w \text{ if } v \sim w\},$$

$$p(k,m) = \#\mathbb{Z}^m \cap k \cdot \{x \in \mathbb{R}^m \mid x_1 \ge x_2 \ge \ldots \ge x_m \ge 0 \text{ and } \sum_{i=1}^{m} x_i = 1\}.$$

We will call the set X the *geometric model* of the counting function ehr_X. The central theme of this exposition is that geometric properties of X often translate into algebraic properties of ehr_X. Ehrhart's theorem is the prime example of this phenomenon. To set the stage, we introduce some terminology and refer to [46,57] for concepts from polyhedral geometry not defined here.

A *polyhedron* is any set of the form $P = \{x \in \mathbb{R}^d \mid Ax \ge b\}$ for a fixed matrix A and vector b. All polyhedra in this article will be *rational*, i.e., we can assume that A and b have only integral entries. A *polytope* is a bounded polyhedron. Any dilate of a polytope contains only a finite number of integer points, whence the Ehrhart function of a polytope is well-defined. A polyhedron is *half-open* if some of its defining inequalities are strict. A *partial polyhedral complex*[2] X is any set that can be written as a disjoint union of half-open polytopes.

Our model of $p(k,m)$ is a polytope. Our model of $\chi_G(k)$ is not, though, as it is non-convex, disconnected and neither closed nor open. It is easily seen to be a partial polytopal complex, though, e.g., by rewriting $x_v \ne x_w$ to $(x_v < x_w) \vee (x_v > x_w)$ and bringing the resulting formula in disjunctive normal form. This makes partial polytopal complexes an extremely flexible modeling framework, as summarized in the following lemma.

---

[2] Classically, a *polyhedral complex* is a collection X of polyhedra that is closed under passing to faces, such that the intersection of any two polyhedra in X is also in X and is a face of both. In contrast, in a partial polyhedral complex some faces are allowed to be open. This means that it is possible to remove an edge from a triangle – including or excluding the incident vertices. It is sometimes useful to regard a partial polyhedral complex as subset of a fixed underlying polyhedral complex, so as to be able to refer to the vertices of the underlying complex, for example. We will disregard these technical issues in this expository paper, however.

Lemma 1. *Let ψ be any boolean formula over homogeneous linear equations and inequalities with rational coefficients in the variables $x_1, \ldots, x_d, k$, such that for every k the set of all x such that $\psi(x)$ is bounded. Then there exists a partial polytopal complex X such that for all $1 \leq k \in \mathbb{Z}$,*

$$\#\{x \in \mathbb{Z}^d \mid \psi(x, k)\} = \mathrm{ehr}_X(k).$$

The generality of Ehrhart functions of partial polytopal complexes underlines the strength of the following famous theorem by Eugène Ehrhart.

Theorem 1 (Ehrhart [31]). *If X is partial polytopal complex[3], then $\mathrm{ehr}_X(k)$ is a quasipolynomial.*

Quasipolynomials are an important class of counting functions which capture both polynomial growth and periodic behavior. A function $p(k)$ is a *quasipolynomial* if there exist polynomials $p_0(k), \ldots, p_{\ell-1}(k)$ such that

$$p(k) = \begin{cases} p_0(k) & \text{if } k \equiv 0 \mod \ell \\ p_1(k) & \text{if } k \equiv 1 \mod \ell \\ \vdots \\ p_{\ell-1}(k) & \text{if } k \equiv \ell - 1 \mod \ell \end{cases}$$

for all $k \in \mathbb{Z}$. The polynomials p_i are called the *constituents* of p and their number is a *period* of p. The period is not uniquely determined, but of course the *minimal period* is; every period is a multiple of the minimal period. The *degree* of p is the maximal degree of the p_i. If d is the degree of p and ℓ a period of p, then p is uniquely determined by $\ell \cdot (d + 1)$ values of p, or more precisely, by $d + 1$ values $p(k' \cdot \ell + i)$ for each $i = 0, \ldots, \ell - 1$. This is why it makes sense to say that ehr_X "is" a quasipolynomial, even though we have defined Ehrhart functions only at positive integers. As an example, the quasipolynomial given by the restricted partition function $p(k, 2)$ is computed by interpolation in Fig. 3.

Given this terminology, we can make our above statement of Ehrhart's theorem more precise. Restricting our attention to polytopes P for the moment, the following hold for ehr_P. First, all constituents of ehr_P have the same degree. That degree is the dimension of P. Second, the leading coefficient of all constituents of P in the monomial basis is the volume of P. Third, the least common multiple of the denominators of all vertices of P is a period of ehr_P. More precisely, if $v_1, \ldots, v_N \in \mathbb{Q}^d$ are the vertices of P and $v_{i,j} = \frac{a_{i,j}}{b_{i,j}} \in \mathbb{Q}$, then

$$\ell = \mathrm{lcm}(\{b_{i,j} \mid i = 1, \ldots, N, \ j = 1, \ldots, d\})$$

is a period of ehr_P. In particular, if the vertices of P are all integral the Ehrhart function is a polynomial.

---

[3] Ehrhart formulated his theorem for polytopes, not for partial polytopal complexes. The generalization follows immediately, however, since for any partial polytopal complex X the Ehrhart function ehr_X is a linear combination of Ehrhart functions of polytopes.

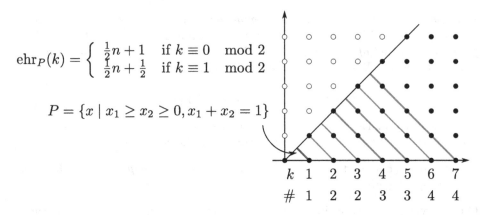

$$\mathrm{ehr}_P(k) = \begin{cases} \frac{1}{2}n + 1 & \text{if } k \equiv 0 \mod 2 \\ \frac{1}{2}n + \frac{1}{2} & \text{if } k \equiv 1 \mod 2 \end{cases}$$

$$P = \{x \mid x_1 \geq x_2 \geq 0, x_1 + x_2 = 1\}$$

k	1	2	3	4	5	6	7
$\#$	1	2	2	3	3	4	4

Fig. 3. By counting the integer points in the dilates of P and interpolating, we can compute the Ehrhart quasipolynomial of P. In this case P is the restricted partition polytope for partitions into at most 2 parts.

Ehrhart's theorem thus provides a very general method for proving that counting functions are (quasi-)polynomials. Simply by virtue of the geometric models from Sect. 2, we immediately obtain that the chromatic function is a polynomial because the vertices of our geometric model are integral. This proof of polynomiality is very different from the standard deletion-contraction method and generalizes to counting functions that do not satisfy such a recurrence, including all scheduling problems. Also we find that the restricted partition function into m parts is a quasipolynomial with period $\mathrm{lcm}(1, 2, \ldots, m)$: The numbers $1, 2, \ldots, m$ appear in the denominators of the vertices, since we intersect the braid arrangement with the simplex $\{x \mid x_i \geq 0, \sum x_i = 1\}$ instead of the cube. For more on the restricted partition function from an Ehrhart perspective, see [19].

Lemma 1 can be generalized even further, for example by allowing quantifiers via Presburger arithmetic [56] or by considering the multivariate case [55]. As a great introductory textbook on Ehrhart theory we recommend [10].

4   Combinatorial Reciprocity Theorems

Now that we know that the Ehrhart function ehr_P of a polytope P is in fact a quasipolynomial we can evaluate it at negative integers. Even though the Ehrhart function itself is defined only at positive integers, it turns out that the values of ehr_P at negative integers have a very elegant geometric interpretation: $\mathrm{ehr}_P(-k)$ counts the number of integer points in the interior of $k \cdot P$.

Theorem 2 (Ehrhart-Macdonald Reciprocity [43]). *If $P \subset \mathbb{R}^n$ is a polytope of dimension d and $0 < k \in \mathbb{Z}$ then*

$$\mathrm{ehr}_P(-k) = (-1)^d \cdot (\#\mathbb{Z}^n \cap k \cdot P^\circ).$$

Here P° denotes the *relative interior* of P, which means the interior of P taken with respect to the affine hull[4] of P. If P is given in terms of a system of linear equations and inequalities, the relative interior is often easy to determine. For example, if P is defined by $Ax \geq b$ and $A'x = b'$, and A' contains all equalities of the system[5], then P° is given by $Ax > b$ and $A'x = b'$. In short, all we need to do is make weak inequalities strict.

Ehrhart-Macdonald reciprocity provides us with a powerful framework for finding combinatorial reciprocity theorems, i.e., combinatorial interpretations of the values of counting functions at negative integers. We start with a counting function f defined in the language of combinatorics and translate this counting function into the language of geometry by constructing a linear model. In the world of geometry, we apply Ehrhart-Macdonald reciprocity to find a geometric interpretation of the values of f at negative integers. Translating this geometric interpretation back into the language of combinatorics, a process which can be quite subtle, we then arrive at a combinatorial reciprocity theorem.

Let us start with the example of the restricted partition function $p(k, m)$. Applying Theorem 2 it follows that, up to sign, $p(-k, m)$ counts vectors x such that $x_1 > x_2 > \ldots > x_m > 0$ and $\sum x_i = k$ for any positive integer k. Interpreting this geometric statement combinatorially, we find:

Theorem 3. *Up to sign, $p(-k, m)$ counts partitions of k into exactly m distinct parts.*

This result seems to be less well-known in partition theory than one would expect, even though it is an immediate consequence of Ehrhart-Macdonald reciprocity; see also [19]. A very similar geometric construction, however, is the basis of Stanley's work on P-partitions and the order polynomial [49] which has many nice extensions, e.g., [39].

Next, we consider the chromatic polynomial $\chi_G(k)$. The model X_G we use here is slightly different from (1) in that we work with the open cube $(0, k+1)^V$. This introduces a shift $\mathrm{ehr}_{X_G}(k) = \chi_G(k-1)$. The advantage is that X_G is now a disjoint union of open polytopes $P_1, \ldots, P_N$. As already motivated by Fig. 1, it turns out that the P_i are in one-to-one correspondence with the acyclic orientations[6] of the graph G [35]. Applying Theorem 2 to each component individually, we find that $\chi_G(-k)$ counts all integer vectors x in the closed cube such that points on the hyperplanes $x_v = x_w$ have a multiplicity equal to the number of closed components $\bar{P}_i$ they are contained in. To interpret this combinatorially, we define an orientation o and a coloring x of G to be *compatible* if, when moving along directed edges, the colors of the vertices always increase or stay the same. Putting everything together and taking the shift into account we obtain

---

[4] The affine hull of P is the smallest affine space containing P. Affine spaces are the translates of linear spaces.

[5] More precisely, we require that the affine hull of P is $\{x \mid A'x = b'\}$ and that for every row a of A the linear functional $\langle a, x \rangle$ is not constant over $x \in P$.

[6] An orientation of a graph G is *acyclic*, if it contains no directed cycles.

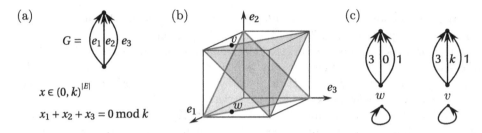

Fig. 4. (a) A directed graph G and the flow problem G defines. (b) The corresponding partial polyhedral complex X. (c) The labelings of G given by the points v, w along with the different totally cyclic orientations of $G/\operatorname{supp}(v) = G/\operatorname{supp}(w)$.

Stanley's reciprocity theorem for the chromatic polynomial below. The geometric proof we described is due to Beck and Zaslavsky [12] and can be generalized to cell-complexes [7].

Theorem 4 (Stanley [47]). *Up to sign, $\chi_G(-k)$ counts pairs (x, o) of (not necessarily proper) k-colorings x and compatible acyclic orientations o of G.*

Next, the modular flow polynomial of a graph provides us with an example of a combinatorial reciprocity theorem that was first discovered via the geometric approach and that makes use of a different construction, unrelated to the braid arrangement. This example is illustrated in Fig. 4. A $\mathbb{Z}_k$-flow on a directed graph G with edge set E is a vector $y \in \mathbb{Z}_k^E$ that assigns to each edge of G a number such that at each vertex v of G the sum of all flows into v equals the sum of all flows out of v, modulo k. The modular flow polynomial $\varphi_G(k)$ of G counts $\mathbb{Z}_k$-flows on G that are nowhere zero. To model this in Euclidean space, we identify the elements of $\mathbb{Z}_k$ with the integers $0, \ldots, k-1$. Nowhere zero vectors $y \in \mathbb{Z}_k^E$ thus correspond to integer points $y \in (0, k)^k$ in the k-th dilate of the open unit cube. If $A \in \mathbb{Z}^{V \times E}$ is the incidence matrix of G, the constraint that flow has to be conserved at each vertex can be expressed simply by requiring $Ay \equiv 0$ mod k, or, equivalently, by $\exists b \in \mathbb{Z}^V : Ay = kb$. Note that for only finitely many $b \in \mathbb{Z}^V$ the hyperplane $Ay = b$ intersects the unit cube $(0,1)^E$. Let $P_1, \ldots, P_N$ denote these sections and let X be their union. Then $\varphi_G(k) = \operatorname{ehr}_X(k)$.

Applying Ehrhart-Macdonald reciprocity, we obtain that, up to sign, $\varphi_G(-k)$ counts integer points in the k-th dilate of the union of the closures $\bar{P}_i$. In particular, we now count vectors that may have both entries 0 and k, which are both congruent zero mod k, but which we have to count as different as Fig. 4 shows. This observation suggests that to find a combinatorial interpretation, we may want to consider assigning two different kinds of labels to the edges with zero flow. Pursuing this line of thought eventually leads to the following combinatorial reciprocity theorem, which again can be generalized to cell complexes [7], see also [13, 15].

Theorem 5 (Breuer-Sanyal [22]). *Up to sign, $\varphi_G(-k)$ counts pairs (y, o) of a $\mathbb{Z}_k$-flow y on G and a totally cyclic reorientation of $G/\operatorname{supp}(y)$.*

Here a reorientation is a labeling of the edges of a directed graph with $+$ or $-$, indicating whether the direction of the edge should be reversed or not. Such a reorientation is *totally cyclic*, if every edge of the resulting directed graph lies on a directed cycle. $\text{supp}(y)$ denotes the set of edges where y is non-zero and $G/\text{supp}(y)$ denotes the graph where $\text{supp}(y)$ has been contracted.

For more on combinatorial reciprocity theorems we recommend the forthcoming book [11].

5   Cones and Fundamental Parallelepipeds

A polyhedral cone or *cone*, for short, is the set of all linear combinations with non-negative real coefficients of a finite set of *generators* $v_1, \ldots, v_d \in \mathbb{Q}^n$. If the generators are linearly independent, the cone is *simplicial*. The cone is *pointed* or *line-free* if it does not contain a line $\{u + \lambda v \mid \lambda \in \mathbb{R}\}$.

Cones are the basic building blocks of Ehrhart theory, because the sets of integer points in simplicial cones have a very elegant description, which is illustrated in Fig. 5. Let $v_1, \ldots, v_d \in \mathbb{Z}^n$ be linearly independent, and consider the simplicial cone $\text{cone}_{\mathbb{R}}(v_1, \ldots, v_d)$ generated by them. The *discrete cone* or *semigroup* $\text{cone}_{\mathbb{Z}}(v_1, \ldots, v_d)$ of all non-negative *integral* combinations of the v_i reaches only those integer points in $\text{cone}_{\mathbb{R}}$ that lie on the lattice $\mathbb{Z}v_1 + \ldots + \mathbb{Z}v_d$ generated by the v_i. However, by shifting the discrete cone to all integer points in the *fundamental parallelepiped* $\Pi(v_1, \ldots, v_d)$ we can not only capture all integer points in C, but we moreover partition them into $\#\mathbb{Z}^{n+1} \cap \Pi(v_1, \ldots, v_d) = |\det(v_1, \ldots, v_d)|$ disjoint classes. This number of integer points in the fundamental parallelepiped is called the *index* of C. Define

$$\text{cone}_{\mathbb{R}}(v_1, \ldots, v_d) = \left\{ \sum_{i=1}^{d} \lambda_i v_i \,\middle|\, 0 \le \lambda_i \in \mathbb{R} \right\},$$

$$\text{cone}_{\mathbb{Z}}(v_1, \ldots, v_d) = \left\{ \sum_{i=1}^{d} \lambda_i v_i \,\middle|\, 0 \le \lambda_i \in \mathbb{Z} \right\},$$

$$\Pi(v_1, \ldots, v_d) = \left\{ \sum_{i=1}^{d} \lambda_i v_i \,\middle|\, 0 \le \lambda_i < 1 \right\}.$$

Lemma 2. *Let $v_1, \ldots, v_d \in \mathbb{Z}^n$ be linearly independent. Then*

$$\mathbb{Z}^n \cap \text{cone}_{\mathbb{R}}(v_1, \ldots, v_d) = (\mathbb{Z}^n \cap \Pi(v_1, \ldots, v_d)) + \text{cone}_{\mathbb{Z}}(v_1, \ldots, v_d).$$

The main benefit of this decomposition is that it splits the problem of describing the integer points in a cone to into two parts: The finite problem of enumerating the integer points in the fundamental parallelepiped, and the problem of describing the discrete cone, which is easy as we shall see below. As an application of this result, we will now prove Ehrhart's theorem for polytopes.

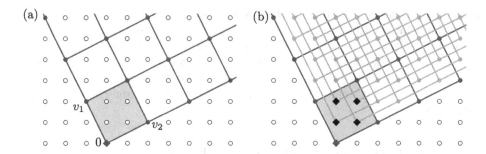

Fig. 5. (a) The discrete cone $\text{cone}_{\mathbb{Z}}(v_1, v_2)$ generated by v_1 and v_2. Its fundamental parallelepiped is shaded. (b) To generate all points in $\mathbb{Z}^2 \cap \text{cone}_{\mathbb{R}}(v_1, v_2)$ the discrete cone has to be translated by every integer point in the fundamental parallelepiped (shown as diamonds).

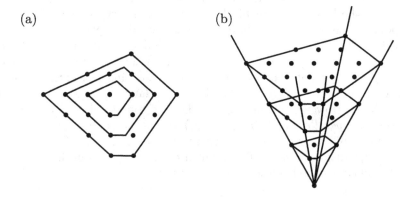

Fig. 6. (a) The three first dilates of a polytope P' in the plane. (b) The polytope $P = P' \times \{1\}$ embedded in 3-space and the cone $C = \text{cone}(P)$ over P. Sections $H_k \cap C$ are lattice equivalent to the dilates of P'.

Suppose we want to compute the Ehrhart function of a polytope $P' \subset \mathbb{R}^n$. We embed P' at height 1 in $\mathbb{R}^{n+1}$, i.e., we pass to $P = P' \times \{1\} \subset \mathbb{R}^{n+1}$. Then, we consider the set $\text{cone}(P)$ of all finite linear combinations of elements in P with non-negative real coefficients as shown in Fig. 6. The intersections of $\text{cone}(P)$ with the hyperplanes $H_k := \{x \mid x_{n+1} = k\}$ are lattice equivalent[7] to the dilates $k \cdot P$ we are interested in. If we can describe the number of integer points in such sections of polyhedral cones, we will have a handle on computing Ehrhart functions.

Before we continue, we observe that we can make two more simplifications. First, we can restrict our attention to simplicial cones. While in general $\text{cone}(P)$ will of course not be simplicial, we can always reduce the problem to simplicial

---

[7] Two sets $X, Y \subset \mathbb{Z}^n$ are *lattice equivalent* if there exists an affine isomorphism $x \mapsto Ax + b$ that maps X to Y and which induces a bijection on $\mathbb{Z}^n$.

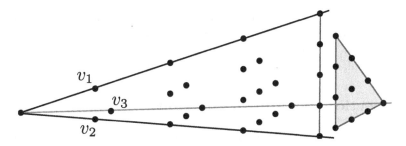

Fig. 7. The 4ℓ-th level of $C = \mathrm{cone}_{\mathbb{Z}}(v_1, v_2, v_3)$ decomposes naturally into the 4ℓ-th level of $\mathrm{cone}_{\mathbb{Z}}(v_1, v_2)$ and a shift of the 3ℓ-th level of C.

cones by triangulating P. Second, we note that while $\mathrm{cone}(P)$ is indeed finitely generated by the vertices $w_1, \ldots, w_N$ of P, these w_i may be rational vectors. Instead, we would like to work with generators v_i that are all integer and all at the same height wrt. the last coordinate. This can be achieved by letting ℓ denote the smallest integer such that $\ell \cdot w_i \in \mathbb{Z}^{n+1}$ for all i and setting $v_i := \ell \cdot w_i$.

We have thus reduced the problem of computing the Ehrhart function of P' to computing $\mathbb{Z}^{n+1} \cap H_k \cap C_{\mathbb{R}}$, the number of integer points at height k in a simplicial cone $C_{\mathbb{R}} := \mathrm{cone}_{\mathbb{R}}(v_1, \ldots, v_d)$ given by integral generators with last coordinate equal to a constant ℓ. Following Lemma 2 we concentrate on $C_{\mathbb{Z}} := \mathrm{cone}_{\mathbb{Z}}(v_1, \ldots, v_d)$ first. Since the last coordinate of all v_i is ℓ, $C_{\mathbb{Z}} \cap H_k$ is empty if $k \not\equiv 0 \mod \ell$. On the other hand, if $k \equiv 0 \mod \ell$ and $k > 0$, then

$$H_k \cap \mathrm{cone}_{\mathbb{Z}}(v_1, \ldots, v_d) = v_d + (H_{k-\ell} \cap \mathrm{cone}_{\mathbb{Z}}(v_1, \ldots, v_d)) \cup (H_k \cap \mathrm{cone}_{\mathbb{Z}}(v_1, \ldots, v_{d-1})).$$

This is an instance of Pascal's recurrence for the binomial coefficients, illustrated in Fig. 7, which yields for all integers $k \geq 0$,

$$\#H_{\ell \cdot k} \cap \mathrm{cone}_{\mathbb{Z}}(v_1, \ldots, v_d) = \binom{k+d-1}{d-1}$$

and $\#H_k \cap \mathrm{cone}_{\mathbb{Z}}(v_1, \ldots, v_d) = 0$ if $k \not\equiv 0 \mod \ell$.

Applying Lemma 2 we see that to get the counting function for $C_{\mathbb{R}}$, we need to shift the discrete cone $C_{\mathbb{Z}}$ by all the integer points in the fundamental parallelepiped, which allows us to reach lattice points at heights k which are not a multiple of ℓ. Organizing these shifts according to the last coordinate, we obtain for any $k \geq 0$ and $0 \leq r < \ell$

$$\# \left(\mathbb{Z}^{n+1} \cap H_{\ell \cdot k + r} \cap C_{\mathbb{R}} \right)$$
$$= h_r^* \binom{k+d-1}{d-1} + h_{\ell+r}^* \binom{k+d-2}{d-1} + \ldots + h_{(d-1) \cdot \ell + r}^* \binom{k}{d-1} \qquad (3)$$

where h_i^* denotes the number of integer points at height i in $\Pi(v_1, \ldots, v_d)$.

Note that if we are interested in $\# \left(\mathbb{Z}^{n+1} \cap H_m \cap C_{\mathbb{R}} \right)$ for an arbitrary nonnegative m then we can always write $m = \ell k + r$ such that $k \geq 0$ and $0 \leq r < \ell$ simply by doing division with remainder. Also note that (3) is a polynomial of degree $d - 1$ in k for each fixed r. Since r changes periodically with m, the counting function $m \mapsto \# \left(\mathbb{Z}^{n+1} \cap H_m \cap C_{\mathbb{R}} \right)$ is a quasipolynomial of period ℓ. By construction, $\mathrm{ehr}_P(k)$ is a sum of such expressions and therefore itself a quasipolynomial, which completes the proof of Theorem 1.

6   Connection to Rational Functions

The results and constructions of the previous section translate immediately into the language of generating functions, formal power series and rational functions. When we represent an integer point $v \in \mathbb{Z}^n$ by a multivariate monomial $z^v := z_1^{v_1} \cdot \ldots \cdot z_n^{v_n}$, the set of integer vectors in any given set $S \subset \mathbb{Z}^n$ can be written as a multivariate generating function

$$\phi_S(z) = \sum_{v \in \mathbb{Z}^n \cap S} z^v.$$

Using the familiar geometric series expansion $\frac{1}{1-z^v} = \sum_{i=0}^{\infty} z^{iv}$ we see that generating functions of "discrete rays" of integer vectors can be represented as rational functions. Indeed, both discrete cones and Lemma 2 can be expressed succinctly in terms of rational functions.

$$\phi_{\mathrm{cone}_{\mathbb{Z}}(v_1,\ldots,v_d)}(z) = \frac{1}{(1 - z^{v_1}) \cdot \ldots \cdot (1 - z^{v_d})}. \tag{4}$$

$$\phi_{\mathrm{cone}_{\mathbb{R}}(v_1,\ldots,v_d)}(z) = \frac{\sum_{v \in \mathbb{Z}^n \cap \Pi(v_1,\ldots,v_d)} z^v}{(1 - z^{v_1}) \cdot \ldots \cdot (1 - z^{v_d})}. \tag{5}$$

If we specialize by substituting $z_i = q$ for each i, then we obtain $(1 - q^\ell)^d$ in the denominator, since, by construction, all generators v_i have coordinate sum ℓ. This explains the appearance of binomial coefficients, since

$$\frac{1}{(1 - q^\ell)^d} = \sum_{k=0}^{\infty} \binom{k + d - 1}{d - 1} q^{\ell \cdot k}$$

which turns (3) into

$$\sum_{k=0}^{\infty} \mathrm{ehr}_P(k) q^k = \frac{h_0^* q^0 + \ldots + h_{d \cdot \ell - 1}^* q^{d \cdot \ell - 1}}{(1 - q^\ell)^d}. \tag{6}$$

In this way, many arithmetic calculations on the level q-series can be viewed as the projection of a geometric construction, via multivariate generating functions. The richer multivariate picture can be of use, for example, when converting arithmetic proofs into a bijective proofs, see [19].

Intuitively, we can think of the generating functions ϕ_S as weighted indicator functions of sets of integer vectors. Starting with generating functions ϕ_P for polyhedra P and taking linear combinations of these, we obtain an algebra $\mathcal{P}$ of polyhedral sets. However, working with rational function representations

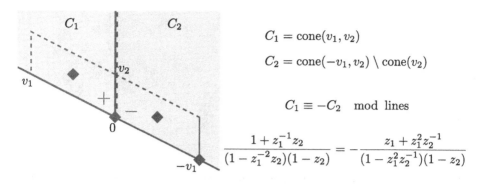

Fig. 8. Let $v_1 = (-2, 1)$ and $v_2 = (0, 1)$. Modulo lines, the closed cone C_1 generated by v_1 and v_2 is equal to the negative of the half-open cone C_2 generated by $-v_1$ and v_2. Integer points in corresponding fundamental parallelepipeds are shown as diamonds.

introduces an equivalence relation on this algebra. For example, we can expand $\frac{1}{1-z}$ either as $\sum_{i=0}^{\infty} z^i$, the indicator function of all non-negative integers, or as $-\sum_{i=1}^{\infty} z^{-i}$, minus the indicator function of all negative integers. This phenomenon generalizes to multivariate generating functions: To determine the formal expansion of a rational function uniquely, we have to fix a "direction of expansion" which can be given for example in terms of a suitable pointed cone. For details we refer the reader to, e.g., [2,6,9,10]. Important for our purposes is that to each rational function there corresponds an equivalence class of indicator functions and the simple example of the geometric series tells us what the equivalence relation is: Two elements in the algebra of polyhedral sets are equivalent if they are equal *modulo lines*, i.e., modulo sets of the form $\{u + \lambda v \mid \lambda \in \mathbb{Z}\}$ for some $u, v \in \mathbb{Z}^n$. We say that a generating function ϕ is represented by some rational function expression ρ if there exists a pointed cone C such that the expansion of ρ in the direction C gives ϕ; for this to be feasible we assume that the support of ϕ does not contain a line. Choosing a different direction C' for the expansion of ρ produces a generating function ϕ' that is equal to ϕ modulo lines.

Working with indicator functions of cones modulo lines does have its advantages. Most importantly, this allows us to "flip" cones by reversing the direction of some (or all) of their generators and opening some of their faces accordingly, as shown in Fig. 8.

One beautiful application of this phenomenon is Brion's theorem, which allows us to represent ϕ_P for any line-free polyhedron P in terms of rational function representations of cones, i.e., as a linear combination of expressions of the form (5). Brion's theorem is motivated in Fig. 9.

For a polyhedron P we define the *vertex cone* $\text{vcone}(v, P)$ at a vertex v of P as the set

$$\text{vcone}(v, P) = v + \text{cone}_{\mathbb{R}}(v_1, \ldots, v_N),$$

where the v_i are the directions of the edges incident to v, oriented away from v. We can easily represent each vertex cone by a rational function: For a simplicial

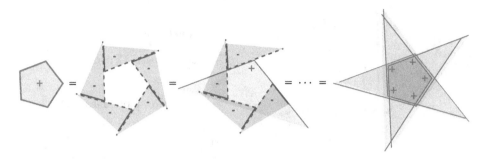

Fig. 9. Modulo lines, a polytope is equal to the sum of its vertex cones. In 2 dimensions, this is easy to see by iteratively flipping cones.

cone C we *define* ρ_C as the rational function expression given in (5).[8] For a non-simplicial cone C we define ρ_C as a linear combination of such expressions, given via a triangulation of C. Then, the generating function of the set of integer points in P is the sum of the rational function representations of the vertex cones.

Theorem 6 (Brion [24]). *Let P be a polyhedron that does not contain any affine line. Then*

$$\phi_P = \sum_{v \text{ vertex of } P} \rho_{\mathrm{vcone}(v,P)}(z).$$

The theorem of Lawrence-Varchenko [41,53] is the corresponding analogue for cases in which it is necessary to work with indicator functions directly, not with equivalence classes modulo lines. It expresses ϕ_P as an inclusion-exclusion of vertex cones which have been "flipped forward" so that their generators all point consistently in one direction of expansion as shown in Fig. 10.

7   Coefficients of (Quasi-)Polynomials

The geometric perspective provides a wide range of methods for establishing bounds on the coefficients of counting (quasi-)polynomials. In this section we will focus on polynomials for simplicity, but the results generalize to quasipolynomials.

The monomial basis is of course the classic choice for computing coefficients of polynomials. Geometrically, the elements of the monomial basis of the space of polynomials are the Ehrhart functions $\mathrm{ehr}_{[0,1)^i}(k) = k^i$ of half-open cubes $[0,1)^i$ of varying dimension. For us, it will be expedient to work with two different binomial bases instead, whose elements are the Ehrhart functions $\mathrm{ehr}_{\Delta_i^d}(k) = \binom{k+d-i}{d}$

---

[8] Here it is important to note that (5) works also for cones with an apex $v \neq 0$: All we have to do is take the fundamental parallelepiped Π to be rooted at v instead of the origin. This simply amounts to translating the fundamental parallelepiped as defined in Sect. 5 by v.

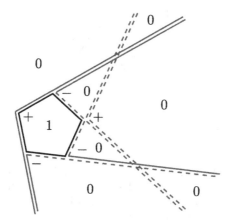

Fig. 10. The Lawrence-Varchenko decomposition of a pentagon. The sign next to the apex of each vertex cone C specifies whether C is to be added or subtracted. The number in each region is the net balance of how often points in the region are counted when the signed vertex cones are summed. All generators point of all vertex cones point to the right, which means all vertex cones are forward.

of unimodular[9] d-dimensional *half-open simplices* Δ_i^d with i open facets. Up to lattice equivalence, such a Δ_i^d has the form

$$\Delta_i^d = \left\{ x \in \mathbb{R}^{d+1} \,\middle|\, x_1 > 0, \dots, x_i > 0, x_{i+1} \geq 0, \dots, x_{d+1} \geq 0, \sum_j x_j = 1 \right\}.$$

These unimodular half-open simplices Δ_i^d form the basic building block of Ehrhart theory. They offer two different ways in which we can use them to construct a basis of the space of polynomials. The first basis, which defines the h^*-coefficients, fixes the dimension d of the simplices and varies the number i of open facets. In contrast, the second basis, which defines the f^*-coefficients, uses only open simplices with $i = d + 1$, but varies their dimension d.

Formally, the h^*-vector $(h_0^*, \dots, h_d^*)$ and the f^*-vector $(f_0^*, \dots, f_d^*)$ of a polynomial $p(k)$ of degree at most d are defined by

$$p(k) = h_0^* \binom{k+d}{d} + h_1^* \binom{k+d-1}{d} + \dots + h_d^* \binom{k}{d}$$
$$= f_0^* \binom{k-1}{0} + f_1^* \binom{k-1}{1} + \dots + f_d^* \binom{k-1}{d}.$$

Let us begin by taking a closer look at the h^*-coefficients. As we have seen in (3) and (6), the h^*-vector of the Ehrhart quasipolynomial of a simplex

---

[9] A simplex Δ with integer vertices is *unimodular* if the fundamental parallelepiped of $\mathrm{cone}(\Delta \times \{1\})$ contains only a single integer vector: the origin. Equivalently $\mathbb{Z}^n \cap \mathrm{cone}_{\mathbb{R}}(\Delta \times \{1\}) = \mathrm{cone}_{\mathbb{Z}}(\Delta \times \{1\})$.

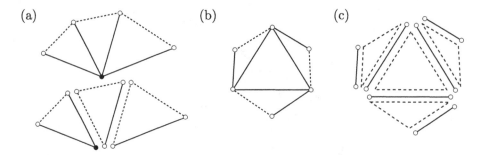

Fig. 11. (a) A half-open 2-dimensional partial polytopal complex and a partition into half-open 2-dimensional simplices. (b) A half-open partial polytopal complex X that is not partitionable. A partition of this complex with half-open 2-dimensional simplices would have to contain a 2-dimensional simplex with at least two edges and, consequently, at least one vertex. However, X does not contain any of its vertices. (c) A decomposition of X into open simplices of various dimension.

Δ counts lattice points at different heights in the fundamental parallelepiped of cone($\Delta \times \{1\}$), which immediately implies $h_i^* \geq 0$. This observation extends to half-open simplices where some facets have been removed. It follows that if a geometric model X can be *partitioned* into half-open simplices that are all of full dimension, as shown in Fig. 11(a), it follows that ehr$_X$ has non-negative h^*-vector as well. As it turns out, all (closed convex) polytopes have such a *partitionable* triangulation, which proves non-negativity of the h^*-vector for all polytopes.

Theorem 7 (Stanley [48]). *If P is an integral polytope, then* ehr$_P$ *has a non-negative h^*-vector.*

However, as the examples of the chromatic polynomial and the flow polynomial from Sects. 2 and 4 show, the geometric models X that appear in combinatorial applications of Ehrhart theory are not simply polytopes: Often they are non-convex, disconnected, half-open or have non-trivial topology. This can lead to geometric models X that are not partitionable and, consequently, to counting polynomials with negative h^*-coefficients.

Figure 11(b) gives an example of a half-open partial polytopal complex that is not partitionable: A partition of the complex in Fig. 11(b), for example, would require 4 half-open simplices of dimension 2 that have, in total, 6 closed edges but contain none of the vertices of the complex, which is impossible. Here it is important to recall that, because we are working with the h^* basis, all half-open simplices participating in a partition are required to have the same dimension (in this example, dimension 2).

Such phenomena appear in practice. One prominent example of natural counting polynomials with negative entries in their h^*-vector are chromatic polynomials of hypergraphs. In this case, it is the non-trivial topology of the geometric models that gives rise to non-partitionability: It is easy to construct hypergraphs whose coloring complexes consists of, say, 2-dimensional spheres that intersect in 0-dimensional subspheres; such complexes are not partitionable and can produce negative h^*-coefficients [18].

As we have seen in Sect. 3, partial polytopal complexes are the right notion to describe combinatorial models in Ehrhart theory. While partial polytopal complexes are not always partitionable, they can always be written as a disjoint union of relatively open simplices of various dimension. The partial polytopal complex in Fig. 11(b) can, for example, be written as a disjoin union of open simplices of dimension 1 and 2 as shown in Fig. 11(c). This motivates the use of the f^*-basis. As it turns out, the f^*-vector of an open simplex Δ has a counting interpretation similar to (3), even though its construction is more subtle [16]. It follows that all partial polytopal complexes with integer vertices have a non-negative f^*-vector. Moreover, this property characterizes Ehrhart polynomials of partial polytopal complexes.

Theorem 8 (Breuer [16]). *If X is an integral partial polytopal complex, then* ehr_X *has a non-negative f^*-vector.*

Conversely, if $p(k)$ is a polynomial with non-negative f^*-vector, then there exists an integral partial polytopal complex X such that $\mathrm{ehr}_X(k) = p(k)$.

While Theorem 8 characterizes Ehrhart polynomials of the kind of geometric objects that appear in many combinatorial applications, the question remains how to characterize Ehrhart polynomials of convex polytopes. This challenge is vastly more difficult, and, even though many constraints on the h^*-vectors of convex polytopes have been proven, is still wide-open even in dimension 3. At least in dimension 2, a complete characterization of the coefficients of Ehrhart polytopes is available. See [8, 36, 37, 51] for more information.

Still, there are a wealth of tools available for proving sharper bounds on the coefficients of counting polynomials ehr_X, by exploiting the particular geometric structure of the partial polytopal complex X, even if X is not convex. One of the most powerful techniques available is the use of convex ear decompositions. A *convex ear decomposition* is a decomposition of a simplicial complex X into "ears" $E_0, \ldots, E_N$ such that E_0 is the boundary complex of a simplicial polytope, the remaining E_i are balls that are subcomplexes of the boundary complex of some simplicial polytope, and E_i is attached to $\bigcup_{j<i} E_j$ along its entire boundary (and not just along some facets), i.e., $E_i \cap \bigcup_{j<i} E_j = \partial E_i$. For example, the complex in Fig. 1, consisting of the boundary of the cube and the two hyperplanes, has a convex ear decomposition: Start with the boundary of the cube as triangulated by the braid arrangement, glue in the triangulated square lying on one of the hyperplanes and then glue in the two triangles on the second hyperplane one after the other. If all simplices in this complex are unimodular (as in many combinatorial applications), this leads to the following bounds, which have been successfully applied to the chromatic polynomial by Hersh and Swartz [38] and to the integral and modular flow and tension polynomials by Breuer and Dall [17].

Theorem 9 (Chari [27], Swartz [52]). *If X is a simplicial complex in which all simplices are unimodular and X has a convex ear decomposition then the h^*-vector of* $\mathrm{ehr}_X(k)$ *satisfies*

(a) $h_0^* \le h_1^* \le \cdots \le h_{\lfloor d/2 \rfloor}^*$,
(b) $h_i^* \le h_{d-i}^*$ for $i \le d/2$, and
(c) $(h_0^*, h_1^* - h_0^*, \ldots, h_{\lceil d/2 \rceil}^* - h_{\lceil d/2 \rceil - 1}^*)$ is an M-vector[10].

8   Quasisymmetric Functions

Polyhedral models are useful for the study of combinatorial objects beyond counting polynomials as well. For example, the simple construction from Sect. 2 of intersecting the cube with a subarrangement of the braid arrangement can serve as a lens into the world of quasisymmetric functions [21].

A *quasisymmetric function* is a formal power series Q of bounded degree in countably many variables $x_1, x_2, \ldots$ such that the coefficients of Q are shift invariant, i.e., for every $(\alpha_1, \ldots, \alpha_m)$ the coefficients of the monomials $x_{i_1}^{\alpha_1} x_{i_2}^{\alpha_2} \cdots x_{i_m}^{\alpha_m}$ for any $i_1 < i_2 < \ldots < i_m$ are equal [50]. Note that a quasisymmetric function can have bounded degree without being a polynomial since we have infinitely many variables at our disposal.

To approach these from a geometric perspective, it is instructive to start with quasisymmetric functions in non-commuting variables or *nc-quasisymmetric functions* for short [14]. Here the variables x_i do not commute multiplicatively and the constraint is that two monomials $x_{i_1} \cdots x_{i_d}$ and $x_{j_1} \cdots x_{j_d}$ have the same coefficient if the tuples $i = (i_1, \ldots, i_d)$ and $j = (j_1, \ldots, j_d)$ induce the same ordered set partition $\Delta(i) = \Delta(j)$. Here $\Delta(i) = (\Delta_1, \ldots, \Delta_m)$ is an ordered partition of the index set $\{1, \ldots, d\}$ such that $i|_{\Delta_l}$ is constant and $i|_{\Delta_l} < i|_{\Delta_{l+1}}$ for all l, e.g., $\Delta(3, 2, 2, 3, 1) = (\{5\}, \{2, 3\}, \{1, 4\}) =: 5|23|14$.

To visualize what is going on here, we need a new way of associating integer vectors with monomials. Classically, we identify monomials in commuting variables with their exponent vector. Here, we identify monomials in non-commuting variables with their vector of indices, i.e., we identify $x_{v_1} \cdots x_{v_d}$ with $(v_1, \ldots, v_d) \in \mathbb{Z}_{\ge 1}^d$. This allows us to picture the map Δ: If ϕ is an ordered set partition of $\{1, \ldots, d\}$, then $\Delta^{-1}(\phi)$ is precisely the set of integer vectors contained in a simplicial cone of the partial polyhedral complex obtained by triangulating the positive orthant by the braid arrangement, as shown in Fig. 12. In other words, the *monomial nc-quasisymmetric functions*

$$\mathcal{M}_\phi = \sum_{v \in \mathbb{Z}^d, \Delta(v) = \phi} x_{v_1} \cdots x_{v_d}$$

form a basis of the space of nc-quasisymmetric functions and these are nothing but cones in the braid arrangement.

---

[10] M-vectors are defined as in Macaulay's theorem, see for example [57, Chap. 8].

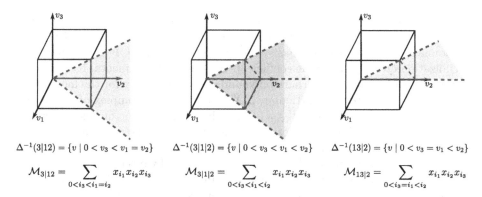

Fig. 12. The "shift-invariant regions" of integer points that are mapped to the same ordered set partition by Δ are the simplicial cones in the braid arrangement and correspond to the monomial quasisymmetric function.

Any nc-quasisymmetric function can be turned into a quasisymmetric function simply by allowing variables to commute. This can be modeled geometrically by taking an integer vector and permuting its entries so that they are in weakly increasing order, i.e., an element of the half-open simplicial cone $C := \{v \mid 0 < v_1 \leq \ldots \leq v_d\}$. This maps $\mathcal{M}_{(\phi_1,\ldots,\phi_m)}$ to the *monomial quasisymmetric function* $M_{(|\phi_1|,\ldots,|\phi_m|)}$ where

$$M_{(\alpha_1,\ldots,\alpha_m)} = \sum_{1 \leq i_1 < \ldots < i_m} x_{i_1}^{\alpha_1} \cdot \ldots \cdot x_{i_m}^{\alpha_m}.$$

The monomial quasisymmetric functions form a basis of the space of quasisymmetric functions. Thus every quasisymmetric function can be visualized as assigning a weight to every face of the cone C. The support of a quasisymmetric function is thus a partial polyhedral subcomplex X of the face lattice of C.

Going one step further it is possible to obtain a polynomial p from a quasisymmetric function Q by substituting 1 into the first k variables and 0 into all other variables, i.e., $Q(1^k) = p(k)$. Geometrically, this substitution eliminates all integer points that contain an entry larger than k. This corresponds to intersecting the complex X of cones with the cube $(0, k]^d$, turning X into a simplicial complex $X \cap (0, k]^d$ and p into the Ehrhart function $\mathrm{ehr}_{X \cap (0,1]^d}(k) = Q(1^k)$.[11] These observations provide a direct translation between Ehrhart functions constructed using the braid arrangement and quasisymmetric functions.

This connection provides fertile ground for future exploration. On the one hand, the geometric approach offers a very flexible framework for defining quasisymmetric functions. Scheduling problems alone capture a wide range of known quasisymmetric functions, such as the chromatic symmetric function, the matroid invariant of Billera-Jia-Reiner, or Ehrenborg's quasisymmetric function

---

[11] This works best if Q is the specialization of an nc-quasisymmetric function with 0-1 coefficients. Otherwise, this would require a linear combination of Ehrhart functions.

for posets, as well as new ones, such as the Bergman and arboricity quasisymmetric functions [21]. On the other hand, many methods for the analysis of Ehrhart polynomials carry over to the quasisymmetric function world. For example, the specialization $Q(1^k)$ collects the coefficients of Q in the fundamental basis in the h^*-vector of the associated Ehrhart-polynomial – and similarly for the monomial basis and the f^*-vector. In particular, if X is a partial subcomplex of the braid arrangement and N, Q and $\mathrm{ehr}_{X \cap (0,1]^d}$ are the associated nc-quasisymmetric, quasisymmetric and Ehrhart functions, then partitionability of X implies nonnegativity of the coefficients in the fundamental basis of N and Q and nonnegativity of the h^*-vector of $\mathrm{ehr}_{X \cap (0,1]^d}$. If X is given by a scheduling problem, partitionability can be guaranteed if the boolean expression defining the scheduling problem takes the form of a certain kind of decision tree [21].

9   Algorithms for Counting Integer Points in Polyhedra

There are many different computational problems associated with polyhedra. The problem of deciding whether there exists a rational vector $v \in \mathbb{Q}^n$ satisfying a linear system of inequalities[12] is polynomial time computable, but when we look for an integer vector $v \in \mathbb{Z}^n$ instead, the problem becomes NP-hard [46]. However, if the dimension of the polyhedron, i.e., the number of variables of the system, is fixed a priori, then there is a polynomial time algorithm for finding an integer solution as Lenstra was able to show in 1983 [42]. While the problem of counting integer solutions is #P-hard as well, the question remained open whether it becomes polynomial time computable if the dimension is fixed. The first algorithm with a polynomial running time in fixed dimension was described by Barvinok in 1994 [5] and it took ten more years until such an algorithm was first implemented by De Loera *et al.* in 2004 [30].

In this section we give an overview over the algorithmic methods for computing the number of integer points in a polyhedron P, and the related problems of computing the Ehrhart polynomial ehr_P and a rational function expression of the multivariate generating function ϕ_P of all integer points in P. Independently of whether the goal is to compute ehr_P by first passing from P to $\mathrm{cone}(P \times \{1\})$ or whether the goal is to compute ϕ_P and $\mathbb{Z}^n \cap P$ directly by using Brion's theorem, the methods employed are similar and consist of three basic steps. First, the polyhedron P is decomposed into simplicial cones. Second, a rational function representation of the integer points in these simplicial cones is computed. We will focus on this step in our exposition since it is crucial with regard to runtime complexity. Third, the obtained rational function expression needs to be specialized if the number of integer points or the Ehrhart (quasi-)polynomial is desired.

To decompose a polyhedron P into simplicial cones, we start by appealing to Brion's theorem and represent P as the sum of its vertex cones, modulo

---

[12] Solving a linear system of *inequalities* over $\mathbb{Z}$ (or, equivalently, solving a linear system of equations over $\mathbb{N}$) is NP-hard. However, solving a linear system of *equations* over $\mathbb{Z}$ is polynomial-time solvable, for example using the Smith normal form, see below.

lines.[13] To achieve this, we need to compute the vertices and edge directions of P. Next, the resulting cones need to be triangulated to make them simplicial. There are sophisticated algorithms available for both tasks [32,33,44,45]. It is also possible to compute a decomposition of P into simplicial cones directly, without computing vertices or triangulating, using the Polyhedral Omega algorithm [23]. Polyhedral Omega is based on simple explicit rules for manipulating simplicial cones formally and is motivated by the symbolic computation framework of partition analysis [1].

In the second step, we use the ideas developed in Sects. 5 and 6 to represent the generating function ϕ_C of integer points in a simplicial cone $C = \mathrm{cone}_{\mathbb{R}}(v_1, \ldots, v_d) \subset \mathbb{Z}^d$ as a rational function. Let V denote the matrix with the generators v_i of C as columns. The straightforward approach is to use (5) and obtain a rational function expression by simply enumerating all integer points in the fundamental parallelepiped $\Pi(V)$ of C. This is both simple and efficient if the index $\mathbb{Z}^d \cap \Pi = |\det(V)|$ of C is sufficiently small. However, in the worst case, the index may be exponential in encoding size of V, as Fig. 14 shows. Thus it is not clear a priori that there exists a rational function expression for ϕ_C whose encoding size is polynomial in the encoding size of the input. Barvinok's key achievement was to find such a representation.

Before we come to Barvinok's short rational function representation, however, it is instructive to take a closer look at how to enumerate the integer points in Π explicitly. There are several well-known approaches to this problem [23,25,40] which are all closely related. We will work with the Smith normal form of the matrix V, which can be computed in polynomial time [46]. The Smith normal form of V is a representation $V = USW$ where U, S, W are integer matrices, U, W have determinant ± 1 and S is a diagonal matrix whose diagonal entries $s_1, \ldots, s_d$ satisfy $s_i | s_{i+1}$. This can be interpreted as shown in Fig. 13. The columns of V form a basis of a sublattice J of the integer lattice $\mathbb{Z}^d$, and V gives the coordinates of this basis with respect to the standard basis of $\mathbb{Z}^d$. The matrices U and W represent changes of basis on *both* lattices such that the new bases B_J of J and $B_{\mathbb{Z}^d}$ of $\mathbb{Z}^d$ line up. Since the elements of B_J are multiples of the elements of $B_{\mathbb{Z}^d}$, the integer points x_i in the fundamental parallelepiped of B_J are easy to enumerate. By computing the coordinates of the x_i wrt. the original basis V of J and taking fractional parts, we translate the x_i into the fundamental parallelepiped $\Pi(V)$ and we are guaranteed that we get every point in $\mathbb{Z}^d \cap \Pi(V)$ exactly once. This process is summarized in the formula

$$\phi_{\mathrm{cone}_{\mathbb{R}}(V)}(z) = \frac{\sum_{k_1=0}^{s_1-1} \cdots \sum_{k_1=0}^{s_1-1} z^{\frac{1}{s_d} V(W^{-1}(s'_1 k_1, \ldots, s'_d k_d)^\top \mod s_d)}}{(1 - z^{v_1}) \cdot \ldots \cdot (1 - z^{v_d})}$$

where $s'_i = \frac{s_d}{s_i}$. This particular expression is taken from [23].

Now we come to Barvinok's central idea. Consider the cone C generated by $(1,0,0)$, $(0,1,0)$ and $(1,1,a)$ for $0 < a \in \mathbb{Z}$. Its fundamental parallelepiped contains a integer points, as shown in Fig. 14, which is exponential in the encoding

---

[13] We can also use the theorem of Lawrence-Varchenko to obtain an exact signed decomposition, without working modulo lines.

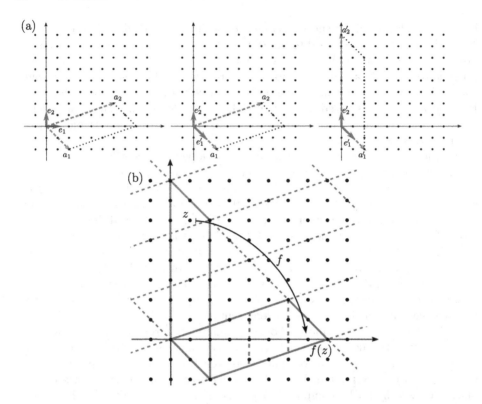

Fig. 13. In order to compute all integer points in the fundamental parallelepiped $\Pi(a_1, a_2)$, shown in the left panel of (a), we proceed as follows. (a) Using the Smith normal form, we first perform a change of basis on the integer lattice $\mathbb{Z}^2$ and then a change of basis on the sublattice generated by a_1, a_2, so that the bases align. (b) Listing all the integer points in the aligned fundamental parallelepiped $\Pi(a_1', a_2')$ is easy. We transform these into integer points in $\Pi(a_1, a_2)$ by modular arithmetic (taking fractional parts of coordinates wrt. the original basis a_1, a_2).

size $\mathcal{O}(\log(a))$ of C. Moreover, there is no way to write C as a union of $\mathcal{O}(\log(a))$ unimodular cones of index 1. Using inclusion-exclusion, however, C can be written as the positive orthant C_1 minus the cone C_2 generated by $(0, 0, 1)$, $(0, 1, 0)$, $(1, 1, a)$ and the cone C_3 generated by $(1, 0, 0)$, $(0, 0, 1)$, $(1, 1, a)$ which all have index 1. This generalizes. Let C denote a simplicial cone in fixed dimension d and let I denote its index. Using the LLL algorithm it is possible to find an integer vector u such that $C = \text{cone}_{\mathbb{R}}(v_1, \ldots, v_d)$ can be written as a signed combination of the cones $C_1 = \text{cone}_{\mathbb{R}}(u, v_2 \ldots, v_d)$, $C_2 = \text{cone}_{\mathbb{R}}(v_1, u, \ldots, v_d)$, $\ldots, C_d = \text{cone}_{\mathbb{R}}(v_1, \ldots, v_{d-1}, u)$, where some facets of the C_i have to be opened according to a few explicit combinatorial rules [40]. The key property of this construction is that indices of the cones C_i decrease quickly. Applying this decomposition recursively, the indices of the cones will eventually reach 1, i.e., the cones will become unimodular. At each node of the recursion tree one cone is split into

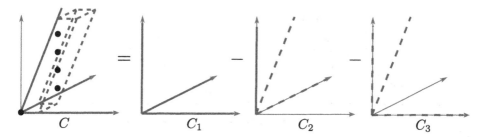

Fig. 14. $C = \mathrm{cone}_{\mathbb{R}}((1,0,0),(0,1,0),(1,1,a))$ has a integer points in its fundamental parallelepiped. This number of integer points is therefore exponential in the encoding size of C which is in $\mathcal{O}(\log(a))$. However C can be written as a signed sum $C = C_1 - C_2 - C_3$ of unimodular cones. Here the facet of C_2 generated by $(0,1,0)$ and $(1,1,a)$ is open and the two facets of C_3 generated by $(1,1,a)$ and one of the other two generators are open.

d-cones, however, the depth of the tree is at most doubly logarithmic in I. Thus the total number of cones obtained is polynomial in the encoding length of C. The result is the following fundamental theorem.

Theorem 10 (Barvinok [5]). *Let $C \subset \mathbb{Z}^d$ be a d-dimensional simplicial cone with integer generators. Then there exists signs ϵ_i and vectors $a_i, b_{i,j}$ such that*

$$\phi_C(z) = \sum_{i=1}^{N} \epsilon_i \frac{z^{a_i}}{(1 - z^{b_{i,1}}) \cdot \ldots \cdot (1 - z^{b_{i,d}})} \tag{7}$$

and for fixed d the number of summands N is bounded by a polynomial in the encoding length of C.

The third step is to specialize the representation of ϕ_P in terms of multivariate rational functions we have obtained thus far, in order to get the Ehrhart polynomial ehr_P or the number $\#\mathbb{Z}^n \cap P$. This specialization is non-trivial, especially if Barvinok decompositions are used, since typically the desired specialization is a pole of the rational function representation. However, using an exponential substitution and limit arguments it is possible to compute this specialization in polynomial time.

The toolbox of algorithms we have described here has many more applications and extensions. For example, it is possible to extend these methods to handle multivariate Ehrhart polynomials [55], to compute intersections $\phi_{P \cap Q}$ given ϕ_P and ϕ_Q [4], to compute Pareto optima in multi-criteria optimization over integer points in polyhedra [28], to integrate and sum polynomials over polyhedra [3] and to convert between rational function representations and piecewise quasi-polynomial representations of counting functions [54] – all in polynomial time if the dimension is fixed. As starting points for further reading we recommend the textbooks [6, 29].

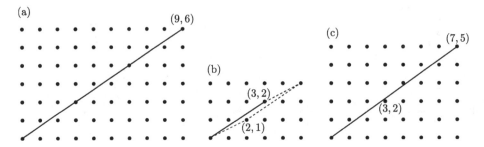

Fig. 15. (a) gcd(9, 6) = 3. (b) $-1 \cdot 3 + 2 \cdot 2 = 1$ since (2, 1) is closest to the line through (3, 2) which means $\mathbb{Z}^2 \cap \Pi((3, 2), (2, 1))$ contains no integer point except the origin. (c) gcd(7, 5) = 1 and $-2 \cdot 7 + 3 \cdot 5 = 1$.

10   Lattice Point Sets and the Euclidean Algorithm

After these very general considerations, we end this exposition on a playful note by taking a closer look at integer point geometry in dimension 2 and discussing several different ways in which the Euclidean algorithm makes an appearance.

The integer lattice in the plane is a great stage for visualizing the greatest common divisor, as Fig. 15 shows. For two integers $a, b \in \mathbb{Z}$, the line segment in the plane from the origin to the point (a, b) contains precisely gcd$(a, b) + 1$ integer points. Let (p, q) denote the coordinates of a lattice point closest to but not on the line L through $(0, 0)$ and (a, b). By construction, the fundamental parallelepiped spanned by (a, b) and (p, q) contains precisely gcd(a, b) lattice points on the line and no lattice points off the line.

$$\gcd(a, b) = \Pi((a, b), (p, q)) = \det \begin{pmatrix} a & p \\ b & q \end{pmatrix} = ap - bq.$$

Thus the coordinates of the closest points give precisely the coefficients produced by the extended Euclidean algorithm.

From the above observation it immediately follows that the value of the GCD increases linearly along any such line L. If (a, b) is the integer point closest to the origin on such a line L, then gcd$(a, b) = 1$. The next values of the GCD on L are thus gcd$(2a, 2b) = 2$, gcd$(3a, 3b) = 3$. The graph of the function gcd : $\mathbb{Z}^2_{>0} \to \mathbb{Z}_{>0}$ is thus contained in a countable collection of rays from the origin through all points (a, b) with gcd$(a, b) = 1$. This "graph" of the GCD is shown in Fig. 16.

A closer look at the graph in Fig. 16 immediately reveals a recursive tree-like structure. It turns out that this tree corresponds precisely to the recursive operation of the Euclidean algorithm. The Euclidean algorithm as described by Euclid moves from (a, b) to $(a - b, a)$ if $a > b$, it moves from (a, b) to $(a, b - a)$ if $a < b$ and it terminates if $a = b$. The perceptive reader will note that this immediately gives a way to enumerate all positive rational numbers as nodes of an infinite binary tree [26]. However, tracing out these paths of the Euclidean algorithm in

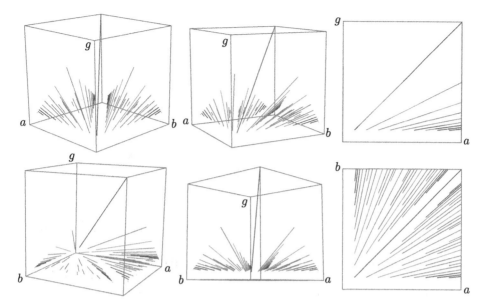

Fig. 16. The graph of the gcd $g = \gcd(a, b)$ as described in the text. Shown are the rays from the origin through (a, b) starting at (a, b) for all a, b with $\gcd(a, b) = 1$. The color of a ray is given by its depth in the recursion tree of the Euclidean algorithm. The plots in the rightmost column show parallel projections of the graph onto the (g, a) and (a, b) planes, respectively (Color figure online).

the plane does not yet reveal the connection to the graph of the GCD. To that end, we turn the Euclidean algorithm on its head.

We fix the point $p = (a, b)$ whose gcd we wish to compute and run the Euclidean algorithm by changing the basis v_1, v_2 of $\mathbb{Z}^2$ in each step. We define the *center* of the current basis as the sum $c = v_1 + v_2$. If p lies below the line through c we change our basis to $v_1' = v_1$ and $v_2' = c$. If p lies above the line through c we change our basis to $v_1' = c$ and $v_2' = v_2$. If p lies on the line through c we are done since $\gcd(a, b) = \frac{a}{c_1} = \frac{b}{c_2}$. Tracing out all the paths the center can take throughout this recursion, we obtain Fig. 17 which reveals the tree structure of the base points of the rays in Fig. 16 and which gives a very natural (and novel) embedding of the Stern-Brocot tree [34, p. 116–117] in the plane.

To conclude, we follow [20] and examine the structure of the integer points below the line L in more detail, going beyond the closest point (p, q). Define $T_{a,b}$ to be the triangle with vertices $(0, 0)$, $(a, 0)$ and (a, b). As we can see in Fig. 18, the "staircase" of integer points in $T_{a,b}$ is irregular: the possible steps as we move from one column to the next are of two different heights, and it is not clear a priori what the underlying pattern is. It turns out, however, that the triangles $T_{a,b}$ have a very nice recursive structure. The key observation is that triangles of the form $T_{c,c}$ are very easy to describe as we always go exactly one step higher as we move from one column to the next. However, if $a > b$, then $T_{a,b}$

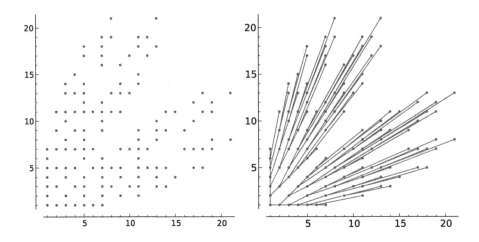

Fig. 17. The left panel shows all lattice points (a, b) in the plane with $\gcd(a, b) = 1$, up to a recursion depth of 5 in the Eulidean algorithm. The right panel also shows the tree structure induced by the inverted Euclidean algorithm as described in the text.

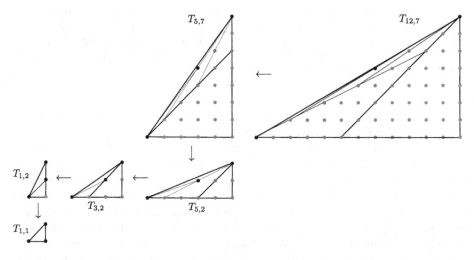

Fig. 18. Following the Euclidean algorithm, we reduce the triangle $T_{12,7}$ recursively, by removing triangles with integral slope and applying shearing lattice transformations. In this way we can decompose the "staircase" of integer points below the line from the origin to $(12, 7)$ into "simple" triangles.

contains a triangle of the form $T_{b,b}$, sitting in the lower right corner. Removing this translate of the half-open triangle $T'_{b,b}$, we are left with a triangle $\tilde{T}$ with vertices $(0, 0), (a - b, 0)$ and (a, b). Shearing $\tilde{T}$ using the linear transformation $A : (x, y) \mapsto (x - y, y)$, we see that the integer points in $\tilde{T}$ have the same structure as those in $T_{a-b,b}$. Here it is crucial that the linear transformation A maps $\mathbb{Z}^2$

bijectively onto itself. Now, if $a < b$ we can apply the same procedure in the other direction. Just like in the Euclidean algorithm we continue recursively until we reach a triangle of the form $T_{c,c}$ at which point we stop. We can thus decompose any triangle $T_{a,b}$ into simple triangles of the form $T_{c,c}$. This process is illustrated in Fig. 18.

This basic approach can yield much more information as detailed in [20]. For example, an analysis of how exactly the big and small steps in the staircase are distributed leads to several characterizations of Sturmian sequences of rational numbers. Moreover, using a recursive procedure similar in spirit to Fig. 18, it is possible to show that the sets of lattice points in 2-dimensional fundamental parallelepipeds always have a short positive description as a union of Minkowski sums of discrete line segments – this short description yields short rational function expressions for 2-dimensional fundamental parallelepipeds, which are quite distinct from those obtained via Barvinok's algorithm.

Acknowledgements. I would like to thank Benjamin Nill, Peter Paule, Manuel Kauers, Christoph Koutschan and an anonymous referee for their helpful comments on earlier versions of this article. I would also like to thank Matthias Beck whose lectures and book [10] were my very own invitation to Ehrhart theory.

References

1. Andrews, G., Paule, P., Riese, A.: MacMahon's partition analysis VI: a new reduction algorithm. Ann. Comb. **5**(3), 251–270 (2001)
2. Monforte, A.A., Kauers, M.: Formal Laurent series in several variables. Expositiones Mathematicae **31**(4), 350–367 (2013)
3. Baldoni, V., Berline, N., De Loera, J.A., Köppe, M., Vergne, M.: How to integrate a polynomial over a simplex. Math. Comput. **80**, 297–325 (2011)
4. Barvinok, A., Woods, K.: Short rational generating functions for lattice point problems. J. Am. Math. Soc. **16**(4), 957–979 (2003)
5. Barvinok, A.I.: A polynomial time algorithm for counting integral points in polyhedra when the dimension is fixed. Math. Oper. Res. **19**(4), 769–779 (1994)
6. Barvinok, A.I.: Integer Points in Polyhedra. European Mathematical Society (2008)
7. Beck, M., Breuer, F., Godkin, L., Martin, J.L.: Enumerating colorings, tensions and flows in cell complexes. J. Comb. Theory, Ser. A **122**, 82–106 (2014)
8. Beck, M., De Loera, J.A., Develin, M., Pfeifle, J., Stanley, R.P.: Coefficients and roots of Ehrhart polynomials. In: Barvinok, A.I. (ed.) Integer Points in Polyhedra, Proceedings of an AMS-IMS-SIAM Joint Summer Research Conference, Snowbird, Utah, 2003, pp. 1–24. AMS (2005)
9. Beck, M., Haase, C., Sottile, F.: Formulas of Brion, Lawrence, and Varchenko on rational generating functions for cones. Math. Intell. **31**(1), 9–17 (2009)
10. Beck, M., Robins, S.: Computing the Continuous Discretely: Integer-Point Enumeration in Polyhedra. Springer, New York (2007)
11. Beck, M., Sanyal, R.: Combinatorial reciprocity theorems (2014, to appear). http://math.sfsu.edu/beck/crt.html
12. Beck, M., Zaslavsky, T.: Inside-out polytopes. Adv. Math. **205**(1), 134–162 (2006)
13. Beck, M., Zaslavsky, T.: The number of nowhere-zero flows on graphs and signed graphs. J. Comb. Theory, Ser. B **96**(6), 901–918 (2006)

14. Bergeron, N., Zabrocki, M.: The Hopf algebras of symmetric functions and quasi-symmetric functions in non-commutative variables are free and co-free. J. Algebra Appl. **8**(4), 581–600 (2009)
15. Breuer, F.: Ham sandwiches, staircases and counting polynomials. Ph.D. thesis, Freie Universität Berlin (2009)
16. Breuer, F.: Ehrhart f^*-coefficients of polytopal complexes are non-negative integers. Electron. J. Comb. **19**(4), P16 (2012)
17. Breuer, F., Dall, A.: Bounds on the coefficients of tension and flow polynomials. J. Algebr. Comb. **33**(3), 465–482 (2011)
18. Breuer, F., Dall, A., Kubitzke, M.: Hypergraph coloring complexes. Discrete Math. **312**(16), 2407–2420 (2012)
19. Breuer, F., Eichhorn, D., Kronholm, B.: Cranks and the geometry of combinatorial witnesses for the divisibility and periodicity of the restricted partition function (2014, in preparation)
20. Breuer, F., von Heymann, F.: Staircases in $\mathbb{Z}^2$. Integers **10**(6), 807–847 (2010)
21. Breuer, F., Klivans, C.J.: Scheduling problems (2014, submitted). arXiv:1401.2978v1
22. Breuer, F., Sanyal, R.: Ehrhart theory, modular flow reciprocity, and the Tutte polynomial. Mathematische Zeitschrift **270**(1), 1–18 (2012)
23. Breuer, F., Zafeirakopoulos, Z.: Polyhedral omega: a new algorithm for solving linear diophantine systems (2014, in preparation)
24. Brion, M.: Points entiers dans les polyèdres convexes. Annales scientifiques de l'École Normale Supérieure **21**(4), 653–663 (1988)
25. Bruns, W., Ichim, B., Söger, C.: The power of pyramid decomposition in normaliz (2012). arXiv:1206.1916v1
26. Calkin, N., Wilf, H.S.: Recounting the rationals. Am. Math. Mon. **107**(4), 360–363 (2000)
27. Chari, M.K.: Two decompositions in topological combinatorics with applications to matroid complexes. Trans. Am. Math. Soc. **349**(10), 3925–3943 (1997)
28. De Loera, J.A., Hemmecke, R., Köppe, M.: Pareto optima of multicriteria integer linear programs. INFORMS J. Comput. **21**(1), 39–48 (2009)
29. De Loera, J.A., Hemmecke, R., Köppe, M.: Algebraic and geometric ideas in the theory of discrete optimization, vol. 14. SIAM (2012)
30. De Loera, J.A., Hemmecke, R., Tauzer, J., Yoshida, R.: Effective lattice point counting in rational convex polytopes. J. Symb. Comput. **38**(4), 1273–1302 (2004)
31. Ehrhart, E.: Sur les polyèdres rationnels homothétiques à n dimensions. C. R. Acad. Sci. Paris **254**, 616–618 (1962)
32. Fukuda, K., Prodon, A.: Double description method revisited. In: Deza, M., Euler, R., Manoussakis, I. (eds.) CCS 1995. LNCS, vol. 1120, pp. 91–111. Springer, Heidelberg (1996)
33. Fukuda, K., Rosta, V.: Combinatorial face enumeration in convex polytopes. Comput. Geom. **4**(4), 191–198 (1994)
34. Graham, R.L., Knuth, D.E., Patashnik, O.: Concrete Mathematics: A Foundation for Computer Science, 2nd edn. Addison-Wesley Longman Publishing Co. Inc., Boston (1994)
35. Greene, C.: Acyclic orientations (notes). In: Aigner, M. (ed.) Higher Combinatorics, pp. 65–68. Reidel, Dordrecht (1977)
36. Haase, C., Schicho, J.: Lattice polygons and the number 2i+7. Am. Math. Mon. **116**(2), 151–165 (2009)
37. Henk, M., Tagami, M.: Lower bounds on the coefficients of Ehrhart polynomials. Eur. J. Comb. **30**(1), 70–83 (2009)

38. Hersh, P., Swartz, E.: Coloring complexes and arrangements. J. Algebr. Comb. **27**(2), 205–214 (2007)
39. Jochemko, K., Sanyal, R.: Arithmetic of marked order polytopes, monotone triangle reciprocity, and partial colorings, pp. 1–16 (2013). arXiv:1206.4066v2
40. Köppe, M., Verdoolaege, S.: Computing parametric rational generating functions with a primal Barvinok algorithm. Electron. J. Comb. **15**, R16 (2008)
41. Lawrence, J.: Valuations and polarity. Discrete Comput. Geom. **3**(1), 307–324 (1988)
42. Lenstra, H.W.: Integer programming with a fixed number of variables. Math. Oper. Res. **8**(4), 538–548 (1983)
43. Macdonald, I.G.: Polynomials associated to finite cell complexes. J. London Math. Soc. **2**(4), 181–192 (1971)
44. Pfeifle, J., Rambau, J.: Computing triangulations using oriented matroids. In: Joswig, M., Takayama, N. (eds.) Algebra, Geometry, and Software Systems, pp. 49–75. Springer, Berlin (2003)
45. De Loera, J.A., Rambau, J., Santos, F.: Triangulations: Structures for Algorithms and Applications. Springer, Berlin (2010)
46. Schrijver, A.: Theory of Linear and Integer Programming. Wiley, New York (1986)
47. Stanley, R.P.: Acyclic orientations of graphs. Discrete Math. **5**, 171–178 (1973)
48. Stanley, R.P.: Decompositions of rational convex polytopes. Ann. Discrete Math. **6**, 333–342 (1980)
49. Stanley, R.P.: Two poset polytopes. Discrete Comput. Geom. **1**, 9–23 (1986)
50. Stanley, R.P.: Enumerative Combinatorics. Cambridge Studies in Advanced Mathematics, vol. 2. Cambridge University Press, Cambridge (2001)
51. Stapledon, A.: Inequalities and Ehrhart δ-vectors. Trans. Am. Math. Soc. **361**, 5615–5626 (2009)
52. Swartz, E.: g-Elements, finite buildings and higher Cohen-Macaulay connectivity. J. Combin. Theory, Ser. A **113**, 1305–1320 (2006)
53. Varchenko, A.N.: Combinatorics and topology of the disposition of affine hyperplanes in real space. Funct. Anal. Appl. **21**(1), 9–19 (1987)
54. Verdoolaege, S., Woods, K.: Counting with rational generating functions. J. Symb. Comput. **43**(2), 75–91 (2008)
55. Verdoolaege, S., Seghir, R., Beyls, K., Loechner, V., Bruynooghe, M.: Counting integer points in parametric polytopes using Barvinok's rational functions. Algorithmica **48**(1), 37–66 (2007)
56. Woods, K.: Presburger arithmetic, rational generating functions, and quasipolynomials. In: Fomin, F.V., Freivalds, R., Kwiatkowska, M., Peleg, D. (eds.) ICALP 2013, Part II. LNCS, vol. 7966, pp. 410–421. Springer, Heidelberg (2013). http://arxiv.org/abs/1211.0020
57. Ziegler, G.M.: Lectures on Polytopes. Graduate Texts in Mathematics. Springer, New York (1995)

Moving Curve Ideals of Rational Plane Parametrizations

Carlos D'Andrea[(✉)]

Facultat de Matemàtiques, Universitat de Barcelona,
Gran Via 585, 08007 Barcelona, Spain
cdandrea@ub.edu
http://atlas.mat.ub.es/personals/dandrea

Abstract. In the nineties, several methods for dealing in a more efficient way with the implicitization of rational parametrizations were explored in the Computer Aided Geometric Design Community. The analysis of the validity of these techniques has been a fruitful ground for Commutative Algebraists and Algebraic Geometers, and several results have been obtained so far. Yet, a lot of research is still being done currently around this topic. In this note we present these methods, show their mathematical formulation, and survey current results and open questions.

1 Rational Plane Curves

Rational curves are fundamental tools in Computer Aided Geometric Design. They are used to trace the boundary of any kind of shape via transforming a parameter (a number) via some simple algebraic operations into a point of the cartesian plane or three-dimensional space. Precision and esthetics in Computer Graphics demands more and more sophisticated calculations, and hence any kind of simplification of the very large list of tasks that need to be performed between the input and the output is highly appreciated in this world. In this survey, we will focus on a simplification of a method for implicitization rational curves and surfaces defined parametrically. This method was developed in the 90's by Thomas Sederberg and his collaborators (see [STD94, SC95, SGD97]), and turned out to become a very rich and fruitful area of interaction among mathematicians, engineers and computer scientist. As we will see at the end of the survey, it is still a very active of research these days.

To ease the presentation of the topic, we will work here only with plane curves and point to the reader to the references for the general cases (spatial curves and rational hypersurfaces).

Let $\mathbb{K}$ be a field, which we will suppose to be algebraically closed so our geometric statements are easier to describe. Here, when we mean "geometric"

Partially supported by the Research Project MTM2010–20279 from the Ministerio de Ciencia e Innovación, Spain.

J. Gutierrez et al. (Eds.): Computer Algebra and Polynomials, LNCS 8942, pp. 30–49, 2015.
DOI: 10.1007/978-3-319-15081-9_2

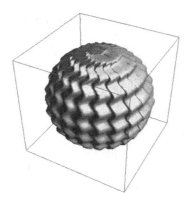

Fig. 1. The shape of an "orange" plotted with `Mathematica` 8.0 [Wol10].

we refer to Algebraic Geometry and not Euclidean Geometry which is the natural domain in Computer Design. Our assumption on $\mathbb{K}$ may look somehow strange in this context, but we do this for the ease of our presentation. We assume the reader also to be familiar with projective lines and planes over $\mathbb{K}$, which will be denoted with $\mathbb{P}^1$ and $\mathbb{P}^2$ respectively. A *rational plane parametrization* is a map

$$
\phi: \quad \mathbb{P}^1 \quad \longrightarrow \quad \mathbb{P}^2
$$
$$
(t_0 : t_1) \longmapsto \big(u_0(t_0, t_1) : u_1(t_0, t_1) : u_2(t_0, t_1)\big), \tag{1}
$$

where $u_0(t_0, t_1)$, $u_1(t_0, t_1)$, $u_2(t_0, t_1)$ are polynomials in $\mathbb{K}[T_0, T_1]$, homogeneous, of the same degree $d \geq 1$, and without common factors. We will call $\mathcal{C}$ to the image of ϕ, and refer to it as *the rational plane curve parametrized by ϕ*.

This definition may sound a bit artificial for the reader who may be used to look at maps of the form

$$
\mathbb{K} \dashrightarrow \quad \mathbb{K}^2
$$
$$
t \longmapsto \left(\frac{a(t)}{c(t)}, \frac{b(t)}{c(t)}\right), \tag{2}
$$

with $a(t), b(t), c(t) \in \mathbb{K}[t]$ without common factors, but it is easy to translate this situation to (1) by extending this "map" (which actually is not defined on all points of $\mathbb{K}$) to one from $\mathbb{P}^1 \to \mathbb{P}^2$, in a sort of *continuous* way. To speak about continuous maps, we need to have a topology on $\mathbb{K}^n$ and/or in $\mathbb{P}^n$, for $n = 1, 2$. We will endow all these sets with the so-called *Zariski topology*, which is the coarsest topology that make polynomial maps as in (2) continuous.

Now it should be clear that there is actually an advantage in working with projective spaces instead of parametrizations as in (2): our rational map defined in (1) is *actually* a map, and the translation from $a(t), b(t), c(t)$ to $u_0(t_0, t_1)$, $u_1(t_0, t_1)$, $u_2(t_0, t_1)$ is very straightforward. The fact that $\mathbb{K}$ is algebraically closed also comes in our favor, as it can be shown that for parametrizations defined over algebraically closed fields (see [CLO07] for instance), the curve $\mathcal{C}$ is actually an *algebraic set* of $\mathbb{P}^2$, i.e. it can be described as the zero set of a finite system of homogeneous polynomial equations in $\mathbb{K}[X_0, X_1, X_2]$.

More can be said on the case of $\mathcal{C}$, the *Implicitization's Theorem* in [CLO07] states essentially that there exists $F(X_0, X_1, X_2) \in \mathbb{K}[X_0, X_1, X_2]$, homogeneous of degree $D \geq 1$, irreducible, such that $\mathcal{C}$ is actually the zero set of $F(X_0, X_1, X_2)$ in $\mathbb{P}^2$, i.e. the system of polynomials equations in this case reduces to one single equation. It can be shown that $F(X_0, X_1, X_2)$ is well-defined up to a nonzero constant in $\mathbb{K}$, and it is called *the defining polynomial* of $\mathcal{C}$. The *implicitization problem* consists in computing F having as input data the polynomials u_0, u_1, u_2 which are the components of ϕ as in (1).

Example 1. Let $\mathcal{C}$ be the unit circle with center in the origin $(0, 0)$ of $\mathbb{K}^2$. A well-known parametrization of this curve by using a pencil of lines centered in $(-1, 0)$ is given in *affine* format (2) as follows:

$$\begin{aligned} \mathbb{K} &\dashrightarrow \mathbb{K}^2 \\ t &\longmapsto \left(\frac{1-t^2}{1+t^2}, \frac{2t}{1+t^2}, \right). \end{aligned} \tag{3}$$

Note that if $\mathbb{K}$ has square roots of -1, these values do not belong to the field of definition of the parametrization above. Moreover, it is straightforward to check that the point $(-1, 0)$ is not in the image of (3). However, by converting (3) into the homogeneous version (1), we obtain the parametrization

$$\begin{aligned} \phi: \quad \mathbb{P}^1 &\longrightarrow \mathbb{P}^2 \\ (t_0 : t_1) &\longmapsto \left(t_0^2 + t_1^2 : t_0^2 - t_1^2 : 2t_0 t_1 \right), \end{aligned} \tag{4}$$

which is well defined on all $\mathbb{P}^1$. Moreover, every point of the circle (in projective coordinates) is in the image of ϕ, for instance $(1 : -1 : 0) = \phi(0 : 1)$, which is the point in $\mathcal{C}$ we were "missing" from the parametrization (3). The defining polynomial of $\mathcal{C}$ in this case is clearly $F(X_0, X_1, X_2) = X_1^2 + X_2^2 - X_0^2$.

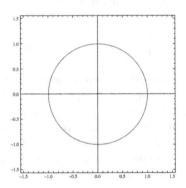

Fig. 2. The unit circle.

In general, the solution to the implicitization problem involves tools from *Elimination Theory*, as explained in [CLO07]: from the equation

$$(X_0 : X_1 : X_2) = \left(u_0(t_0 : t_1) : u_1(t_0 : t_1) : u_2(t_0 : t_1) \right),$$

one "eliminates" the variables t_0 and t_1 to get an expression involving only the X's variables.

The elimination process can be done with several tools. The most popular and general is provided by *Gröbner bases*, as explained in [AL94] (see also [CLO07]). In the case of a rational parametrization like the one we are handling here, we can consider a more efficient and suitable tool: the *Sylvester resultant* of two homogeneous polynomials in t_0, t_1, as defined in [AJ06] (see also [CLO05]). We will denote this resultant with $\text{Res}_{t_0,t_1}(\cdot,\cdot)$. The following result can be deduced straightforwardly from the section of Elimination and Implicitization in [CLO07].

Proposition 1. *There exist α, $\beta \in \mathbb{N}$ such that -up to a nonzero constant-*

$$\text{Res}_{t_0,t_1}\big(X_2 u_0(t_0,t_1) - X_0 u_2(t_0,t_1), X_2 u_1(t_0,t_1) - X_1 u_2(t_0,t_1)\big) = X_2^\alpha F(X_0, X_1, X_2)^\beta. \tag{5}$$

Note that as the polynomial $F(X_0, X_1, X_2)$ is well-defined up to a nonzero constant, all formulae involving it must also hold this way. For instance, an explicit computation of (3) in Example 1 shows that this resultant is equal to

$$- 4X_2^2\big(X_0^2 - X_1^2 - X_2^2\big). \tag{6}$$

One may think that the number -4 which appears above is just a random constant, but indeed it is indicating us something very important: if the characteristic of $\mathbb{K}$ is 2, then it is easy to verify that (3) does not describe a circle, but the line $X_2 = 0$. What is even worse, (4) is not the parametrization of a curve, as its image is just the point $(1 : 1 : 0)$.

To compute the Sylvester Resultant one can use the well-known *Sylvester matrix* (see [AJ06, CLO07]), whose nonzero entries contain coefficients of the two polynomials $X_2 u_0(t_0,t_1) - X_0 u_2(t_0,t_1)$ and $X_2 u_1(t_0,t_1) - X_1 u_2(t_0,t_1)$, regarded as polynomials in the variables t_0 and t_1. The resultant is then the determinant of that (square) matrix.

For instance, in Example 1, we have

$$X_2 u_0(t_0,t_1) - X_0 u_2(t_0,t_1) = X_2 t_0^2 - 2X_0 t_0 t_1 + X_2 t_1^2$$
$$X_2 u_1(t_0,t_1) - X_1 u_2(t_0,t_1) = X_2 t_0^2 - 2X_1 t_0 t_1 - X_2 t_1^2,$$

and (6) is obtained as the determinant of the Sylvester matrix

$$\begin{pmatrix} X_2 & -2X_0 & X_2 & 0 \\ 0 & X_2 & -2X_0 & X_2 \\ X_2 & -2X_1 & -X_2 & 0 \\ 0 & X_2 & -2X_1 & -X_2 \end{pmatrix}. \tag{7}$$

Having X_2 as a factor in (5) is explained by the fact that the polynomials whose resultant is being computed in (3) are not completely symmetric in the X's parameters, and indeed X_2 is the only X-monomial appearing in both expansions.

The exponent β in (5) has a more subtle explanation, it is the *tracing index* of the map ϕ, or the cardinality of its *generic fiber*. Geometrically, for all but a finite number of points $(p_0 : p_1 : p_2) \in \mathcal{C}$, β is the cardinality of the set $\phi^{-1}(p_0 : p_1 : p_2)$. Algebraically, it is defined as the degree of the extension

$$\left[\mathbb{K}\big(u_0(t_0,t_1)/u_2(t_0,t_1), u_1(t_0,t_1)/u_2(t_0,t_1)\big) : \mathbb{K}(t_0/t_1)\right].$$

In the applications, one already starts with a map ϕ as in (1) which is *generically injective*, i.e. with $\beta = 1$. This assumption is not a big one, due to the fact that generic parametrizations are generically injective, and moreover, thanks to *Lüroth's theorem* (see [vdW66]), every parametrization ϕ as in (1) can be factorized as $\phi = \overline{\phi} \circ \mathcal{P}$, with $\overline{\phi} : \mathbb{P}^1 \to \mathbb{P}^2$ generically injective, and $\mathcal{P} : \mathbb{P}^1 \to \mathbb{P}^1$ being a map defined by a pair of coprime homogeneous polynomial both of them having degree β. One can then regard $\overline{\phi}$ as a "reparametrization" of $\mathcal{C}$, and there are very efficient algorithms to deal with this problem, see for instance [SWP08].

In closing this section, we should mention the difference between "algebraic (plane) curves" and the rational curves introduced above. An algebraic plane curve is a subset of $\mathbb{P}^2$ defined by the zero set of a homogeneous polynomial $G(X_0, X_1, X_2)$. In this sense, any rational plane curve is algebraic, as we can find its defining equation via the implicitization described above. But not all algebraic curve is rational, and moreover, if the curve has degree 3 or more, a generic algebraic curve will not be rational. Being rational or not is actually a geometric property of the curve, and one should not expect to detect it from the form of the defining polynomial, see [SWP08] for algorithms to decide whether a given polynomial $G(X_0, X_1, X_2)$ defines a rational curve or not. For instance, the Folium of Descartes (see Fig. 3) is a rational curve with parametrization

$$(t_0 : t_1) \mapsto (t_0^3 + t_1^3 : 3t_0^2 t_1 : 3t_0 t_1^2),$$

and implicit equation given by the polynomial $F(X_0, X_1, X_2) = X_1^3 + X_2^3 - 3X_0 X_1 X_2$. On the other hand, Fermat's cubic plotted in Fig. 4 is defined by the vanishing of $G(X_0, X_1, X_2) = X_1^3 + X_2^3 - X_0^3$ but it is not rational.

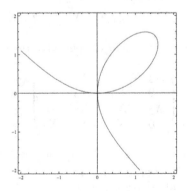

Fig. 3. The Folium of Descartes.

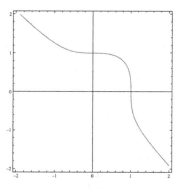

Fig. 4. Fermat's cubic.

The reason why rational curves play a central role in Visualization and Computer Design should be easy to get, as they are

- easy to "manipulate" and plot,
- enough to describe all possible kind of shape by using patches (so-called spline curves).

2   Moving Lines and μ-Bases

Moving lines were introduced by Thomas W. Sederberg and his collaborators in the nineties, [STD94, SC95, SGD97, CSC98]. The idea is the following: in each row of the Sylvester matrix appearing in (7) one can find the coefficients as a polynomial in t_0, t_1 of a form $\mathcal{L}(t_0, t_1, X_0, X_1, X_2) \in \mathbb{K}[t_0, t_1, X_0, X_1, X_2]$ of degree 3 in the variables t's, and satisfying:

$$\mathcal{L}\big(t_0, t_1, u_0(t_0, t_1), u_1(t_0, t_1), u_2(t_0, t_1)\big) = 0. \tag{8}$$

The first row of (7) for instance, contains the coefficients of

$$t_0(X_2\, u_0(t_0, t_1) - X_0\, u_2(t_0, t_1)) = \mathbf{X_2}t_0^3 - \mathbf{2X_0}t_0^2 t_1 + \mathbf{X_2}t_0 t_1^2 + \mathbf{0}t_1^3,$$

which clearly vanishes if we set $X_i \mapsto u_i(t_0, t_1)$. Note that all the elements in (7) are linear in the X's variables.

With this interpretation in mind, we can regard any such $\mathcal{L}(t_0, t_1, X_0, X_1, X_2)$ as a family of lines in $\mathbb{P}^2$ in such a way that for any $(t_0 : t_1) \in \mathbb{P}^1$, this line passes through the point $\phi(t_0 : t_1) \in \mathcal{C}$. Motivated by this idea, the following central object in this story has been defined.

Definition 1. *A moving line of degree δ which follows the parametrization ϕ is a polynomial*

$\mathcal{L}_\delta(t_0, t_1, X_0, X_1, X_2) = v_0(t_0, t_1)X_0 + v_1(t_0, t_1)X_1 + v_2(t_0, t_1)X_2 \in \mathbb{K}[t_0, t_1, X_0, X_1, X_2]$,

with each v_i homogeneous of degree δ, $i = 0, 1, 2$, such that

$$\mathcal{L}_\delta(t_0, t_1, u_0(t_0, t_1), u_1(t_0, t_1), u_2(t_0, t_1)) = 0,$$

i.e.

$$v_0(t_0, t_1)u_0(t_0, t_1) + v_1(t_0, t_1)u_1(t_0, t_1) + v_2(t_0, t_1)u_2(t_0, t_1) = 0. \tag{9}$$

Note that both $X_2 u_0(t_0, t_1) - X_0 u_2(t_0, t_1)$ and $X_2 u_1(t_0, t_1) - X_1 u_2(t_0, t_1)$ are always moving lines following ϕ. Moreover, note that if we multiply any given moving line by a homogeneous polynomial in $\mathbb{K}[t_0, t_1]$, we obtain another moving line of higher degree. The set of moving lines following a given parametrization has an algebraic structure of a *module* over the ring $\mathbb{K}[t_0, t_1]$. Indeed, another way of saying that $\mathcal{L}_\delta(t_0, t_1, X_0, X_1, X_2)$ is a moving line which follows ϕ is that the vector $(v_0(t_0, t_1), v_1(t_0, t_1), v_2(t_0, t_1))$ is a homogeneous element of the *syzygy module* of the ideal generated by the sequence $\{u_0(t_0, t_1), u_1(t_0, t_1), u_2(t_0, t_1)\}$ -the coordinates of ϕ- in the ring of polynomials $\mathbb{K}[t_0, t_1]$.

We will not go further in this direction yet, as the definition of moving lines does not require understanding concepts like syzygies or modules. Note that computing moving lines is very easy from an equality like (9). Indeed, one first fixes δ as small as possible, and then sets $v_0(t_0, t_1)$, $v_1(t_0, t_1)$, $v_2(t_0, t_1)$ as homogeneous polynomials of degree δ and unknown coefficients, which can be solved via the linear system of equations determined by (9).

With this very simple but useful object, the *method of implitization by moving lines* as stated in [STD94] says essentially the following: look for a set of moving lines of the same degree δ, with δ as small as possible, which are "independent" in the sense that the matrix of their coefficients (as polynomials in t_0, t_1) has maximal rank. If you are lucky enough, you will find $\delta + 1$ of these forms, and hence the matrix will be square. Compute then the determinant of this matrix, and you will get a non-trivial multiple of the implicit equation. If your are even luckier, your determinant will be equal to $F(X_0, X_1, X_2)^\beta$.

Example 2. Let us go back to the parametrization of the unit circle given in Example 1. We check straightforwardly that both

$$\mathcal{L}_1(t_0, t_1, X_0, X_1, X_2) = -t_1 X_0 - t_1 X_1 + t_0 X_2 = X_2 t_0 - (X_0 + X_1) t_1$$
$$\mathcal{L}_2(t_0, t_1, X_0, X_1, X_2) = -t_0 X_0 + t_0 X_1 + t_1 X_2 = (-X_0 + X_1) t_0 + X_2 t_1.$$

satisfy (8). Hence, they are moving lines of degree 1 which follow the parametrization of the unit circle. Here, $\delta = 1$. We compute the matrix of their coefficients as polynomials (actually, linear forms) in t_0, t_1, and get

$$\begin{pmatrix} X_2 & -X_0 - X_1 \\ -X_0 + X_1 & X_2 \end{pmatrix}. \tag{10}$$

It is easy to check that the determinant of this matrix is equal to

$$F(X_0, X_1, X_2) = X_1^2 + X_2^2 - X_0^2.$$

Note that the size of (10) is actually half of the size of (7), and also that the determinant of this matrix gives the implicit equation without any extraneous factor.

Of course, in order to convince the reader that this method is actually better than just performing (5), we must shed some light on how to compute algorithmically a matrix of moving lines. The following result was somehow discovered by Hilbert more than a hundred years ago, and rediscovered in the CAGD community in the late nineties (see [CSC98]).

Theorem 1. *For ϕ as in (1), there exist a unique $\mu \leq \frac{d}{2}$ and two moving lines following ϕ which we will denote as $\mathcal{P}_\mu(t_0, t_1, X_0, X_1, X_2)$, $\mathcal{Q}_{d-\mu}(t_0, t_1, X_0, X_1, X_2)$ of degrees μ and $d-\mu$ respectively such that any other moving line following ϕ is a polynomial combination of these two, i.e. every $\mathcal{L}_\delta(t_0, t_1, X_0, X_1, X_2)$ as in the Definition 1 can be written as*

$$\mathcal{L}_\delta(t_0, t_1, X_0, X_1, X_2) = p(t_0, t_1)\mathcal{P}_\mu(t_0, t_1, X_0, X_1, X_2) + q(t_0, t_1)\mathcal{P}_{d-\mu}(t_0, t_1, X_0, X_1, X_2),$$

with $p(t_0, t_1)$, $q(t_0, t_1) \in \mathbb{K}[t_0, t_1]$ homogeneous of degrees $\delta - \mu$ and $\delta - d + \mu$ respectively.

This statement is consequence of a stronger one, which essentially says that a parametrization ϕ as in (1), can be "factorized" as follows:

Theorem 2 (Hilbert-Burch). *For ϕ as in (1), there exist a unique $\mu \leq \frac{d}{2}$ and two parametrizations φ_μ, $\psi_{d-\mu} : \mathbb{P}^1 \to \mathbb{P}^2$ of degrees μ and $d-\mu$ respectively such that*

$$\phi(t_0 : t_1) = \varphi_\mu(t_0 : t_1) \times \psi_{d-\mu}(t_0 : t_1), \tag{11}$$

where $\times$ denotes the usual cross product of vectors.

Note that we made an abuse of notation in the statement of (11), as $\varphi_\mu(t_0 : t_1)$ and $\psi_{d-\mu}(t_0 : t_1)$ are elements in $\mathbb{P}^2$ and the cross product is not defined in

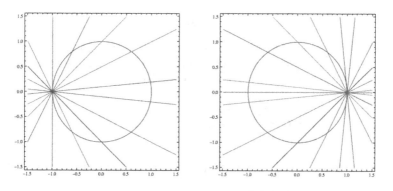

Fig. 5. Moving lines $\mathcal{L}_1$ (left) and $\mathcal{L}_2$ (right).

this space. The meaning of $\times$ in (11) should be understood as follows: *pick representatives in $\mathbb{K}^3$ of both $\varphi_\mu(t_0 : t_1)$ and $\psi_{d-\mu}(t_0 : t_1)$, compute the cross product of these two representatives, and then "projectivize" the result to $\mathbb{P}^2$ again.*

The parametrizations φ_μ and $\psi_{d-\mu}$ can be explicited by computing a *free resolution* of the ideal $\langle u_0(t_0, t_1), u_1(t_0, t_1), u_2(t_0, t_1)\rangle \subset \mathbb{K}[t_0, t_1]$, and there are algorithms to do that, see for instance [CDNR97]. Note that even though general algorithms for computing free resolutions are based on computations of Gröbner bases, which have in general bad complexity time, the advantage here is that we are working with a graded resolution, and also that the resolution of an ideal like the one we deal with here is of *Hilbert-Burch* type in the sense of [Eis95]. This means that the coordinates of both φ_d and $\psi_{d-\mu}$ appear in the columns of the 2×3 matrix of the first syzygies in the resolution. We refer the reader to [CSC98] for more details on the proofs of Theorems 1 and 2.

The connection between the moving lines $\mathcal{P}_\mu(t_0, t_1, X_0, X_1, X_2)$, $\mathcal{Q}_{d-\mu}(t_0, t_1, X_0, X_1, X_2)$ of Theorem 1 and the parametrizations φ_μ, $\psi_{d-\mu}$ in (11) is the obvious one: the coordinates of φ_μ (resp. $\psi_{d-\mu}$) are the coefficients of $\mathcal{P}_\mu(t_0, t_1, X_0, X_1, X_2)$ (resp. $\mathcal{Q}_{d-\mu}(t_0, t_1, X_0, X_1, X_2)$) as a polynomial in X_0, X_1, X_2.

Definition 2. *A sequence $\{\mathcal{P}_\mu(t_0, t_1, X_0, X_1, X_2), \mathcal{Q}_{d-\mu}(t_0, t_1, X_0, X_1, X_2)\}$ as in Theorem 1, is called a μ-basis of ϕ.*

Note that both Theorems 1 and 2 only state the uniquenes of the value of μ, and not of $\mathcal{P}_\mu(t_0, t_1, X_0, X_1, X_2)$ and $\mathcal{Q}_{d-\mu}(t_0, t_1, X_0, X_1, X_2)$. Indeed, if $\mu = d - \mu$ (which happens generically if d is even), then any two generic linear combinations of the elements of a μ-basis is again another μ-basis. If $\mu < d - \mu$, then any polynomial multiple of $\mathcal{P}_\mu(t_0, t_1, X_0, X_1, X_2)$ of the proper degree can be added to $\mathcal{Q}_{d-\mu}(t_0, t_1, X_0, X_1, X_2)$ to produce a different μ-basis of the same parametrization.

Example 3. For the parametrization of the unit circle given in Example 1, one can easily check that

$$\varphi_1(t_0 : t_1) = (-t_1 : -t_1 : t_0),$$
$$\psi_1(t_0 : t_1) = (-t_0 : t_0 : t_1)$$

is a μ-basis of ϕ defined in (4), i.e. this parametrization has $\mu = d - \mu = 1$. Indeed, we compute the cross product in (11) as follows: denote with e_0, e_1, e_2 the vectors of the canonical basis of $\mathbb{K}^3$. Then, we get

$$\begin{vmatrix} e_0 & e_1 & e_2 \\ -t_1 & t_1 & t_0 \\ -t_0 & t_0 & t_1 \end{vmatrix} = \left(-t_0^2 - t_1^2, \, t_1^2 - t_0^2, \, -2t_0t_1 \right),$$

which shows that the $\varphi_1(t_0 : t_1) \times \psi_1(t_0 : t_1) = \phi(t_0 : t_1)$, according to (11).

The reason the computation with μ-bases are important, not only because with them we can generate all the moving lines which follow a given parametrization, but also because they will allow us to produce small matrices of moving lines whose determinant give the implicit equation. Indeed, the following result has been proven in [CSC98, Theorem 1].

Theorem 3. *With notation as above, let β be the tracing index of ϕ. Then, up to a nonzero constant in $\mathbb{K}$, we have*

$$\text{Res}_{t_0,t_1}\big(\mathcal{P}_\mu(t_0,t_1,X_0,X_1,X_2),\mathcal{Q}_{d-\mu}(t_0,t_1,X_0,X_1,X_2)\big) = F(X_0,X_1,X_2)^\beta. \tag{12}$$

As shown in [SGD97] if you use any kind of matrix formulation for computing the Sylvester resultant, in each row of these matrices, when applied to formulas (5) and (12), you will find the coefficients (as a polynomial in t_0, t_1) of a moving line following the parametrization. Note that the formula given by Theorem 3 always involves a smaller matrix than the one in (5), as the t-degrees of the polynomials $\mathcal{P}_\mu(t_0,t_1,X_0,X_1,X_2)$ and $\mathcal{Q}_{d-\mu}(t_0,t_1,X_0,X_1,X_2)$ are roughly half of the degrees of those in (5).

There is, of course, a connection between these two formulas. Indeed, denote with $\text{Syl}_{t_0,t_1}(G,H)$ (resp. $\text{Bez}_{t_0,t_1}(G,H)$) the Sylvester (resp. *Bézout*) matrix for computing the resultant of the two homogeneous polynomials $G, H \in \mathbb{K}[t_0,t_1]$. For more about definitions and properties of these matrices, see [AJ06]. In [BD12, Proposition 6.1], we proved with Laurent Busé the following:

Theorem 4. *There exists an invertible matrix $M \in \mathbb{K}^{d\times d}$ such that*

$$X_2 \cdot \text{Sylv}_{t_0,t_1}\big(\mathcal{P}_\mu(t_0,t_1,X_0,X_1,X_2),\mathcal{Q}_{d-\mu}(t_0,t_1,X_0,X_1,X_2)\big)$$
$$= M \cdot \text{Bez}_{t_0,t_1}\big(X_2 u_0(t_0,t_1) - X_0 u_2(t_0,t_1), X_2 u_1(t_0,t_1) - X_1 u_2(t_0,t_1)\big).$$

From the identity above, one can easily deduce that it is possible to compute the implicit equation (or a power of it) of a rational parametrization with a determinant of a matrix of coefficients of d moving lines, where d is the degree of ϕ. Can you do it with less? Unfortunately, the answer is *no*, as each row or column of a matrix of moving lines is linear in X_0, X_1, X_2, and the implicit equation has typically degree d. So, the method will work optimally with a matrix of size $d \times d$, and essentially you will be computing the Sylvester matrix of a μ-basis of ϕ.

3   Moving Conics, Moving Cubics...

One can actually take advantage of the resultant formulation given in (12) and get a determinantal formula for the implicit equation by using the square matrix

$$\text{Bez}_{t_0,t_1}\big(\mathcal{P}_\mu(t_0,t_1,X_0,X_1,X_2),\mathcal{Q}_{d-\mu}(t_0,t_1,X_0,X_1,X_2)\big),$$

which has smaller size (it will have $d - \mu$ rows and columns) than the Sylvester matrix of these polynomials. But this will not be a matrix of coefficients of moving lines anymore, as the input coefficients of the Bézout matrix will be quadratic in X_0, X_1, X_2. Yet, due to the way the Bézout matrix is being built (see for instance [SGD97], one can find in the rows of this matrix the coefficients of a polynomial which also vanishes on the parametrization ϕ. This motivates the following definition:

Definition 3. A moving curve *of bidegree* (ν, δ) *which follows the parametrization* ϕ *is a polynomial* $\mathcal{L}_{\nu,\delta}(t_0, t_1, X_0, X_1, X_2) \in \mathbb{K}[t_0, t_1, X_0, X_1, X_2]$ *homogeneous in* X_0, X_1, X_2 *of degree* ν *and in* t_0, t_1 *of degree* δ, *such that*

$$\mathcal{L}(t_0, t_1, u_0(t_0, t_1), u_1(t_0, t_1), u_2(t_0, t_1)) = 0.$$

If $\nu = 1$ we recover the definition of moving lines given in 1. For $\nu = 2$, the polynomial $\mathcal{L}(t_0, t_1, X_0, X_1, X_2)$ is called a *moving conic* which follows ϕ [ZCG99]. *Moving cubics* will be curves with $\nu = 3$, and so on.

A series of experiments made by Sederberg and his collaborators showed something interesting: one can compute the defining polynomial of $\mathcal{C}$ as a determinant of a matrix of coefficients of moving curves following the parametrization, but the more singular the curve is (i.e. the more singular points it has), the smaller the determinant of moving curves becomes. For instance, the following result appears in [SC95]:

Theorem 5. *The implicit equation of a quartic curve with no base points can be written as a* 2×2 *determinant. If the curve doesn't have a triple point, then each element of the determinant is quadratic; otherwise one row is linear and the other is cubic.*

To illustrate this, we consider the following examples.

Example 4. Set $u_0(t_0, t_1) = t_0^4 - t_1^4$, $u_1(t_0, t_1) = -t_0^2 t_1^2$, $u_2(t_0, t_1) = t_0 t_1^3$. These polynomials define a parametrization ϕ as in (1) with implicit equation given by the polynomial $F(X_0, X_1, X_2) = X_2^4 - X_1^4 - X_0 X_1 X_2^2$. From the shape of this polynomial, it is easy to show that $(1 : 0 : 0) \in \mathbb{P}^2$ is a point of multiplicity 3 of this curve, see Fig. 6. In this case, we have $\mu = 1$, and it is also easy to verify that

$$\mathcal{L}_{1,1}(t_0, t_1, X_0, X_1, X_2) = t_0 X_2 + t_1 X_1$$

is a moving line which follows ϕ. The reader will now easily check that the following moving curve of bidegree $(3, 1)$ also follows ϕ:

$$\mathcal{L}_{1,3}(t_0, t_1, X_0, X_1, X_2) = t_0(X_1^3 + X_0 X_2^2) + t_1 X_2^3.$$

And the 2×2 matrix claimed in Theorem 5 for this case is made with the coefficients of both $\mathcal{L}_{1,1}(t_0, t_1, X_0, X_1, X_2)$ and $\mathcal{L}_{1,3}(t_0, t_1, X_0, X_1, X_2)$ as polynomials in t_0, t_1:

$$\begin{pmatrix} X_2 & X_1 \\ X_1^3 + X_0 X_2^2 & X_2^3 \end{pmatrix}.$$

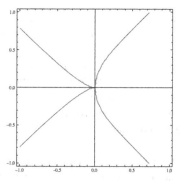

Fig. 6. The curve of Example 4.

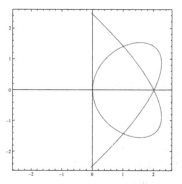

Fig. 7. The curve of Example 5.

Example 5. We reproduce here Example 2.7 in [Cox08]. Consider

$$u_0(t_0, t_1) = t_0^4, \ u_1(t_0, t_1) = 6t_0^2 t_1^2 - 4t_1^4, \ u_2(t_0, t_1) = 4t_0^3 t_1 - 4t_0 t_1^3.$$

This input defines a quartic curve with three nodes, with implicit equation given by $F(X_0, X_1, X_2) = X_2^4 + 4X_0 X_1^3 + 2X_0 X_1 X_2^2 - 16X_0^2 X_1^2 - 6X_0^2 X_2^2 + 16X_0^3 X_1$, see Fig. 7.

The following two moving conics of degree 1 in t_0, t_1 follow the parametrization:

$$\mathcal{L}_{1,2}(t_0, t_1, X_0, X_1, X_2) = t_0(X_1 X_2 - X_0 X_2) + t_1(-X_2^2 - 2X_0 X_1 + 4X_0^2)$$

$$\tilde{\mathcal{L}}_{1,2}(t_0, t_1, X_0, X_1, X_2) = t_0(X_1^2 + \frac{1}{2}X_2^2 - 2X_0 X_1) + t_1(X_0 X_2 - X_1 X_2).$$

As in the previous example, the 2×2 matrix of the coefficients of these moving conics is the matrix claimed in Theorem 5.

4   The Moving Curve Ideal of ϕ

Now it is time to introduce some tools from Commutative Algebra which will help us understand all the geometric constructions defined above. The set of all moving curves following a given parametrization generates a *bi-homogeneous* ideal in $\mathbb{K}[t_0, t_1, X_0, X_1, X_2]$, which we will call the *moving curve ideal* of this parametrization.

As explained above, the method of moving curves for implicitization of a rational parametrization looks for small determinants made with coefficients of moving curves which follow the parametrization of low degree in t_0, t_1. To do this, one would like to have a description as in Theorem 1, of a set of "minimal" moving curves from which we can describe in an easy way all the other elements of the moving curve ideal.

Fortunately, Commutative Algebra provides the adequate language and tools for dealing with this problem. As it was shown by David Cox in [Cox08], all we have to do is look for minimal generators of the kernel $\mathcal{K}$ of the following morphism of rings:

$$
\begin{array}{rcll}
\mathbb{K}[t_0, t_1, X_0, X_1, X_2] & \longrightarrow & \mathbb{K}[t_0, t_1, z] & \\
t_i & \longmapsto & t_i & i = 0, 1, \\
X_j & \longmapsto & u_j(t_0, t_1)\, z & j = 0, 1, 2.
\end{array} \tag{13}
$$

Here, z is a new variable. The following result appears in [Cox08, Nice Fact 2.4] (see also [BJ03] for the case when ϕ is not generically injective):

Theorem 6. $\mathcal{K}$ *is the moving curve ideal of* ϕ.

Let us say some words about the map (13). Denote with $I \subset \mathbb{K}[t_0, t_1]$ the ideal generated by $u_0(t_0, t_1)$, $u_1(t_0, t_1)$, $u_2(t_0, t_1)$. The image of (13) is actually isomorphic to $\mathbb{K}[t_0, t_1][z\, I]$, which is called the *Rees Algebra* of I. By the Isomorphism Theorem, we then get that $\mathbb{K}[t_0, t_1, X_0, X_1, X_2]/\mathcal{K}$ is isomorphic to the Rees Algebra of I. This is why the generators of $\mathcal{K}$ are called the *defining equations* of the Rees Algebra of I. The Rees Algebra that appears in the moving lines method corresponds to the blow-up of $V(I)$, the variety defined by I. Geometrically, it is just the blow-up of the empty space (the effect of this blow-up is just to introduce torsion), but yet the construction should explain somehow why moving curves are sensitive to the presence of complicated singularities. It is somehow strange that the fact that the description of $\mathcal{K}$ actually gets much simpler if the singularities of $\mathcal{C}$ are more entangled.

Let us show this with an example. It has been shown in [Bus09] that, by unravelling some duality theory developed by Jouanolou in [Jou97], that for any proper parametrization of a curve of degree d having $\mu = 2$ and only cusps as singular points, the kernel $\mathcal{K}$ has $\frac{(d+1)(d-4)}{2} + 5$ minimal generators. On the other hand, in a joint work with Teresa Cortadellas [CD13b] (see also [KPU13]), we have shown that if $\mu = 2$ and there is a point of very high multiplicity (it can be proven that if the multiplicity of a point is larger than 3 when $\mu = 2$, then it must be equal to $d - 2$), then the number of generators drops to $\lfloor \frac{d+6}{2} \rfloor$, i.e. the description of $\mathcal{K}$ is simpler in this case. In both cases, these generators can be made explicit, see [Bus09, CD13b, KPU13].

Further evidence supporting this claim is what is already known for the case $\mu = 1$, which was one of the first one being worked out by several authors:

[HSV08, CHW08, Bus09, CD10]. It turns out (cf. [CD10, Corollary 2.2]) that $\mu = 1$ if and only if the parametrization is proper (i.e. generically injective), and there is a point on $\mathcal{C}$ which has multiplicity $d - 1$, which is the maximal multiplicity a point can have on a curve of degree d. If this is the case, a set of minimal generators of the kernel has exactly $d + 1$ elements.

In both cases ($\mu = 1$ and $\mu = 2$), explicit elements of a set of minimal generators of $\mathcal{K}$ can be given in terms of the input parametrization. But in general, very little is known about how many are them and which are their bidegrees. Let $n_0(\mathcal{K})$ be the 0-th *Betti number* of $\mathcal{K}$ (i.e. the cardinal of any minimal set of generators of $\mathcal{K}$). We propose the following problem which is the subject of attention of several researchers at the moment.

Problem 1. Describe *all* the possible values of $n_0(\mathcal{K})$ and the parameters that this function depends on, for a proper parametrization ϕ as in (1).

Recall that "proper" here means "generically injective". For instance, we just have shown above that, for $\mu = 1$, $n_0(\mu) = d + 1$. If $\mu = 2$, the value of $n_0(\mathcal{K})$ depends on whether there is a very singular point or not. Is n_0 a function of only d, μ and the multiplicity structure of $\mathcal{C}$?

A more ambitious problem of course is the following. Let $\mathcal{B}(\mathcal{K}) \subset \mathbb{N}^2$ be the (multi)-set of bidegrees of a minimal set of generators of $\mathcal{K}$.

Problem 2. Describe *all* the possible values of $\mathcal{B}(\mathcal{K})$.

For instance, if $\mu = 1$, we have that (see [CD10, Theorem 2.9])

$$\mathcal{B}(\mathcal{K}) = \{(0, d), (1, 1), (1, d - 1), (2, d - 2), \dots, (d - 1, 1)\}.$$

Explicit descriptions of $\mathcal{B}(\mathcal{K})$ have been done also for $\mu = 2$ in [Bus09, CD13b, KPU13]. In this case, the value of $\mathcal{B}(\mathcal{K})$ depends on whether the parametrization has singular point of multiplicity $d - 2$ or not.

For $\mu = 3$ the situation gets a bit more complicated as we have found in [CD13b]: consider the parametrizations ϕ_1 and ϕ_2 whose μ-bases are respectively:

$$P_{3,1}(t_0, t_1, X_0, X_1, X_2) = t_0^3 X_0 + (t_1^3 - t_0 t_1^2) X_1$$
$$Q_{7,1}(t_0, t_1, X_0, X_1, X_2) = (t_0^6 t_1 - t_0^2 t_1^5) X_0 + (t_0^4 t_1^3 + t_0^2 t_1^5) X_1 + (t_0^7 + t_1^7) X_2,$$

$$P_{3,2}(t_0, t_1, X_0, X_1, X_2) = (t_0^3 - t_0^2 t_1) X_0 + (t_1^3 + t_0 t_1^2 - t_0^2 t_1) X_1$$
$$Q_{7,2}(t_0, t_1, X_0, X_1, X_2) = (t_0^6 t_1 - t_0^2 t_1^5) X_0 + (t_0^4 t_1^3 + t_0^2 t_1^5) X_1 + (t_0^7 + t_1^7) X_2.$$

Each of them parametrizes properly a rational plane curve of degree 10 having the point $(0 : 0 : 1)$ with multiplicity 7. The rest of them are either double or triple points. Set $\mathcal{K}_1$ and $\mathcal{K}_2$ for the respective kernels, we have then

$$\mathcal{B}(\mathcal{K}_1) = \{(3, 1), (7, 1), (2, 3), (2, 3), (4, 2), (2, 4), (1, 6), (1, 6), (1, 6), (0, 10)\},$$
$$\mathcal{B}(\mathcal{K}_2) = \{(3, 1), (7, 1), (2, 3), (2, 3), (4, 2), (2, 4), (\mathbf{1, 5}), (1, 6), (1, 6), (0, 10)\}.$$

The parameters to find in the description of $n_0(\mathcal{K})$ proposed in Problem 1 may be more than μ and the multiplicities of the curve. For instance, in [CD13], we have shown that if there is a minimal generator of bidegree $(1, 2)$ in $\mathcal{K}$, then the whole set $\mathcal{B}(\mathcal{K})$ is constant, and equal to

$$\begin{cases} \{(0, d), (1, 2), (1, d - 2), (2, d - 4), \ldots, (\frac{d-1}{2}, 1), (\frac{d+1}{2}, 1)\} & \text{if } d \text{ is odd} \\ \{(0, d), (1, 2), (1, d - 2), (2, d - 4), \ldots, (\frac{d}{2}, 1), (\frac{d}{2}, 1)\} & \text{if } d \text{ is even.} \end{cases}$$

To put the two problems above in a more formal context, we proceed as in [CSC98, Sect. 3]: For $d \geq 1$, denote with $\mathbb{V}_d \subset \mathbb{K}[t_0, t_1]_d^3$ the set of triples of homogeneous polynomials $(u_0(t_0, t_1), u_1(t_0, t_1), u_2(t_0, t_1))$ defining a proper parametrization ϕ as in (1). Note that one can regard $\mathbb{V}_d$ as an open set in an algebraic variety in the space of parameters. Moreover, $\mathbb{V}_d$ could actually be taken as a quotient of $\mathbb{K}[t_0, t_1]_d^3$ via the action of $\mathrm{SL}(2, \mathbb{K})$ acting on the monomials t_0, t_1.

Problem 3. Describe the subsets of $\mathbb{V}_d$ where $\mathcal{B}(\mathcal{K})$ is constant.

Note that, naturally the μ-basis is contained in $\mathcal{K}$, and moreover, we have (see [BJ03, Proposition 3.6]):

$$\mathcal{K} = \langle \mathcal{P}_\mu(t_0, t_1, X_0, X_1, X_2), \mathcal{Q}_{d-\mu}(t_0, t_1, X_0, X_1, X_2) \rangle : \langle t_0, t_1 \rangle^\infty,$$

so the role of the μ-basis is crucial to understand $\mathcal{K}$. Indeed, any minimal set of generators of $\mathcal{K}$ contains a μ-basis, so the pairs $(1, \mu)$, $(1, d - \mu)$ are always elements of $B(\mathcal{K})$. The study of the geometry of $\mathbb{V}_d$ according to the stratification done by μ has been done in [CSC98, Sect. 3] (see also [DAn04, Iar13]). Also, in [CKPU13], a very interesting study of how the μ-basis of a parametrization having generic μ ($\mu = \lfloor d/2 \rfloor$) and very singular points looks like has been made. It would be interesting to have similar results for $\mathcal{K}$.

In this context, one could give a positive answer to the experimental evidence provided by Sederberg and his collaborators about the fact that "the more singular the curve, the simpler the description of K" as follows. For $\mathbb{W} \subset \mathbb{V}_d$, we denote by $\overline{\mathbb{W}}$ the closure of $\mathbb{W}$ with respect to the Zariski topology.

Conjecture 1. If $\mathbb{W}_1, \mathbb{W}_2 \subset \mathbb{V}_d$ are such that $n_0|_{\mathbb{W}_i}$ is constant for $i = 1, 2$, and $\overline{\mathbb{W}}_1 \subset \overline{\mathbb{W}}_2$, then

$$n_0(\mathbb{W}_1) \leq n_0(\mathbb{W}_2).$$

Note that this condition is equivalent to the fact that $n_0(\mathcal{K})$ is *upper semicontinuous* on $\mathbb{V}_d$ with its Zariski topology. Very related to this conjecture is the following claim, which essentially asserts that in the "generic" case, we obtain the largest value of $n_0(\mathcal{K})$:

Conjecture 2. Let $\mathbb{W}_d$ be open set of $\mathbb{V}_d$ parametrizing all the curves with $\mu = \lfloor d/2 \rfloor$, and having all its singular points being ordinary double points. Then, $n_0(\mathcal{K})$ is constant on $\mathbb{W}_d$, and attains its maximal value on $\mathbb{V}_d$ in this component.

Note that a "refinement" of Conjecture 1 with $\mathcal{B}(\mathcal{K}_1) \subset \mathcal{B}(\mathcal{K}_2)$ will not hold in general, as in the examples computed for $\mu = 2$ in [Bus09, CD13b, KPU13] show. Indeed, we have in this case that the Zariski closure of those parametrizations with a point of multiplicity $d - 2$ is contained in the case where all the points are only cusps, but the bidegrees of the minimal generators of $\mathcal{K}$ in the case of parametrizations with points of multiplicity $d - 2$ appear at lower values than the more general case (only cusps).

5 Why Rational Plane Curves Only?

All along this text we were working with the parametrization of a rational plane curve, but most of the concepts, methods and properties worked out here can be extended in two different directions. The obvious one is to consider "surface" parametrizations, that is maps of the form

$$\phi_S : \quad \mathbb{P}^2 \quad \dashrightarrow \quad \mathbb{P}^3$$
$$(t_0 : t_1 : t_2) \longmapsto \left(u_0(t_0, t_1, t_2) : u_1(t_0, t_1, t_2) : u_2(t_0, t_1, t_2) : u_3(t_0, t_1, t_2) \right)$$
$$(14)$$

where $u_i(t_0, t_1, t_2) \in \mathbb{K}[t_0, t_1, t_2]$, $i = 0, 1, 2, 3$, are homogeneous of the same degree, and without common factors. Obviously, one can do this in higher dimensions also, but we will restrict the presentation just to this case. The reason we have now a dashed arrow in (14) is because even with the conditions imposed upon the u_i's, the map may not be defined on all points of $\mathbb{P}^2$. For instance, if

$$u_0(t_0, t_1, t_2) = t_1 t_2, \, u_1(t_0, t_1, t_2) = t_0 t_2, \, u_2(t_0, t_1, t_2) = t_0 t_1, \, u_3(t_0, t_1, t_2) = t_0 t_1 + t_1 t_2,$$

ϕ_S will not be defined on the set $\{(1 : 0 : 0), (0 : 1 : 0), (0 : 0 : 1)\}$.

In this context, there are methods to deal with the implicitization analogues to those presented here for plane curves. For instance, one can use a *multivariate resultant* or a *sparse resultant* (as defined in [CLO05]) to compute the implicit equation of the Zariski closure of the image of ϕ_S. Other tools from Elimination Theory such as determinants of complexes can be also used to produce matrices whose determinant (or quotient or gcd of some determinants) can also be applied to compute the implicit equation, see for instance [BJ03, BCJ09].

The method of moving lines and curves presented before gets translated into a *method of moving planes and surfaces* which follows ϕ_S, and its description and validity is much more complicated, as both the Algebra and the Geometry involved have more subtleties, see [SC95, CGZ00, Cox01, BCD03, KD06]. Even though, it has been shown in [CCL05] that there exists an equivalent of a μ-basis in this context, its computation is not as easy as in the planar case. Part of the reason is that the syzygy module of general $u_i(t_0, t_1, t_2)$, $i = 0, 1, 2, 3$ is not free anymore (i.e. it does not have sense the meaning of a "basis" as we defined in the case of curves), but if one set $t_0 = 1$ and regards these polynomials as affine bivariate forms, a nicer situation appears but without control on the degrees of the elements of the μ-basis, see [CCL05, Proposition 2.1] for more on this. Some explicit descriptions have been done for either low degree parametrizations, and also for surfaces having some additional geometric features (see [CSD07, WC12, SG12, SWG12]), but the general case remains yet to be explored.

A generalization of a map like (13) to this situation is straightforward, and one can then consider the defining ideal of the Rees Algebra associated to ϕ_S. Very little seems to be known about the minimal generators of $\mathcal{K}$ in this situation. In [CD10] we studied the case of *monoid* surfaces, which are rational parametrizations with a point of the highest possible multiplicity. This situation can be regarded as a possible generalization of the case $\mu = 1$ for plane curves, and has been actually generalized to *de Jonquières* parametrizations in [HS12].

We also dealt in [CD10] (see also [HW10]) with the case where there are two linearly independent moving planes of degree 1 following the parametrization plus some geometric conditions, this may be regarded of a generalization of the "$\mu = 1$" situation for plane curves. But the general description of the defining ideal of the Rees Algebra for the surface situation is still an open an fertile area for research.

The other direction where we can go after consider rational plane parametrizations is to look at spatial curves, that is maps

$$\phi_C : \quad \mathbb{P}^1 \quad \longrightarrow \quad \mathbb{P}^3$$
$$(t_0 : t_1) \longmapsto (u_0(t_0, t_1) : u_1(t_0, t_1) : u_2(t_0, t_1) : u_3(t_0, t_1)),$$

where $u_i \in \mathbb{K}[t_0, t_1]$, homogeneous of the same degree $d \geq 1$ in $\mathbb{K}[t_0, t_1]$ without any common factor. In this case, the image of ϕ_C is a curve in $\mathbb{P}^3$, and one has to replace "an" implicit equation with "the set of" implicit equations, as there will be more than one in the same way that the implicit equations of the line joining $(1 : 0 : 0 : 1)$ and $(0 : 0 : 0 : 1)$ in $\mathbb{P}^3$ are given by the vanishing of the equations $X_1 = X_2 = 0$.

As explained in [CSC98], both Theorems 1 and 2 carry on to this situation, so there is more ground to play and theoretical tools to help with the computations. In [CKPU13], for instance, the singularities of the spatial curve are studied as a function of the shape of the μ-basis. Further computations have been done in [KPU09] to explore the generalization of the case $\mu = 1$ and produce generators for $\mathcal{K}$ in this case. These generators, however, are far from being minimal. More explorations have been done in [JG09, HWJG10, JWG10], for some specific values of the degrees of the generators of the μ-basis.

It should be also mentioned that in the recently paper [Iar13], an attempt of the stratification proposed in Problem 2 for this kind of curves is done, but only with respect to the value of μ and no further parameters.

As the reader can see, there are lots of recent work in this area, and many challenges yet to solve. We hope that in the near future we can get more and deeper insight in all these matters, and also to be able to apply these results in the Computer Aided and Visualization community.

Acknowledgments. I am grateful to Laurent Busé, Eduardo Casas-Alvero and Teresa Cortadellas Benitez for their careful reading of a preliminary version of this manuscript, and very helpful comments. Also, I thank the anonymous referee for her further comments and suggestions for improvements, and to Marta Narváez Clauss for her help with the computations of some examples. All the plots in this text have been done with `Mathematica 8.0` [Wol10].

References

[AL94] Adams, W.W., Loustaunau, P.: An Introduction to Gröbner Bases. Graduate Studies in Mathematics, vol. 3. American Mathematical Society, Providence (1994)

[AJ06] Apéry, F., Jouanolou, J.-P.: Élimination: le cas d'une variable. Collection Méthodes, Hermann Paris (2006)

[Bus09] Busé, L.: On the equations of the moving curve ideal of a rational algebraic plane curve. J. Algebra **321**(8), 2317–2344 (2009)

[BCJ09] Busé, L., Chardin, M., Jouanolou, J.-P.: Torsion of the symmetric algebra and implicitization. Proc. Amer. Math. Soc. **137**(6), 1855–1865 (2009)

[BCD03] Busé, L., Cox, D., D'Andrea, C.: Implicitization of surfaces in $\mathbb{P}^3$ in the presence of base points. J. Algebra Appl. **2**(2), 189–214 (2003)

[BD12] Busé, L., D'Andrea, C.: Singular factors of rational plane curves. J. Algebra **357**, 322–346 (2012)

[BJ03] Busé, L., Jouanolou, J.-P.: On the closed image of a rational map and the implicitization problem. J. Algebra **265**(1), 312–357 (2003)

[CDNR97] Capani, A., De Dominicis, G., Niesi, G., Robbiano, L.: Computing minimal finite free resolutions. Algorithms for algebra (Eindhoven, 1996). J. Pure Appl. Algebra **117**(118), 105–117 (1997)

[CCL05] Chen, F., Cox, D., Liu, Y.: The μ-basis and implicitization of a rational parametric surface. J. Symbolic Comput. **39**(6), 689–706 (2005)

[CSD07] Chen, F., Shen, L., Deng, J.: Implicitization and parametrization of quadratic and cubic surfaces by μ-bases. Computing **79**(2–4), 131–142 (2007)

[CD10] Cortadellas Benitez, T., D'Andrea, C.: Minimal generators of the defining ideal of the Rees Algebra associated to monoid parametrizations. Comput. Aided Geom. Des. **27**(6), 461–473 (2010)

[Cox01] Cox, D.A.: Equations of parametric curves and surfaces via syzygies. In: Symbolic computation: solving equations in algebra, geometry, and engineering (South Hadley, MA, 2000). Contemporary Mathematics, vol. 286, pp. 1–20. American Mathematical Society, Providence (2001)

[CD13] Cortadellas Benitez, T., D'Andrea, C.: Rational plane curves parametrizable by conics. J. Algebra **373**, 453–480 (2013)

[CD13b] Cortadellas Benitez, T., D'Andrea, C.: Minimal generators of the defining ideal of the Rees Algebra associated to a rational plane parameterization with $\mu = 2$. Canad. J. Math. **66**(6), 1225–1249 (2014)

[Cox08] Cox, D.A.: The moving curve ideal and the Rees algebra. Theoret. Comput. Sci. **392**(1–3), 23–36 (2008)

[CGZ00] Cox, D., Goldman, R., Zhang, M.: On the validity of implicitization by moving quadrics of rational surfaces with no base points. J. Symbolic Comput. **29**(3), 419–440 (2000)

[CHW08] Cox, D., Hoffman, J.W., Wang, H.: Syzygies and the Rees algebra. J. Pure Appl. Algebra **212**(7), 1787–1796 (2008)

[CKPU13] Cox, D., Kustin, A., Polini, C., Ulrich, B.: A study of singularities on rational curves via syzygies. vol. 222, Number 1045. American Mathematical Society, Providence (2013)

[CLO05] Cox, D.A., Little, J., O'Shea, D.: Using Algebraic Geometry. Graduate Texts in Mathematics, vol. 185, 2nd edn. Springer, New York (2005)

[CLO07] Cox, D., Little, J., O'Shea, D.: Ideals, Varieties, and Algorithms: An intro-duction to Computational Algebraic Geometry and Commutative Algebra. Undergraduate Texts in Mathematics, 3rd edn. Springer, New York (2007)

[CSC98] Cox, D.A., Sederberg, T.W., Chen, F.: The moving line ideal basis of planar rational curves. Comput. Aided Geom. D. **15**(8), 803–827 (1998)

[DAn04] D'Andrea, C.: On the structure of μ-classes. Comm. Algebra **32**(1), 159–165 (2004)

[Eis95] Eisenbud, D.: Commutative Algebra: With a View Toward Algebraic Geometry. Graduate Texts in Mathematics, vol. 150. Springer, New York (1995)

[HS12] Hassanzadeh, S.H., Simis, A.: Implicitization of the Jonquières parame-trizations. J. Commut. Algebra **6**(2), 149–172 (2014)

[HW10] Hoffman, J.W., Wang, H.: Defining equations of the Rees algebra of certain parametric surfaces. J. Algebra Appl. **9**(6), 1033–1049 (2010)

[HWJG10] Hoffman, J.W., Wang, H., Jia, X., Goldman, R.: Minimal generators for the Rees algebra of rational space curves of type $(1, 1, d - 2)$.. Eur. J. Pure Appl. Math. **3**(4), 602–632 (2010)

[HSV08] Hong, J., Simis, A., Vasconcelos, W.V.: On the homology of two-dimensional elimination. J. Symbolic Comput. **43**(4), 275–292 (2008)

[Iar13] Iarrobino, A.: Strata of vector spaces of forms in $k[x, y]$ and of rational curves in $\mathbb{P}^k$. arXiv:1306.1282

[JG09] Jia, X., Goldman, R.: μ-bases and singularities of rational planar curves. Comput. Aided Geom. Des. **26**(9), 970–988 (2009)

[JWG10] Jia, X., Wang, H., Goldman, R.: Set-theoretic generators of rational space curves. J. Symbolic Comput. **45**(4), 414–433 (2010)

[Jou97] Jouanolou, J.P.: Formes d'inertie et résultant: un formulaire. Adv. Math. **126**(2), 119–250 (1997)

[KD06] Khetan, A., D'Andrea, C.: Implicitization of rational surfaces using toric varieties. J. Algebra **303**(2), 543–565 (2006)

[KPU09] Kustin, A.R., Polini, C., Ulrich, B.: Rational normal scrolls and the defining equations of Rees Algebras. J. Reine Angew. Math. **650**, 23–65 (2011)

[KPU13] Kustin, A., Polini, C., Ulrich, B.: The bi-graded structure of Symmetric Algebras with applications to Rees rings. arXiv:1301.7106

[SC95] Sederberg, T., Chen, F.: Implicitization using moving curves and surfaces. In: Proceedings of SIGGRAPH, pp. 301–308 (1995)

[SGD97] Sederberg, T., Goldman, R., Du, H.: Implicitizing rational curves by the method of moving algebraic curves. J. Symbolic Comput. **23**(2–3), 153–175 (1997). Parametric algebraic curves and applications (Albuquerque, NM, 1995)

[STD94] Sederberg, T.W., Saito, T., Qi, D., Klimaszewski, K.S.: Curve implicitiza-tion using moving lines. Comput. Aided Geom. Des. **11**, 687–706 (1994)

[SWP08] Sendra, J.R., Winkler, F., Pérez-Díaz, S.: Rational Algebraic Curves: A Computer Algebra Approach. Algorithms and Computation in Mathemat-ics, vol. 22. Springer, Berlin (2008)

[SG12] Shi, X., Goldman, R.: Implicitizing rational surfaces of revolution using μ-bases. Comput. Aided Geom. Des. **29**(6), 348–362 (2012)

[SWG12] Shi, X., Wang, X., Goldman, R.: Using μ-bases to implicitize rational sur-faces with a pair of orthogonal directrices. Comput. Aided Geom. Des. **29**(7), 541–554 (2012)

[vdW66] van der Waerden, B.L.: Modern Algebra, vol. 1, 2nd edn. Frederick Ungar, New York (1966)

[WC12] Wang, X., Chen, F.: Implicitization, parameterization and singularity computation of Steiner surfaces using moving surfaces. J. Symbolic Comput. **47**(6), 733–750 (2012)

[Wol10] Wolfram Research, Inc. Mathematica, Version 8.0, Champaign, IL (2010)

[ZCG99] Zhang, M., Chionh, E.-W., Goldman, R.N.: On a relationship between the moving line and moving conic coefficient matrices. Comput. Aided Geom. Des. **16**(6), 517–527 (1999)

Survey on Counting Special Types
of Polynomials

Joachim von zur Gathen and Konstantin Ziegler$^{(\boxtimes)}$

B-IT, Universität Bonn, 53113 Bonn, Germany
{gathen,zieglerk}@bit.uni-bonn.de
http://cosec.bit.uni-bonn.de/

Abstract. Most integers are composite and most univariate polynomials over a finite field are reducible. The Prime Number Theorem and a classical result of Gauß count the remaining ones, approximately and exactly. For polynomials in two or more variables, the situation changes dramatically. Most multivariate polynomials are irreducible. This survey presents counting results for some special classes of multivariate polynomials over a finite field, namely the reducible ones, the s-powerful ones (divisible by the sth power of a nonconstant polynomial), the relatively irreducible ones (irreducible but reducible over an extension field), the decomposable ones, and also for reducible space curves. These come as exact formulas and as approximations with relative errors that essentially decrease exponentially in the input size.

Furthermore, a univariate polynomial f is decomposable if $f = g \circ h$ for some nonlinear polynomials g and h. It is intuitively clear that the decomposable polynomials form a small minority among all polynomials.

The tame case, where the characteristic p of $\mathbb{F}_q$ does not divide $n = \deg f$, is fairly well-understood, and we obtain closely matching upper and lower bounds on the number of decomposable polynomials. In the wild case, where p does divide n, the bounds are less satisfactory, in particular when p is the smallest prime divisor of n and divides n exactly twice. The crux of the matter is to count the number of collisions, where essentially different (g, h) yield the same f. We present a classification of all collisions at degree $n = p^2$ which yields an exact count of those decomposable polynomials.

Keywords: Counting special polynomials · Finite fields · Combinatorics on polynomials · Generating functions · Analytic combinatorics · Asymptotic behavior · Multivariate polynomials · Polynomial decomposition · Ritt's Second Theorem

2010 Mathematics Subject Classification. 00B25, 11T06, 12Y05

1   Introduction

Most integers are composite and most univariate polynomials over a finite field are reducible. The classical results of the Prime Number Theorem and a theorem

J. Gutierrez et al. (Eds.): Computer Algebra and Polynomials, LNCS 8942, pp. 50–75, 2015.
DOI: 10.1007/978-3-319-15081-9_3

of Gauß present approximations saying that randomly chosen integers up to x or polynomials of degree up to n are prime or irreducible with probability about $1/\ln x$ or $1/n$, respectively.

Concerning special classes of univariate polynomials over a finite field, Zsigmondy (1894) counts those with a given number of distinct roots or without irreducible factors of a given degree. In the same situation, Artin (1924) counts the irreducible ones in an arithmetic progression and Hayes (1965) generalizes these results. Cohen (1969) and Car (1987) count polynomials with certain factorization patterns and Williams (1969) those with irreducible factors of given degree. Polynomials that occur as a norm in field extensions are studied by Gogia and Luthar (1981).

In two or more variables, the situation changes dramatically. Most multivariate polynomials are irreducible. Carlitz (1963) provides the first count of irreducible multivariate polynomials. In Carlitz (1965), he goes on to study the fraction of irreducibles when bounds on the degrees in each variable are prescribed; see also Cohen (1968). In this survey, we opt for bounding the total degree because it has the charm of being invariant under invertible linear transformations. Gao and Lauder (2002) consider the counting problem in yet another model, namely where one variable occurs with maximal degree. The natural generating function (or zeta function) for the irreducible polynomials in two or more variables does not converge anywhere outside of the origin. Wan (1992) notes that this explains the lack of a simple combinatorial formula for the number of irreducible polynomials. But he gives a p-adic formula, and also a (somewhat complicated) combinatorial formula. For further references, see Mullen and Panario (2013, Sect. 3.6).

In the bivariate case, von zur Gathen (2008) proves precise approximations with an exponentially decreasing relative error. Von zur Gathen et al. (2013) extend those results to multivariate polynomials and give further information such as exact formulas and generating functions. Bodin (2008) gives a recursive formula for the number of irreducible bivariate polynomials and remarks on a generalization for more than two variables; he follows up with Bodin (2010).

We present exact formulas for the numbers of reducible (Sects. 2.1–2.3), s-powerful (Sect. 2.4), and relatively irreducible polynomials (Sect. 2.5). The formulas also yield simple, yet precise, approximations to these numbers, with rapidly decaying relative errors.

Geometrically, a single polynomial corresponds to a hypersurface, that is, to a cycle in affine or projective space of codimension 1. This correspondence preserves the respective notions of reducibility. Thus, Sects. 2.1–2.3 can also be viewed as counting reducible hypersurfaces, in particular planar curves, and Sect. 2.4 those with an s-fold component. From a geometric perspective, these results say that almost all hypersurfaces are irreducible. Can we say something similar for other types of varieties? Cesaratto et al. (2013) give an affirmative answer for curves in $\mathbb{P}^r$ for arbitrary r. A first question is how to parametrize the curves. Moduli spaces only include irreducible curves, and systems of defining equations do not work except for complete intersections. The natural parametrization is by the Chow variety $C_{r,n}$ of curves of degree n in $\mathbb{P}^r$, for some fixed r and n. The foundation of this approach is a result by Eisenbud and Harris (1992), who identified the irreducible components of $C_{r,n}$ of maximal dimension. We present the counting results in Sect. 2.6.

It is intuitively clear that the decomposable polynomials form a small minority among all multivariate polynomials over a field. Von zur Gathen (2011) gives a quantitative version of this intuition (see Sect. 2.7). The number of multivariate decomposable polynomials is also studied by Bodin et al. (2009).

This concludes the first half (Sect. 2) of our survey, dealing with multivariate polynomials. The second half (Sect. 3) is devoted to counting univariate decomposable polynomials.

Some of the results in this survey are from joint work with Raoul Blankertz, Eda Cesaratto, Mark Giesbrecht, Guillermo Matera, and Alfredo Viola.

2   Counting Multivariate Polynomials

We work in the polynomial ring $F[x_1, \ldots, x_r]$ in $r \geq 1$ variables over a field F and consider polynomials with total degree equal to some nonnegative integer n:

$$P_{r,n}^{\text{all}}(F) = \{f \in F[x_1, \ldots, x_r] \colon \deg f = n\}.$$

The polynomials of degree at most n form an F-vector space of dimension $\binom{r+n}{r}$.

The property of a certain polynomial to be reducible, squareful, relatively irreducible, or decomposable is shared with all polynomials associated to the given one. For counting them, it is sufficient to take one representative. We choose an arbitrary monomial order, say, the degree-lexicographic one, so that the monic polynomials are those with leading coefficient 1, and write

$$P_{r,n}(F) = \{f \in P_{r,n}^{\text{all}}(F) \colon f \text{ is monic}\}.$$

We use two different methodologies to obtain such bounds: generating functions and combinatorial counting. The usual approach, see Flajolet and Sedgewick (2009), of analytic combinatorics on series with integer coefficients leads, in our case, to power series that diverge everywhere (except at 0). We have not found a way to make this work. Instead, we use power series with symbolic coefficients, namely rational functions in a variable representing the field size. Several useful relations from standard analytic combinatorics carry over to this new scenario. In a first step, this yields in a straightforward manner an exact formula for the number under consideration (Theorem 5). This formula is, however, not very transparent. Even the leading term is not immediately visible.

In a second step, coefficient comparisons yield easy-to-use approximations to our number (Theorem 7). The relative error is exponentially decreasing in the bit size of the data. Thus, Theorem 7 gives a "third order" approximation for the number of reducible polynomials, and thus a "fourth order" approximation for the irreducible ones. The error term is in the big-Oh form and thus contains an unspecified constant.

In a third step, a different method, namely some combinatorial counting, yields "second order" approximations with explicit constants in the error term (Theorem 9).

The results of Sects. 2.1–2.5 are from von zur Gathen et al. (2013) unless otherwise attributed, those of Sect. 2.6 are from Cesaratto et al. (2013), and those of Sect. 2.7 are from von zur Gathen (2011).

2.1 Exact Formula for Reducible Polynomials

To study reducible polynomials, we consider the following subsets of $P_{r,n}(F)$:

$$I_{r,n}(F) = \{f \in P_{r,n}(F): f \text{ is irreducible}\},$$
$$R_{r,n}(F) = P_{r,n}(F) \setminus I_{r,n}(F).$$

In the usual notions, the polynomial 1 is neither reducible nor irreducible. In our context, it is natural to have $R_{r,0}(F) = \{1\}$ and $I_{r,0}(F) = \varnothing$.

The sets of polynomials

$$\mathcal{P}_r = \bigcup_{n \geq 0} P_{r,n}(\mathbb{F}_q),$$

$$\mathcal{I}_r = \bigcup_{n \geq 0} I_{r,n}(\mathbb{F}_q),$$

$$\mathcal{R}_r = \mathcal{P}_r \setminus \mathcal{I}_r,$$

are combinatorial classes with the total degree as size functions and we denote the corresponding generating functions by $\mathrm{P}_r, \mathrm{I}_r, \mathrm{R}_r \in \mathbb{Z}_{\geq 0}[\![z]\!]$, respectively. Their coefficients are

$$\mathrm{P}_{r,n} = \#P_{r,n}(\mathbb{F}_q) = q^{\binom{r+n}{r}-1} \frac{1 - q^{-\binom{r+n-1}{r-1}}}{1 - q^{-1}}, \tag{1}$$

$$\mathrm{R}_{r,n} = \#R_{r,n}(\mathbb{F}_q),$$
$$\mathrm{I}_{r,n} = \#I_{r,n}(\mathbb{F}_q), \tag{2}$$

respectively, dropping the finite field $\mathbb{F}_q$ with q elements from the notation. By definition, $\mathcal{P}_r$ equals the disjoint union of $\mathcal{R}_r$ and $\mathcal{I}_r$, and therefore

$$\mathrm{R}_r = \mathrm{P}_r - \mathrm{I}_r.$$

By unique factorization, every element in $\mathcal{P}_r$ corresponds to an unordered finite sequence of elements in $\mathcal{I}_r$, where repetition is allowed, and therefore

$$\mathrm{I}_r = \sum_{k \geq 1} \frac{\mu(k)}{k} \log \mathrm{P}_r(z^k) \tag{3}$$

by Flajolet and Sedgewick (2009, Theorem I.5), where μ is the number-theoretic Möbius function. A resulting algorithm is easy to program and returns exact results with lightning speed.

This approach quickly leads to explicit formulas. A *composition* of a positive integer n is a sequence $j = (j_1, j_2, \ldots, j_{|j|})$ of positive integers $j_1, j_2, \ldots, j_{|j|}$ with $j_1 + j_2 + \cdots + j_{|j|} = n$, where $|j|$ denotes the length of the sequence. We define the set

$$M_n = \{\text{compositions of } n\}. \tag{4}$$

This standard combinatorial notion is not to be confused with the composition of polynomials, which we discuss in Sects. 2.7 and 3.

Theorem 5 (Exact counting). *Let $r \geq 1$, $q \geq 2$, $P_{r,n}$ as in (1), and $I_{r,n}$ the number of irreducible monic r-variate polynomials of degree n over $\mathbb{F}_q$. Then we have*

$$I_{r,0} = 0,$$

$$I_{r,n} = -\sum_{k \mid n} \frac{\mu(k)}{k} \sum_{j \in M_{n/k}} \frac{(-1)^{|j|}}{|j|} P_{r,j_1} P_{r,j_2} \cdots P_{r,j_{|j|}},$$

for $n \geq 1$, and therefore for the number $R_{r,n}$ of reducible monic r-variate polynomials of degree n over $\mathbb{F}_q$

$$R_{r,0} = 1,$$

$$R_{r,n} = P_{r,n} + \sum_{k \mid n} \frac{\mu(k)}{k} \sum_{j \in M_{n/k}} \frac{(-1)^{|j|}}{|j|} P_{r,j_1} P_{r,j_2} \cdots P_{r,j_{|j|}},$$

for $n \geq 1$.

The formula of Theorem 5 is exact but somewhat cumbersome. The following two sections provide simple yet precise approximations with rapidly decaying error terms.

2.2 Symbolic Approximation for Reducible Polynomials

For $r \geq 2$, the power series P_r, I_r, and R_r do not converge anywhere except at 0, and the standard asymptotic arguments of analytic combinatorics are inapplicable. We now deviate from this approach and move from power series in $\mathbb{Q}[\![z]\!]$ to power series in $\mathbb{Q}(\mathbf{q})[\![z]\!]$, where $\mathbf{q}$ is a symbolic variable representing the field size. For $r \geq 2$ and $n \geq 0$ we let

$$P_{r,n}(\mathbf{q}) = \mathbf{q}^{\binom{r+n}{r}-1} \frac{1 - \mathbf{q}^{-\binom{r+n-1}{r-1}}}{1 - \mathbf{q}^{-1}} \in \mathbb{Z}[\mathbf{q}]$$

in analogy to (1). We define the power series $P_r, I_r, R_r \in \mathbb{Q}(\mathbf{q})[\![z]\!]$ by

$$P_r(\mathbf{q}, z) = \sum_{n \geq 0} P_{r,n}(\mathbf{q}) z^n, \tag{6}$$

$$I_r(\mathbf{q}, z) = \sum_{k \geq 1} \frac{\mu(k)}{k} \log P_r(\mathbf{q}, z^k),$$

$$R_r(\mathbf{q}, z) = P_r(\mathbf{q}, z) - I_r(\mathbf{q}, z).$$

Then $R_{r,n}(\mathbf{q})$ denotes the coefficient of z^n in R_r and counts symbolically the reducible monic r-variate polynomials of degree n.

For nonzero $f \in \mathbb{Q}(\mathbf{q})$, $\deg_{\mathbf{q}} f$ is the degree of f, that is, the numerator degree minus the denominator degree. The appearance of $O(\mathbf{q}^{-m})$ with a positive integer m in an equation means the existence of some f with degree at most $-m$ that makes the equation valid. If a term $O(\mathbf{q}^{-m})$ appears, then we may conclude a numerical asymptotic result for growing prime powers q.

Theorem 7 (Symbolic approximation). *Let $r \geq 2$ and*

$$\rho_{r,n}(\mathbf{q}) = \mathbf{q}^{\binom{r+n-1}{r}+r-1}\frac{1-\mathbf{q}^{-r}}{(1-\mathbf{q}^{-1})^2} \in \mathbb{Q}(\mathbf{q}).$$

Then the symbolic formula $R_{r,n}(\mathbf{q})$ for the number of reducible monic r-variate polynomials of degree n over $\mathbb{F}_q$ satisfies

$$R_{r,0}(\mathbf{q}) = 1, \quad R_{r,1}(\mathbf{q}) = 0, \quad R_{r,2}(\mathbf{q}) = \frac{\rho_{r,2}(\mathbf{q})}{2} \cdot (1-\mathbf{q}^{-r-1}),$$

$$R_{r,3}(\mathbf{q}) = \rho_{r,3}(\mathbf{q})\left(1-\mathbf{q}^{-r(r+1)/2} + \mathbf{q}^{-r(r-1)/2}\frac{1-2\mathbf{q}^{-r}+2\mathbf{q}^{-2r-1}-\mathbf{q}^{-2r-2}}{3(1-\mathbf{q}^{-1})}\right),$$

$$R_{r,4}(\mathbf{q}) = \rho_{r,4}(\mathbf{q}) \cdot \left(1+\mathbf{q}^{-\binom{r+1}{3}}\right) \cdot \frac{1+O(\mathbf{q}^{-r(r-1)/2})}{2(1-\mathbf{q}^{-r})},$$

and for $n \geq 5$

$$R_{r,n}(\mathbf{q}) = \rho_{r,n}(\mathbf{q}) \cdot \left(1+\mathbf{q}^{-\binom{r+n-2}{r-1}+r(r+1)/2} \cdot \frac{1+O(\mathbf{q}^{-r(r-1)/2})}{1-\mathbf{q}^{-r}}\right). \tag{8}$$

Alekseyev (2006) lists $(\#I_{r,n}(\mathbb{F}_q))_{n\geq 0}$ as A115457–A115472 in The On-Line Encyclopedia of Integer Sequences, for $2 \leq r \leq 6$ and prime $q \leq 7$. Bodin (2008, Theorem 7) states (in our notation)

$$1 - \frac{\#I_{r,n}}{\#P_{r,n}} \sim q^{-\binom{n+r-1}{r-1}-r}\frac{1-q^{-r}}{1-q^{-1}}.$$

Hou and Mullen (2009) provide results for $\#I_{r,n}(\mathbb{F}_q)$. These do not yield error bounds for the approximation of $\#R_{r,n}(\mathbb{F}_q)$. Bodin (2010) also uses (3) to claim a result similar to (8).

2.3 Explicit Bounds for Reducible Polynomials

The third approach by "combinatorial counting" is somewhat more involved. The payoff of this additional effort is an explicit relative error bound. However, the calculations are sufficiently complicated for us to stop at the first error term. Thus we replace the asymptotic $1 + O(\mathbf{q}^{-r(r-1)/2})$ in (8) by $1/(1-q^{-1})$.

Theorem 9 (Explicit approximation). *Let* $r, q \geq 2$, *and* $\rho_{r,n}$ *as in Theorem 7. For the number* $\#R_{r,n}(\mathbb{F}_q)$ *of reducible monic* r-*variate polynomials of degree* n *over* $\mathbb{F}_q$ *we have*

$$\#R_{r,0}(\mathbb{F}_q) = 1, \quad \#R_{r,1}(\mathbb{F}_q) = 0, \quad \#R_{r,2}(\mathbb{F}_q) = \frac{\rho_{r,2}(q)}{2} \cdot (1 - q^{-r-1}),$$

$$|\#R_{r,3}(\mathbb{F}_q) - \rho_{r,3}(q)| = \rho_{r,3}(q) \cdot q^{-r(r-1)/2} \frac{1 - 2q^{-r} + 2q^{-2r-1} - q^{-2r-2}}{3(1 - q^{-1})}$$

$$\leq \rho_{r,3}(q) \cdot q^{-r(r-1)/2},$$

and for $n \geq 4$

$$|\#R_{r,n}(\mathbb{F}_q) - \rho_{r,n}(q)| \leq \rho_{r,n}(q) \cdot \frac{q^{-\binom{r+n-2}{r-1}+r(r+1)/2}}{(1 - q^{-1})(1 - q^{-r})}$$

$$\leq \rho_{r,n}(q) \cdot 3q^{-\binom{r+n-2}{r-1}+r(r+1)/2}.$$

Remark 10. How close is our relative error estimate to being exponentially decaying in the input size? The usual dense representation of a polynomial in r variables and of degree n requires $b_{r,n} = \binom{r+n}{r}$ monomials, each of them equipped with a coefficient from $\mathbb{F}_q$, using about $\log_2 q$ bits. Thus the total input size is about $\log_2 q \cdot b_{r,n}$ bits. This differs from $\log_2 q \cdot (b_{r-1,n-1} - b_{r-1,2})$ by a factor of

$$\frac{b_{r,n}}{b_{r-1,n-1} - b_{r-1,2}} < \frac{b_{r,n}}{\frac{1}{2}b_{r-1,n-1}} = \frac{2(n+r)(n+r-1)}{nr}.$$

Up to this polynomial difference (in the exponent), the relative error is exponentially decaying in the bit size of the input, that is, $\log q$ times the number of coefficients in the usual dense representation. In particular, it is exponentially decaying in any of the parameters r, n, and $\log_2 q$, when the other two are fixed.

2.4   Powerful Polynomials

For an integer $s \geq 2$, a polynomial is called s-*powerful* if it is divisible by the sth power of some nonconstant polynomial, and s-*powerfree* otherwise; it is *squarefree* if $s = 2$. Let

$$Q_{r,n,s}(F) = \{f \in P_{r,n}(F) \colon f \text{ is } s\text{-powerful}\},$$
$$S_{r,n,s}(F) = P_{r,n}(F) \setminus Q_{r,n,s}(F).$$

As in the previous section, we restrict our attention to a finite field $F = \mathbb{F}_q$, which we omit from the notation.

For the approach by generating functions, we consider the combinatorial classes $\mathcal{Q}_{r,s} = \bigcup_{n \geq 0} Q_{r,n,s}$ and $\mathcal{S}_{r,s} = \mathcal{P}_r \setminus \mathcal{Q}_{r,s}$. Any monic polynomial f factors uniquely as $f = g \cdot h^s$ where g is a monic s-powerfree polynomial and h an arbitrary monic polynomial, hence

$$P_r = S_{r,s} \cdot P_r(z^s) \tag{11}$$

and by definition $Q_{r,s} = P_r - S_{r,s}$ for the generating functions of $\mathcal{S}_{r,s}$ and $\mathcal{Q}_{r,s}$, respectively. For univariate polynomials, Carlitz (1932) derives (11) directly from generating functions to prove the counting formula (13) for $r = 1$. Flajolet et al. (2001, Sect. 1.1) use (11) for $s = 2$ to count univariate squarefree polynomials, see also Flajolet and Sedgewick (2009, Note I.66).

As in Theorem 5, this approach quickly leads to explicit formulas.

Theorem 12 (Exact counting). *For $r \geq 1$, $q, s \geq 2$, $P_{r,n}$ as in (1), and M_n as in (4), we have for the number $Q_{r,n,s} = \#\mathcal{Q}_{r,n,s}(\mathbb{F}_q)$ of s-powerful monic r-variate polynomials of degree n over $\mathbb{F}_q$*

$$Q_{r,n,s} = - \sum_{\substack{1 \leq i \leq n/s \\ j \in M_i}} (-1)^{|j|} P_{r,j_1} P_{r,j_2} \cdots P_{r,j_{|j|}} P_{r,n-is}. \tag{13}$$

To study the asymptotic behavior of $Q_{r,n,s}$ for $r \geq 2$ we again deviate from the standard approach and move to power series in $\mathbb{Q}(\mathbf{q})\, [\![z]\!]$. With P_r from (6), we define $S_{r,s}, Q_{r,s} \in \mathbb{Q}(\mathbf{q})\, [\![z]\!]$ by

$$P_r = S_{r,s} \cdot P_r(z^s),$$
$$Q_{r,s} = P_r - S_{r,s}.$$

The approach by generating functions now yields the following result. Its "general" case is (iv). We give exact expressions in special cases, namely for $n < 3s$ in (ii) and for $(n, s) = (6, 2)$ in (iii), which also apply when we substitute the size q of a finite field $\mathbb{F}_q$ for $\mathbf{q}$.

Theorem 14 (Symbolic approximation). *Let $r, s \geq 2$, $n \geq 0$, and*

$$\eta_{r,n,s}(\mathbf{q}) = \mathbf{q}^{\binom{r+n-s}{r}+r-1} \frac{(1-\mathbf{q}^{-r})(1-\mathbf{q}^{-\binom{r+n-s-1}{r-1}})}{(1-\mathbf{q}^{-1})^2} \in \mathbb{Q}(\mathbf{q}),$$

$$\delta = \binom{r+n-s}{r} - \binom{r+n-2s}{r} - \frac{r(r+1)}{2}.$$

Then the symbolic formula $Q_{r,n,s}(\mathbf{q})$ for the number of s-powerful monic r-variate polynomials of degree n over $\mathbb{F}_q$ satisfies the following.

(i) If $n \geq 2s$, then $\delta \geq r$.
(ii)

$$Q_{r,n,s}(\mathbf{q}) = \begin{cases} 0 & \text{for } n < s, \\ \eta_{r,n,s}(\mathbf{q}) & \text{for } s \leq n < 2s, \\ \eta_{r,n,s}(\mathbf{q}) \left(1 + \mathbf{q}^{-\delta} \cdot \dfrac{1-\mathbf{q}^{-\binom{n+r-2s-1}{r-1}}}{1-\mathbf{q}^{-\binom{n+r-s-1}{r-1}}} \right. \\ \left. \quad \cdot \left(\dfrac{1-\mathbf{q}^{-r(r+1)/2}}{1-\mathbf{q}^{-r}} - \mathbf{q}^{-r(r-1)/2}\dfrac{1-\mathbf{q}^{-r}}{1-\mathbf{q}^{-1}}\right)\right) & \text{for } 2s \leq n < 3s. \end{cases}$$

$$\tag{15}$$

(iii) For $(n, s) = (6, 2)$, we have

$$Q_{r,6,2}(\mathbf{q}) = \eta_{r,6,2}(\mathbf{q})\big(1 + \mathbf{q}^{-\delta + (r-2)(r-1)(r+3)/6}(1 + O(\mathbf{q}^{-1}))\big). \tag{16}$$

(iv) For $n \geq 2s$ and $(n, s) \neq (6, 2)$, we have

$$Q_{r,n,s}(\mathbf{q}) = \eta_{r,n,s}(\mathbf{q})\big(1 + \mathbf{q}^{-\delta}(1 + O(\mathbf{q}^{-1}))\big).$$

For $r \geq 3$, we can replace $1 + O(\mathbf{q}^{-1})$ in (16) by $\mathbf{q}^{-1} + O(\mathbf{q}^{-2})$. The combinatorial approach replaces the asymptotic $1 + O(\mathbf{q}^{-1})$ for $n \geq 3s$ with an explicit bound. For $n < 3s$ the exact formula (15) of Theorem 14 (ii) applies.

Theorem 17 (Explicit approximation). *Let $r, s, q \geq 2$, $\#Q_{r,n,s}(\mathbb{F}_q)$ the number of s-powerful monic r-variate polynomials of degree n over $\mathbb{F}_q$, and $\eta_{r,n,s}$ and δ as in Theorem 14.*

(i) For $(n, s) = (6, 2)$, we have $\delta = r(r+1)(r^2 + 9r + 2)/24$ and

$$|\#Q_{r,6,2}(\mathbb{F}_q) - \eta_{r,6,2}(q)| \leq \eta_{r,6,2}(q) \cdot 2q^{-\delta + (r-2)(r-1)(r+3)/6}.$$

(ii) For $n \geq 3s$ and $(n, s) \neq (6, 2)$, we have

$$|\#Q_{r,n,s}(\mathbb{F}_q) - \eta_{r,n,s}(q)| \leq \eta_{r,n,s}(q) \cdot 6q^{-\delta}.$$

As noted in Remark 10 for reducible polynomials, the relative error term is (essentially) exponentially decreasing in the input size, and exponentially decaying in any of the parameters r, n, s, and $\log_2 q$, when the other three are fixed.

2.5 Relatively Irreducible Polynomials

A polynomial over F is *absolutely irreducible* if it is irreducible over an algebraic closure of F, and *relatively irreducible* (or *exceptional*) if it is irreducible over F but factors over some extension field of F. We define

$$A_{r,n}(F) = \{f \in P_{r,n}(F) \colon f \text{ is absolutely irreducible}\} \subseteq I_{r,n}(F),$$
$$E_{r,n}(F) = I_{r,n}(F) \setminus A_{r,n}(F).$$

As before, we restrict ourselves to finite fields and recall that all our polynomials are monic. We relate the generating function $A_r(\mathbb{F}_q)$ of $\#A_{r,n}(\mathbb{F}_q)$ to the generating function $I_r(\mathbb{F}_q)$ of irreducible polynomials as introduced in Sect. 2.1 and obtain

$$[z^n] I_r(\mathbb{F}_q) = \sum_{k \mid n} \frac{1}{k} \sum_{s \mid k} \mu(k/s) \cdot [z^{n/k}] A_r(\mathbb{F}_{q^s}), \tag{18}$$

$$[z^n] A_r(\mathbb{F}_q) = \sum_{k \mid n} \frac{1}{k} \sum_{s \mid k} \mu(s) \cdot [z^{n/k}] I_r(\mathbb{F}_{q^s})$$

with Möbius inversion. For an explicit formula, we combine the expression for $I_{r,n}(\mathbb{F}_q)$ from Theorem 5 with (18).

Theorem 19 (Exact counting). *For $r, n \geq 1$, $q \geq 2$, M_n as in (4), $P_{r,n}$ as in (1), and $I_{r,n}$ as in (2), we have for the number $E_{r,n}$ of relatively irreducible monic r-variate polynomials of degree n over $\mathbb{F}_q$*

$$E_{r,0}(\mathbb{F}_q) = 0,$$

$$E_{r,n}(\mathbb{F}_q) = - \sum_{1 < k \mid n} \frac{1}{k} \sum_{s \mid k} \mu(s) I_{r,n/k}(\mathbb{F}_{q^s})$$

$$= \sum_{1 < k \mid n} \frac{1}{k} \sum_{\substack{s \mid k \\ m \mid n/k}} \frac{\mu(s)\mu(m)}{m}$$

$$\cdot \sum_{j \in M_{n/(km)}} \frac{(-1)^{|j|}}{|j|} P_{r,j_1}(\mathbb{F}_{q^s}) P_{r,j_2}(\mathbb{F}_{q^s}) \cdots P_{r,j_{|j|}}(\mathbb{F}_{q^s}).$$

The approach by generating functions gives the following result.

Theorem 20 (Symbolic approximation). *Let $r, n \geq 2$, let ℓ be the smallest prime divisor of n, and*

$$\epsilon_{r,n}(\mathbf{q}) = \frac{\mathbf{q}^{\ell\left(\binom{r+n/\ell}{r}-1\right)}}{\ell(1 - \mathbf{q}^{-\ell})} \in \mathbb{Q}(\mathbf{q}),$$

$$\kappa = (\ell - 1)\left(\binom{r-1+n/\ell}{r-1} - r\right) + 1.$$

Then the symbolic formula $E_{r,n}(\mathbf{q})$ for the number of relatively irreducible monic r-variate polynomials of degree n over $\mathbb{F}_q$ satisfies the following.

(i) $E_{r,1}(\mathbf{q}) = 0$.
(ii) *If n is prime, then*

$$E_{r,n}(\mathbf{q}) = \epsilon_{r,n}(\mathbf{q})(1 - \mathbf{q}^{-nr})\left(1 - \mathbf{q}^{-r(n-1)}\frac{(1 - \mathbf{q}^{-r})(1 - \mathbf{q}^{-n})}{(1 - \mathbf{q}^{-1})(1 - \mathbf{q}^{-nr})}\right).$$

(iii) *If n is composite, then $\kappa \geq 2$ and*

$$E_{r,n}(\mathbf{q}) = \epsilon_{r,n}(\mathbf{q})(1 + O(\mathbf{q}^{-\kappa})).$$

While (i) and (ii) yield explicit bounds, the combinatorial approach does this for (iii).

Theorem 21 (Explicit approximation). *Let $r, q \geq 2$, and $\epsilon_{r,n}$ and κ as in Theorem 20, and n be composite. Then for the number $\#E_{r,n}(\mathbb{F}_q)$ of relatively irreducible monic r-variate polynomials of degree n over $\mathbb{F}_q$ we have*

$$|\#E_{r,n}(\mathbb{F}_q) - \epsilon_{r,n}(q)| \leq \epsilon_{r,n}(q) \cdot 3q^{-\kappa}.$$

2.6 Reducible Space Curves

The *Chow variety* of curves of degree n in the r-dimensional projective space $\mathbb{P}^r = \mathbb{P}^r(\overline{\mathbb{F}_q})$ over an algebraic closure $\overline{\mathbb{F}_q}$ is denoted by $C_{r,n}$. Each point of the Chow variety $C_{r,n}$ actually corresponds to a unique *effective cycle* in $\mathbb{P}^r$ of dimension 1 and degree n, that is, to a formal linear combination $\sum a_i C_i$, where each C_i is an irreducible curve in $\mathbb{P}^r$, each a_i is a positive integer, and $\sum a_i \deg(C_i) = n$.

For a subfield $F \subseteq \overline{\mathbb{F}_q}$, an effective F-cycle C is called *F-reducible* if there exist $m \geq 2$ and effective F-cycles $C_1, \ldots, C_m$ such that $C = \sum_{i=1}^{m} C_i$ holds. Let $C_{r,n}(\mathbb{F}_q)$ denote the Chow variety of effective $\mathbb{F}_q$-cycles and $R^*_{r,n}(\mathbb{F}_q)$ its closed subvariety of $\mathbb{F}_q$-reducible $\mathbb{F}_q$-cycles. Methods of algebraic geometry yield the following bounds on the probability that a random curve of degree n in $\mathbb{P}^r(\mathbb{F}_q)$ is $\mathbb{F}_q$-reducible.

Theorem 22. *Let $r \geq 3$ and*

$$g_{r,n} = \binom{r+n-2}{n}^2 \cdot \frac{r+n-1}{(r-1)(n+1)},$$

$$c_{r,n} = (2en)^{r(r+1)(n^2+1)+4rg_{r,n}},$$

where e denotes the basis of the natural logarithm. For the number $\#R^*_{r,n}(\mathbb{F}_q)$ of $\mathbb{F}_q$-reducible cycles of degree n we have the following.

(i) If $n \geq \min\{4r - 7, 7\}$, then

$$\frac{1}{4c_{r,n}} q^{-(n-2r+3)} \leq \frac{\#R^*_{r,n}(\mathbb{F}_q)}{\#C_{r,n}(\mathbb{F}_q)} \leq c_{r,n} q^{-(n-2r+3)}.$$

(ii) If $n = 4r - 8$, then

$$\frac{1}{2n!\, c_{r,n}} q^{-r+2} \leq \frac{\#R^*_{r,n}(\mathbb{F}_q)}{\#C_{r,n}(\mathbb{F}_q)} \leq c_{r,n} q^{-r+2}.$$

We call an $\overline{\mathbb{F}_q}$-reducible cycle *absolutely reducible*. An $\mathbb{F}_q$-cycle can be absolutely reducible for two reasons: either it is $\mathbb{F}_q$-reducible, as treated above, or *relatively $\mathbb{F}_q$-irreducible*, that is, is $\mathbb{F}_q$-irreducible and $\overline{\mathbb{F}_q}$-reducible. The set of relatively $\mathbb{F}_q$-irreducible (or *exceptional*) $\mathbb{F}_q$-curves of degree n in $\mathbb{P}^r$ is denoted by $E^*_{r,n}(\mathbb{F}_q)$.

Theorem 23. *Let $r \geq 3$, $n \geq 4r - 8$, let ℓ denote the smallest prime divisor of n, and*

$$b_{r,n} = 3(r-2) + n(n+3)/2,$$

$$d_{\ell,n,r} = (en/\ell)^{r(r+1)(n^2/\ell^2+1)+4rg_{r,n/\ell}}.$$

For the number $\#E^*_{r,n}(\mathbb{F}_q)$ of relatively $\mathbb{F}_q$-irreducible cycles of degree n we have

$$q^{2n(r-1)}(1 - 4q^{2(1-n)(r-1)}) \leq \#E^*_{r,n}(\mathbb{F}_q) \leq 2d_{\ell,n,r} q^{2n(r-1)} \text{ for } n/\ell \leq 4r - 7,$$

$$q^{\ell b_{r,n/\ell}}(1 - 16q^{\ell-n}) \leq \#E^*_{r,n}(\mathbb{F}_q) \leq 3d_{\ell,n,r} q^{\ell b_{r,n/\ell}} \text{ for } n/\ell \geq 4r - 8.$$

2.7 Decomposable Polynomials

For monic univariate $g \in F[y]$ and $h \in P_{r,n}$, we define their *composition*

$$f = g \circ h = g(h) \in P_{r,n}.$$

If $\deg g \geq 2$ and $\deg h \geq 1$, then (g, h) is a *decomposition* of f. A polynomial $f \in P_{r,n}$ is *decomposable* if there exist such g and h. There are other notions of decompositions. The present one is called uni-multivariate in von zur Gathen et al. (2003). Another one is studied in Faugère and Perret (2009) for cryptanalytic purposes. In the context of univariate polynomials $\deg h \geq 2$ is also required, see Sect. 3.

It is sufficient to concentrate on polynomials with vanishing constant term, see Subsect. 3.1, and we denote by $D_{r,n}(F)$ the set of all decomposable polynomials $f \in P_{r,n}(F)$ with $f(0, \ldots, 0) = 0$.

Theorem 24. *Let $\mathbb{F}_q$ be a finite field with q elements, $r \geq 2$, and ℓ the smallest prime divisor of the composite integer $n \geq 2$. Let*

$$m = \begin{cases} n \text{ if } r = 2, n/\ell \text{ is prime, and } n/\ell \leq 2\ell - 5, \\ \ell \text{ otherwise,} \end{cases}$$

$$\alpha_{r,n} = q^{\binom{r+n/m}{r}+m-3} \frac{1 - q^{-\binom{r-1+n/m}{r-1}}}{1 - q^{-1}},$$

$$\beta_{r,n} = \frac{2q^{-\frac{1}{2}\binom{r-1+n/\ell}{r-1}+1}}{1 - q^{-1}}.$$

Then for the number $\#D_{r,n}(\mathbb{F}_q)$ of decomposable monic r-variate polynomials with vanishing constant term of degree n over $\mathbb{F}_q$ we have

$$|\#D_{r,n}(\mathbb{F}_q) - \alpha_{r,n}| \leq \alpha_{r,n} \cdot \beta_{r,n}.$$

3 Counting Univariate Decomposable Polynomials

The *composition* of two univariate polynomials $g, h \in F[x]$ over a field F is denoted as $f = g \circ h = g(h)$, and then (g, h) is a *decomposition* of f, and f is *decomposable* if g and h have degree at least 2. In the 1920s, Ritt, Fatou, and Julia studied structural properties of these decompositions over $\mathbb{C}$, using analytic methods. Particularly important are two theorems by Ritt on the uniqueness, in a suitable sense, of decompositions, the first one for (many) indecomposable components and the second one for two components, as above. Engstrom (1941) and Levi (1942) proved them over arbitrary fields of characteristic zero using algebraic methods.

The theory was extended to arbitrary characteristic by Fried and MacRae (1969), Dorey and Whaples (1974), Schinzel (1982; 2000), Zannier (1993), and others. Its use in a cryptographic context was suggested by Cade (1985). In computer algebra, the decomposition method of Barton and Zippel (1985) requires exponential time. A fundamental dichotomy is between the *tame case*, where the characteristic p does not divide $\deg g$, and the *wild case*, where p divides $\deg g$, see von zur Gathen (1990a; 1990b). (Schinzel (2000), Sect. 1.5, uses *tame* in a different sense.) A breakthrough result of Kozen and Landau (1989) was their polynomial-time algorithm to compute tame decompositions; see also von zur Gathen et al. (1987); Kozen et al. (1996); Gutierrez and Sevilla (2006), and the survey articles of von zur Gathen (2002) and Gutierrez and Kozen (2003) with further references. Schur's conjecture, as proven by Turnwald (1995), offers a natural connection between the tame indecomposable polynomials in this section and certain absolutely irreducible bivariate polynomials, as studied in Sect. 2.5. More precisely, a tame polynomial f is indecomposable if $(f(x) - f(y))/(x - y)$ is absolutely irreducible. Aside from natural exceptions, the converse is also true.

In the wild case, considerably less is known, both mathematically and computationally. Zippel (1991) suggests that the block decompositions of Landau and Miller (1985) for determining subfields of algebraic number fields can be applied to decomposing rational functions even in the wild case. A version of Zippel's algorithm in Blankertz (2014) computes in polynomial time all decompositions of a polynomial that are minimal in a certain sense. Avanzi and Zannier (2003) study ambiguities in the decomposition of rational functions over $\mathbb{C}$. On a different but related topic, Zieve and Müller (2008) found interesting characterizations for Ritt's First Theorem, which deals with complete decompositions, where all components are indecomposable.

We have seen fairly precise estimates for the number of multivariate decomposable polynomials in Sect. 2.7. It is intuitively clear that the univariate decomposable polynomials also form only a small minority among all univariate polynomials over a field, and this second part of our survey confirms this intuition. The task is to approximate the number of decomposables over a finite field, together with a good relative error bound. One readily obtains an upper bound. The challenge then is to find an essentially matching lower bound.

A set of distinct decompositions of f is called a *collision*. The number of decomposable polynomials of degree n is thus the number of all pairs (g, h) with $\deg g \cdot \deg h = n$ reduced by the ambiguities introduced by collisions. An important tool for estimating the number of collisions is Ritt's Second Theorem. The first algebraic versions of this in positive characteristic p required $p > \deg(g \circ h)$. Zannier (1993) reduced this to the milder and more natural requirement $g' \neq 0$ for all g in the collision. His proof works over an algebraic closed field, and Schinzel's (2000) monograph adapts it to finite fields. In Sect. 3.2, we provide a precise quantitative version of Ritt's Second Theorem, by determining exactly the number of such collisions in the tame case, assuming that $p \nmid n/\ell$, where n is the degree of the composition and ℓ is the smallest prime divisor of n. This is based on a unique normal form for the polynomials occurring in Ritt's Second Theorem.

Giesbrecht (1988) was the first to consider this counting problem. He showed that the decomposable polynomials form an exponentially small fraction of all univariate polynomials. General approximations to the number of univariate decomposable polynomials are shown in Sect. 3.3. They come with satisfactory (rapidly decreasing) relative error bounds except when p divides $n = \deg f$ exactly twice. In Sect. 3.4, we determine exactly the number of decomposable polynomials in one of these difficult cases, namely when $n = p^2$.

Zannier (2008) studies a different but related question, namely compositions $f = g \circ h$ in $\mathbb{C}[x]$ with a *sparse* polynomial f, having t terms. The degree is not bounded. He gives bounds, depending only on t, on the degree of g and the number of terms in h. Furthermore, he gives a parametrization of all such f, g, h in terms of varieties (for the coefficients) and lattices (for the exponents). Bodin et al. (2009) also deal with counting.

Unless otherwise attributed, the results of Sect. 3.2 are from von zur Gathen (2014b), those of Sect. 3.3 from von zur Gathen (2014a), and those of Sect. 3.4 from Blankertz et al. (2013).

3.1 Notation

A nonzero polynomial $f \in F[x]$ over a field F of characteristic $p \geq 0$ is *monic* if its leading coefficient $\mathrm{lc}(f)$ equals 1. We call f *original* if its graph contains the origin, that is, $f(0) = 0$. For $g, h \in F[x]$,

$$f = g \circ h = g(h) \in F[x] \tag{25}$$

is their *composition*. If $\deg g, \deg h \geq 2$, then (g, h) is a *decomposition* of f. A polynomial $f \in F[x]$ of degree at least 2 is *decomposable* if there exist such g and h, otherwise f is *indecomposable*. A decomposition (25) is *tame* if $p \nmid \deg g$, and f is *tame* if $p \nmid \deg f$.

Multiplication by a unit or addition of a constant does not change decomposability, since

$$f = g \circ h \iff af + b = (ag + b) \circ h$$

for all f, g, h as above and $a, b \in F$ with $a \neq 0$. In other words, the set of decomposable polynomials is invariant under this action of $F^\times \times F$ on $F[x]$.

Furthermore, any decomposition (g, h) of a monic original f can be normalized by this action, by taking $a = \mathrm{lc}(h)^{-1} \in F^\times$, $b = -a \cdot h(0) \in F$, $g^* = g((x - b)a^{-1}) \in F[x]$, and $h^* = ah + b$. Then $g \circ h = g^* \circ h^*$ and g^* and h^* are monic original.

It is therefore sufficient to consider compositions $f = g \circ h$ where all three polynomials are monic original. In such a tame decomposition, g and h are uniquely determined by f and $\deg g$. For $n \geq 1$ and any proper divisor e of n, we write

$$P_n(F) = \{f \in F[x] : f \text{ is monic original of degree } n\},$$
$$D_n(F) = \{f \in P_n(F) : f \text{ is decomposable}\},$$
$$D_{n,e}(F) = \{f \in P_n(F) : f = g \circ h \text{ for some } (g, h) \in P_e(F) \times P_{n/e}(F)\}.$$

Thus $P_n(F)$ and $D_n(F)$ are the subsets of original polynomials in the sets $P_{1,n}(F)$ and $D_{1,n}(F)$, respectively, as defined in the context of multivariate polynomials (Subsect. 2.7) but with right component h of degree at least 2. We sometimes leave out F from the notation when it is clear from the context and have over a finite field $\mathbb{F}_q$ with q elements

$$\#P_n = q^{n-1},$$
$$\#D_{n,e} \leq q^{e+n/e-2}.$$

The set D_n of all decomposable polynomials in P_n satisfies

$$D_n = \bigcup_{\substack{e|n \\ 1<e<n}} D_{n,e}.$$

In particular, $D_n = \varnothing$ if n is prime and $x \in P_1$ is neither decomposable nor indecomposable. For the resulting inclusion-exclusion formula for $\#D_n$, we have to determine the *collisions* (or nonuniqueness) of decompositions, that is, different components $(g, h) \neq (g^*, h^*)$ with equal composition $g \circ h = g^* \circ h^*$.

It is useful to single out a special case of wild compositions when $p > 0$.

Example 26. We call an $f \in P_n \cap F[x^p]$ a *Frobenius composition*, since then $f = g^* \circ x^p$ for some $g^* \in P_{n/p}$, and any decomposition (g, h) of $f = g \circ h$ is a *Frobenius decomposition*. We denote by $\varphi \colon F \longrightarrow F$ the Frobenius endomorphism over a field F of characteristic p, with $\varphi(a) = a^p$ for all $a \in F$, and extend it to an $\mathbb{F}_p$-linear map $\varphi \colon P_n \longrightarrow P_n$ with $\varphi(x) = x$. For $h \in P_{n/p} \setminus \{x^p\}$, this provides the collision

$$x^p \circ h = \varphi(h) \circ x^p. \tag{27}$$

If F is perfect – in particular if F is finite or algebraically closed – then φ is an automorphism on F and every Frobenius composition except x^{p^2} is a collision as in (27). Over $F = \mathbb{F}_q$, this yields $q^{p-1} - 1$ collisions in D_{p^2} and $q^{n/p-1}$ collisions in D_n for $p \mid n \neq p^2$, called *Frobenius collisions*. This example is noted in Schinzel (1982, Sect. I.5, p. 39).

For $f \in P_n(F)$ and $a \in F$, the *original shift* of f by a is

$$f^{[a]} = (x - f(a)) \circ f \circ (x + a) \in P_n(F).$$

Original shifting defines a group action of the additive group of F on $P_n(F)$. Shifting respects decompositions in the sense that for each decomposition (g, h) of f we have a decomposition $(g^{[h(a)]}, h^{[a]})$ of $f^{[a]}$, and vice versa. We denote $(g^{[h(a)]}, h^{[a]})$ as $(g, h)^{[a]}$.

3.2 Normal Form for Ritt's Second Theorem

Ritt presented two types of essential collisions:

$$x^\ell \circ x^k w(x^\ell) = x^{k\ell} w^\ell(x^\ell) = x^k w^\ell \circ x^\ell, \tag{28}$$
$$T_m(x, z^\ell) \circ T_\ell(x, z) = T_{\ell m}(x, z) = T_\ell(x, z^m) \circ T_m(x, z),$$

where $w \in F[x]$, $z \in F^\times = F \setminus \{0\}$, and T_m is the mth Dickson polynomial of the first kind. And then he proved that these are all possibilities up to composition with linear polynomials. This involved four unspecified linear functions, and it is not clear whether there is a relation between the first and the second type of example.

Von zur Gathen (2014b) presents a normal form for the decompositions in Ritt's Second Theorem under Zannier's assumption $g'(g^*)' \neq 0$ and the standard assumption $\gcd(\ell, m) = 1$, where $m = k + \ell \deg w$ in (28). This normal form is unique unless $p \mid m$.

Theorem 29 (Ritt's Second Theorem, normal form). *Let F be a field of characteristic $p \geq 0$, let $m > \ell \geq 2$ be integers with $\gcd(\ell, m) = 1$ and $n = \ell m$. Furthermore, we have monic original $f, g, h, g^*, h^* \in F[x]$ satisfying*

$$f = g \circ h = g^* \circ h^*, \tag{30}$$

$$f, g, h, g^*, h^* \text{ are monic original,} \tag{31}$$

$$\deg g = \deg h^* = m, \deg h = \deg g^* = \ell, \tag{32}$$

$$g'(g^*)' \neq 0, \tag{33}$$

where $g' = \partial g / \partial x$ is the derivative of g. Then either (i) or (ii) hold, and (iii) is also valid.

(i) *(First Case) There exists a monic polynomial $w \in F[x]$ of degree s and $a \in F$ so that*

$$f = (x^{k\ell} w^\ell (x^\ell))^{[a]},$$

where $m = s\ell + k$ is the division with remainder of m by ℓ, with $1 \leq k < \ell$. Furthermore, we have

$$\begin{aligned}(g, h) &= (x^k w^\ell, x^\ell)^{[a]}, \\ (g^*, h^*) &= (x^\ell, x^k w(x^\ell))^{[a]},\end{aligned} \tag{34}$$

$$kw + \ell x w' \neq 0 \text{ and } p \nmid \ell. \tag{35}$$

Conversely, any (w, a) as above for which (35) holds yields a collision satisfying (30) through (33), via (34). If $p \nmid m$, then (w, a) is uniquely determined by f and ℓ.

(ii) *(Second Case) There exist $z, a \in F$ with $z \neq 0$ so that*

$$f = T_n(x, z)^{[a]}.$$

Now (z, a) is uniquely determined by f. Furthermore, we have

$$\begin{aligned}(g, h) &= (T_m(x, z^\ell), T_\ell(x, z))^{[a]}, \\ (g^*, h^*) &= (T_\ell(x, z^m), T_m(x, z))^{[a]},\end{aligned} \tag{36}$$

$$p \nmid n. \tag{37}$$

Conversely, if (37) holds, then any (z, a) as above yields a collision satisfying (30) through (33), via (36).

(iii) *When $\ell \geq 3$, the First and Second Cases are mutually exclusive. For $\ell = 2$, the Second Case is included in the First Case.*

If $p \nmid n$, then the case where $\gcd(\ell, m) \neq 1$ is reduced to the previous one by a result of Tortrat (1988). This determines $D_{n,\ell} \cap D_{n,m}$ exactly if $p \nmid n = \ell m$.

Theorem 38 (Tame case). *Let $\mathbb{F}_q$ be a finite field of characteristic p, let δ denote Kronecker's delta function, and let $m > \ell \geq 2$ be integers with $p \nmid n = \ell m$, $i = \gcd(\ell, m)$ and $s = \lfloor m/\ell \rfloor$. For the number of monic original polynomials of degree n over $\mathbb{F}_q$ with left components of degree ℓ and m we have*

$$\#(D_{n,\ell}(\mathbb{F}_q) \cap D_{n,m}(\mathbb{F}_q)) = \begin{cases} q^{2\ell+s-3} & \text{if } \ell \mid m, \\ q^{2i}(q^{s-1} + (1 - \delta_{\ell,2})(1 - q^{-1})) & \\ \qquad \leq q^{2\ell+s-3} & \text{otherwise.} \end{cases}$$

In the remaining case where $p \mid n$, the Frobenius collisions are easily counted, see Example 26, and therefore excluded. We have the following upper bounds.

Corollary 39 (Wild case, upper bounds). *Let $\mathbb{F}_q$ be a finite field of characteristic p and ℓ, m, $n \geq 2$ be integers with $p \mid n = \ell m$, and let $c = \#(D_{n,\ell}(\mathbb{F}_q) \cap D_{n,m}(\mathbb{F}_q) \setminus F[x^p])$ be the number of monic original polynomials of degree n over $\mathbb{F}_q$ with left components of degree ℓ and m that are not Frobenius collisions. Then the following hold.*

(i) *If $p \nmid \ell$, then*
$$c \leq q^{m+\lceil \ell/p \rceil - 2}.$$

(ii) *If $p \mid \ell$ and $\ell < m$, we set $b = \lceil (m - \ell + 1)/\ell \rceil$. Then*
$$c \leq q^{m+\ell-b+\lceil b/p \rceil - 2}.$$

For perspective, we also note the following lower bounds on c from von zur Gathen (2013; 2014a). Unlike the exact result of Theorem 38, there is a substantial gap between the upper and lower bounds.

Corollary 40 (Wild case, lower bounds). *Let $\mathbb{F}_q$ be a finite field of characteristic p, ℓ a prime number dividing $m > \ell$, assume that $p \mid n = \ell m$, and let $c = \#(D_{n,\ell}(\mathbb{F}_q) \cap D_{n,m}(\mathbb{F}_q) \setminus F[x^p])$ be the number of monic original polynomials of degree n over $\mathbb{F}_q$ with left components of degree ℓ and m that are not Frobenius collisions. Then the following hold.*

(i) *If $p = \ell \mid m$ and each nontrivial divisor of m/p is larger than p, then*
$$c \geq q^{2p+m/p-3}(1 - q^{-1})(1 - q^{-p+1}).$$

(ii) *If $p \neq \ell$ divides m exactly $d \geq 1$ times, then*
$$c \geq q^{2\ell+m/\ell-3}(1 - q^{-m/\ell})(1 - q^{-1}(1 + q^{-p+2}\frac{(1 - q^{-1})^2}{1 - q^{-p}}))$$

if $\ell \nmid p^d - 1$. Otherwise we set $\mu = \gcd(p^d - 1, \ell)$, $r = (p^d - 1)/\mu$ and have

$$c \geq q^{2\ell + m/\ell - 3}\left((1 - q^{-1}(1 + q^{-p+2}\frac{(1 - q^{-1})^2}{1 - q^{-p}}))(1 - q^{-m/\ell})\right.$$

$$\left. - q^{-m/\ell - r + 2}\frac{(1 - q^{-1})^2(1 - q^{-r(\mu - 1)})}{1 - q^{-r}}(1 + q^{-r(p-2)}))\right).$$

3.3   The Number of Decomposable Univariate Polynomials

The basic statement is that α_n as in (42) is an approximation to the number of monic original decomposable polynomials of degree n, with relative error bounds of varying quality. The following is a condensed version of the more precise bounds in von zur Gathen (2014a).

Theorem 41. *Let $\mathbb{F}_q$ be a finite field with q elements and characteristic p, let ℓ be the smallest prime divisor of the composite integer $n \geq 2$, and*

$$\alpha_n = \begin{cases} 2q^{\ell + n/\ell - 2} & \text{if } n \neq \ell^2, \\ q^{2\ell - 2} & \text{if } n = \ell^2. \end{cases} \tag{42}$$

Then the following hold for the number $\#D_n(\mathbb{F}_q)$ of decomposable monic original polynomials of degree n over $\mathbb{F}_q$, where $p \| n$ means that p divides n exactly twice.

(i) $q^{2\sqrt{n}-2} \leq \alpha_n \leq 2q^{n/2}$.
(ii) $\alpha_n/2 \leq \#D_n(\mathbb{F}_q) \leq \alpha_n(1 + q^{-n/3\ell^2}) < 2\alpha_n \leq 4q^{n/2}$.
(iii) If $n \neq p^2$ and $q > 5$, then $\#D_n(\mathbb{F}_q) \geq (3 - 2q^{-1})\alpha_n/4 \geq q^{2\sqrt{n}-2}/2$.
(iv) Unless $p = \ell \| n$ and, we have $\#D_n(\mathbb{F}_q) \geq \alpha_n(1 - 2q^{-1})$.
(v) If $p \nmid n$, then $|\#D_n(\mathbb{F}_q) - \alpha_n| \leq \alpha_n \cdot q^{-n/3\ell^2}$.

The relative error in (v) is exponentially decreasing in the input size $n \log q$, in the tame case and for growing $n/3\ell^2$. In (iv), the factor is $1 + O(q^{-1})$ over $\mathbb{F}_q$. When $p = \ell \| n$, then we have a factor of about 2 in (ii), which is improved to about $4/3$ in (iii). The case $n = p^2$ is settled in Subsect. 3.4.

Beyond the previous precise bounds, without asymptotics or unspecified constants, we now derive some conclusions about the asymptotic behavior. There are two parameters: the field size q and the degree n. When n is prime, then $\#D_n(\mathbb{F}_q) = 0$, and prime values of n are excepted in the following. We consider the asymptotics in one parameter, where the other one is fixed, and also the special situations where $\gcd(q, n) = 1$. Furthermore, we denote as "$q, n \longrightarrow \infty$" any infinite sequence of pairwise distinct (q, n). The cases $n = 4$ and $p^2 \| n \neq p^2$ for some prime p are the only ones where our methods do not show that $\#D_n(\mathbb{F}_q)/\alpha_n \longrightarrow 1$.

Theorem 43. *Let $\#D_n(\mathbb{F}_q)$ be the number of decomposable monic original polynomials of degree n over $\mathbb{F}_q$, α_n as in (42), and $\nu_{q,n} = \#D_n(\mathbb{F}_q)/\alpha_n$. We only consider composite n.*

(i) *For any q, we have*

$$\lim_{\substack{n\to\infty \\ \gcd(q,n)=1}} \nu_{q,n} = 1,$$

$$\limsup_{n\to\infty} \nu_{q,n} = 1,$$

$$\frac{1}{2} \le \nu_{q,n} \text{ for any } n,$$

$$\frac{3 - 2q^{-1}}{4} \le \nu_{q,n} \text{ for any } n \text{ if } q > 5.$$

(ii) *Let n be a composite integer and ℓ its smallest prime divisor. Then*

$$\lim_{\substack{q\to\infty \\ \gcd(q,n)=1}} \nu_{q,n} = 1,$$

$$\limsup_{q\to\infty} \nu_{q,n} = 1,$$

$$\liminf_{q\to\infty} \nu_{q,n} \begin{cases} = 2/3 & \text{if } n = 4, \\ \ge \frac{1}{4}(3 + \frac{1}{\ell+1}) & \text{if } \ell^2 \,\|\, n \text{ and } n \ne \ell^2, \\ = 1 & \text{otherwise.} \end{cases}$$

(iii) *For any sequence q, n → ∞, we have*

$$\lim_{\substack{q,n\to\infty \\ \gcd(q,n)=1}} \nu_{q,n} = 1,$$

$$\frac{1}{2} \le \liminf_{q,n\to\infty} \nu_{q,n} \le \limsup_{q,n\to\infty} \nu_{q,n} = 1.$$

3.4   Collisions at Degree p^2

The previous section gives satisfactory estimates for the number of decomposable polynomials at degree n unless $p^2 \,\|\, n$. The material of this section determines the number in the easiest of these open cases, namely for $n = p^2$.

First, we present two classes of explicit collisions at degree r^2, where r is a power of the characteristic $p > 0$ of the field F. The collisions of Fact 44 consist of additive and subadditive polynomials. A polynomial A of degree r^k is r-*additive* if it is of the form $A = \sum_{0 \le i \le k} a_i x^{r^i}$ with all $a_i \in F$. We call a polynomial *additive* if it is p-additive. A polynomial is additive if and only if it acts additively on an algebraic closure $\overline{F}$ of F, that is $A(a + b) = A(a) + A(b)$ for all $a, b \in \overline{F}$; see Goss (1996, Corollary 1.1.6). The composition of additive polynomials is additive, see for instance Proposition 1.1.2 of the cited book. The decomposition structure of additive polynomials was first studied by Ore (1933). Dorey and Whaples (1974, Theorem 4) show that all components of an additive polynomial are additive. Giesbrecht (1988) gives lower bounds on the number of decompositions and algorithms to determine them.

For a divisor m of $r - 1$, the (r, m)-*subadditive* polynomial associated with the r-additive polynomial A is $S = x(\sum_{0 \le i < k} a_i x^{(r^i - 1)/m})^m$ of degree r^k. Then A and S are related as $x^m \circ A = S \circ x^m$. Dickson (1897) notes a special case of subadditive polynomials, and Cohen (1985) is concerned with the reducibility of some related polynomials. Cohen (1990a; 1990b) investigates their connection to exceptional polynomials and coins the term "sub-linearized"; see also Cohen and Matthews (1994). Coulter et al. (2004) derive the number of indecomposable subadditive polynomials and present an algorithm to decompose subadditive polynomials.

Ore (1933, Theorem 3) describes exactly the right components of degree p of an additive polynomial. Henderson and Matthews (1999) relate such additive decompositions to subadditive polynomials, and in their Theorems 3.4 and 3.8 describe the collisions of Fact 44 below. Theorem 48 shows that together with those of Theorem 46 and the Frobenius collisions of Example 26, these examples and their shifts comprise all collisions at degree p^2.

Fact 44. *Let r be a power of p, $u, s \in F^{\times}$, $\varepsilon \in \{0, 1\}$, m a positive divisor of $r - 1$, $\ell = (r - 1)/m$, and*

$$f = S(u, s, \varepsilon, m) = x(x^{\ell(r+1)} - \varepsilon u s^r x^{\ell} + u s^{r+1})^m \in P_{r^2}(F),$$
$$T = \{t \in F : t^{r+1} - \varepsilon u t + u = 0\}. \tag{45}$$

For each $t \in T$ and
$$g = x(x^{\ell} - u s^r t^{-1})^m,$$
$$h = x(x^{\ell} - st)^m,$$

both in $P_r(F)$, we have $f = g \circ h$. Moreover, f has at least $\#T$ distinct decompositions.

The polynomials f in (45) are "simply original" in the sense that they have a simple root at 0. This motivates the designation S. The second construction of collisions goes as follows.

Theorem 46. *Let r be a power of p, $b \in F^{\times}$, $a \in F \setminus \{0, b^r\}$, $a^* = b^r - a$, m an integer with $1 < m < r - 1$ and $p \nmid m$, $m^* = r - m$, and*

$$f = M(a, b, m) = x^{mm^*}(x - b)^{mm^*}\left(x^m + a^* b^{-r}((x - b)^m - x^m)\right)^m$$
$$\cdot \left(x^{m^*} + ab^{-r}((x - b)^{m^*} - x^{m^*})\right)^{m^*},$$
$$g = x^m(x - a)^{m^*},$$
$$h = x^r + a^* b^{-r}(x^{m^*}(x - b)^m - x^r), \tag{47}$$
$$g^* = x^{m^*}(x - a^*)^m,$$
$$h^* = x^r + ab^{-r}(x^m(x - b)^{m^*} - x^r).$$

Then $f = g \circ h = g^* \circ h^* \in P_{r^2}(F)$ has at least two distinct decompositions.

The polynomials f in (47) are "multiply original" in the sense that they have a multiple root at 0. This motivates the designation M. The notation is set up so that * acts as an involution on our data, leaving b, f, r, and x invariant.

Zieve (2011) points out that the rational functions of case (4) in Proposition 5.6 of Avanzi and Zannier (2003) can be transformed into (47). Zieve also mentions that this example already occurs in unpublished work of his, joint with Robert Beals.

Theorem 48. *Let F be a perfect field of characteristic p and $f \in P_{p^2}(F)$. Then f has a collision $\{(g,h),(g^*,h^*)\}$ if and only if exactly one of the following holds.*

(F) *The polynomial f is a Frobenius collision as in Example 26.*
(S) *The polynomial f is simply original and there are u, s, ε, and m as in Fact 44 and $w \in F$ such that*

$$f^{[w]} = S(u,s,\varepsilon,m)$$

and the collision $\{(g,h)^{[w]},(g^*,h^*)^{[w]}\}$ is contained in the collision described in Fact 44, with $\#T \geq 2$.
(M) *The polynomial f is multiply original and there are a, b, and m as in Theorem 46 and $w \in F$ such that*

$$f^{[w]} = M(a,b,m)$$

and the collision $\{(g,h)^{[w]},(g^*,h^*)^{[w]}\}$ is as in Theorem 46.

In particular, the collisions in case (S) and case (M) have exactly $\#T$ and 2 distinct decompositions, respectively. Inclusion-exclusion now yields the following exact formula for the number of decomposable polynomials of degree p^2 over $\mathbb{F}_q$.

Theorem 49. *Let $\mathbb{F}_q$ be a finite field of characteristic p, δ Kronecker's delta function, and τ the number of positive divisors of $p-1$. Then for the number $\#D_{p^2}(\mathbb{F}_q)$ of decomposable monic original polynomials of degree p^2 over $\mathbb{F}_q$ we have*

$$\#D_{p^2}(\mathbb{F}_q) = q^{2p-2} - q^{p-1} + 1 - \frac{(\tau q - q + 1)(q-1)(qp - p - 2)}{2(p+1)}$$
$$- (1 - \delta_{p,2})\frac{q(q-1)(q-2)(p-3)}{4}.$$

In particular, we have

$$\#D_4(\mathbb{F}_q) = q^2 \cdot \frac{2 + q^{-2}}{3} \qquad \text{for } p = 2,$$

$$\#D_9(\mathbb{F}_q) = q^4\left(1 - \frac{3}{8}(q^{-1} + q^{-2} - q^{-3} - q^{-4})\right) \qquad \text{for } p = 3,$$

$$\#D_{p^2}(\mathbb{F}_q) = q^{2p-2}\left(1 - q^{-p+1} + O(q^{-2p+5+1/d})\right) \qquad \text{for } q = p^d \text{ and } p \geq 5.$$

We have two independent parameters p and d, and $q = p^d$. For two eventually positive functions $f, g: \mathbb{N}^2 \to \mathbb{R}$, here $g \in O(f)$ means that there are constants

b and c so that $g(p,d) \leq c \cdot f(p,d)$ for all p and d with $p + d \geq b$. We have the following asymptotics.

Corollary 50. *Let $p \geq 5$, $d \geq 1$, $q = p^d$, and $k \geq 1$. Then the number c_k of monic original polynomials of degree p^2 over $\mathbb{F}_q$ with exactly k decompositions is as follows*

$$c_1 = q^{2p-2}(1 - 2q^{-p+1} + O(q^{-2p+5+1/d})),$$
$$c_2 = q^{p-1}(1 + O(q^{-p+4+1/d})),$$
$$c_{p+1} = (\tau - 1)q^{3-3/d}\big(1 + O(q^{-\max\{2/d, 1-1/d\}})\big)$$
$$\subseteq O\big(q^{3-3/d+1/(d \ \log\log \ p)}\big),$$
$$c_k = 0 \ \text{if } k \notin \{1, 2, p+1\}.$$

Theorem 49 leads to $\lim_{q \to \infty} \nu_{q,\ell^2} = 1$ for any prime $\ell > 2$ in Theorem 43 (ii). For $n = 4$, the sequence has no limit, but oscillates between values close to $\liminf_{q \to \infty} \nu_{q,4} = 2/3$ and to $\limsup_{q \to \infty} \nu_{q,4} = 1$, and these are the only two accumulation points of the sequence $\nu_{q,4}$.

4   Open Problems

Further types of multivariate polynomials that are examined from a counting perspective include singular bivariate ones (von zur Gathen 2008) and pairs of coprime polynomials (Hou and Mullen 2009). It remains open to extend the methods of Sect. 2 to singular multivariate ones and achieve exponentially decreasing error bounds for coprime multivariate polynomials.

For univariate decomposable polynomials, the question of good asymptotics for $\nu_{q,n}$ when q is fixed and $n \to \infty$ is still open. More work is needed to understand the case where the characteristic p is the smallest prime divisor of the degree n, divides n exactly twice, and $n \neq p^2$. Ritt's Second Theorem covers distinct-degree collisions, even in the wild case, see Zannier (1993); it would be interesting to see a parametrization even for $p \mid m$ and obtain a similar classification for general equal-degree collisions. Ziegler (2014) provides an exact count of tame univariate polynomials.

Finally, this survey deals with polynomials only and the study of rational functions with the same methods remains open.

Acknowledgments. This work was funded by the B-IT Foundation and the Land Nordrhein-Westfalen.

References

Alekseyev, M.: A115457–A115472. In: The On-Line Encyclopedia of Integer Sequences. OEIS Foundation Inc. (2006). http://oeis.org. Last download 4 Dec 2012

Artin, E.: Quadratische Körper im Gebiete der höheren Kongruenzen. II. (Analytischer Teil.). Math. Z. **19**(1), 207–246 (1924). http://dx.doi.org/10.1007/BF01181075

Avanzi, R.M., Zannier, U.M.: The equation $f(X) = f(Y)$ in rational functions $X = X(t)$, $Y = Y(t)$. Compositio Math. **139**(3), 263–295 (2003). http://dx.doi.org/10.1023/B:COMP.0000018136.23898.65

Barton, D.R., Zippel, R.: Polynomial Decomposition Algorithms. J. Symbolic Comput. **1**, 159–168 (1985)

Blankertz, R.: A polynomial time algorithm for computing all minimal decompositions of a polynomial. ACM Commun. Comput. Algebra **48**(1, Issue 187), 13–23 (2014)

Blankertz, R., von zur Gathen, J., Ziegler, K.: Compositions and collisions at degree p^2. J. Symbolic Comput. **59**, 113–145 (2013). ISSN 0747-7171. http://dx.doi.org/10.1016/j.jsc.2013.06.001. http://arxiv.org/abs/1202.5810. Extended Abstract in: Proceedings of the 2012 International Symposium on Symbolic and Algebraic Computation, ISSAC 2012, Grenoble, France, pp. 91–98 (2012)

Bodin, A.: Number of irreducible polynomials in several variables over finite fields. Am. Math. Monthly **115**(7), 653–660 (2008). ISSN 0002-9890

Bodin, A.: Generating series for irreducible polynomials over finite fields. Finite Fields Their Appl. **16**(2), 116–125 (2010). http://dx.doi.org/10.1016/j.ffa.2009.11.002

Bodin, A., Dèbes, P., Najib, S.: Indecomposable polynomials and their spectrum. Acta Arith. **139**(1), 79–100 (2009)

Cade, J.J.: A new public-key cipher which allows signatures. In: Proceedings of the 2nd SIAM Conference on Applied Linear Algebra. SIAM, Raleigh (1985)

Car, M.: Théorèmes de densité dans $\mathbb{F}_q[X]$. Acta Arith. **48**, 145–165 (1987)

Carlitz, L.: The arithmetic of polynomials in a Galois field. Am. J. Math. **54**, 39–50 (1932)

Carlitz, L.: The distribution of irreducible polynomials in several indeterminates. Ill. J. Math. **7**, 371–375 (1963)

Carlitz, L.: The distribution of irreducible polynomials in several indeterminates II. Can. J. Math. **17**, 261–266 (1965)

Cesaratto, E., von zur Gathen, J., Matera, G.: The number of reducible space curves over a finite field. J. Number Theory **133**, 1409–1434 (2013). http://dx.doi.org/10.1016/j.jnt.2012.08.027

Cohen, S.: The distribution of irreducible polynomials in several indeterminates over a finite field. Proc. Edinburgh Math. Soc. **16**, 1–17 (1968)

Cohen, S.: Some arithmetical functions in finite fields. Glasgow Math. Soc. **11**, 21–36 (1969)

Cohen, S.D.: Reducibility of sub-linear polynomials over a finite field. Bull. Korean Math. Soc. **22**, 53–56 (1985)

Cohen, S.D.: Exceptional polynomials and the reducibility of substitution polynomials. Enseign. Math. (2) **36**(1–2), 53–65 (1990a). ISSN 0013-8584

Cohen, S.D.: The factorable core of polynomials over finite fields. J. Australian Math. Soc. Ser. A **49**(02), 309–318 (1990b). http://dx.doi.org/10.1017/S1446788700030585

Cohen, S.D., Matthews, R.W.: A class of exceptional polynomials. Trans. Am. Math. Soc. **345**(2), 897–909 (1994). ISSN 0002-9947. http://www.jstor.org/stable/2155005

Coulter, R.S., Havas, G., Henderson, M.: On decomposition of sub-linearised polynomials. J. Australian Math. Soc. **76**(3), 317–328 (2004). ISSN 1446-7887. http://dx.doi.org/10.1017/S1446788700009885

Dickson, L.E.: The analytic representation of substitutions on a power of a prime number of letters with a discussion of the linear group. Part I & Part II. Ann. Math. **11**, 65–120, 161–183 (1897). http://www.jstor.org/stable/1967217, http://www.jstor.org/stable/1967224

Dorey, F., Whaples, G.: Prime and Composite Polynomials. J. Algebra **28**, 88–101 (1974). http://dx.doi.org/10.1016/0021-8693(74)90023-4

Eisenbud, D., Harris, J.: The dimension of the Chow variety of curves. Compositio Math. **83**(3), 291–310 (1992)

Engstrom, H.T.: Polynomial substitutions. Am. J. Math. **63**, 249–255 (1941). http://www.jstor.org/stable/2371520

Faugère, J.-C., Perret, L.: High Order Derivatives and Decomposition of Multivariate Polynomials. In: May, J.P. (ed.) Proceedings of the 2009 International Symposium on Symbolic and Algebraic Computation ISSAC '09, Seoul, Korea, pp. 207–214. ACM Press (2009). http://dx.doi.org/10.1145/1576702.1576732. ISBN 978-1-60558-609-0. Extended Abstract in: Extended Abstracts of the Second Workshop on Mathematical Cryptology, WMC 2008, pp. 15–19 (2008)

Flajolet, P., Gourdon, X., Panario, D.: The complete analysis of a polynomial factorization algorithm over finite fields. J. Algorithms **40**(1), 37–81 (2001). Meyer auf der Heide, F., Monien, B. (eds.) Extended Abstract in: Proceedings of the 23rd International Colloquium on Automata, Languages and Programming, ICALP 1996. LNCS, vol. 1099, pp. 232–243. Springer, Heidelberg (1996)

Flajolet, P., Sedgewick, R.: Analytic Combinatorics, 824 pp. Cambridge University Press, Cambridge (2009). ISBN 0521898064

Fried, M.D., MacRae, R.E.: On the invariance of chains of fields. Ill. J. Math. **13**, 165–171 (1969)

Gao, S., Lauder, A.G.B.: Hensel lifting and bivariate polynomial factorisation over finite fields. Math. Comput. **71**(240), 1663–1676 (2002). http://dx.doi.org/10.1090/S0025-5718-01-01393-X

von zur Gathen, J.: Functional Decomposition of Polynomials: the Tame Case. J. Symbolic Comput. **9**, 281–299 (1990a). http://dx.doi.org/10.1016/S0747-7171(08)80014-4. Extended Abstract in: Proceedings of the 28th Annual IEEE Symposium on Foundations of Computer Science, Los Angeles, CA (1987)

von zur Gathen, J.: Functional Decomposition of Polynomials: the Wild Case. J. Symbolic Comput. **10**, 437–452 (1990b). http://dx.doi.org/10.1016/S0747-7171(08)80054-5

von zur Gathen, J.: Factorization and Decomposition of Polynomials. In: Mikhalev, A.V., Pilz, G.F. (eds.) The Concise Handbook of Algebra, pp. 159–161. Kluwer Academic Publishers, Dordrecht (2002). ISBN 0-7923-7072-4

von zur Gathen, J.: Counting reducible and singular bivariate polynomials. Finite Fields Their Appl. **14**(4), 944–978 (2008). http://dx.doi.org/10.1016/j.ffa.2008.05.005. Extended Abstract in: Proceedings of the 2007 International Symposium on Symbolic and Algebraic Computation, ISSAC 2007, Waterloo, ON, Canada, pp. 369–376 (2007)

von zur Gathen, J.: Counting decomposable multivariate polynomials. Appl. Algebra Eng. Commun. Comput. **22**(3), 165–185 (2011). http://dx.doi.org/10.1007/s00200-011-0141-9. Abstract in: Abstracts of the Ninth International Conference on Finite Fields and their Applications, pp. 21–22. Claude Shannon Institute, Dublin, July 2009. http://www.shannoninstitute.ie/fq9/AllFq9Abstracts.pdf

von zur Gathen, J.: Lower bounds for decomposable univariate wild polynomials. J. Symbolic Comput. **50**, 409–430 (2013). http://dx.doi.org/10.1016/j.jsc.2011.01.008. Extended Abstract in: Proceedings of the 2009 International Symposium on Symbolic and Algebraic Computation, ISSAC 2009, Seoul, Korea (2009)

von zur Gathen, J.: Counting decomposable univariate polynomials. Combin. Probab. Comput. Special Issue 01 **24**, 294–328 (2014a).

http://dx.doi.org/10.1017/S0963548314000388. Extended Abstract in: Proceedings of the 2009 International Symposium on Symbolic and Algebraic Computation, ISSAC 2009, Seoul, Korea (2009). Preprint (2008). http://arxiv.org/abs/0901.0054

von zur Gathen, J.: Normal form for Ritt's Second Theorem. Finite Fields Their Appl. **27**, 41–71 (2014b). ISSN 1071-5797. http://dx.doi.org/10.1016/j.ffa.2013.12. 004. http://arxiv.org/abs/1308.1135

von zur Gathen, J., Gutierrez, J., Rubio, R.: Multivariate polynomial decomposition. Appl. Algebra Eng. Commun. Comput. **14**(1), 11–31 (2003). http://www.springerlink.com/content/0a5c0vvp82xbx5je/. Extended Abstract in: Proceedings of the Second Workshop on Computer Algebra in Scientific Computing, CASC 1999, München, Germany (1999)

von zur Gathen, J., Kozen, D., Landau, S.: Functional decomposition of polynomials. In: Proceedings of the 28th Annual IEEE Symposium on Foundations of Computer Science, Los Angeles, CA, pp. 127–131. IEEE Computer Society Press, Washington, DC (1987). http://dx.doi.org/10.1109/SFCS.1987.29

von zur Gathen, J., Viola, A., Ziegler, K.: Counting reducible, powerful, and relatively irreducible multivariate polynomials over finite fields. SIAM J. Discrete Math. **27**(2), 855–891 (2013). http://dx.doi.org/10.1137/110854680. http://arxiv.org/abs/0912. 3312. Extended Abstract in: Proceedings of LATIN 2010, Oaxaca, Mexico (2010)

Giesbrecht, M.W.: Some Results on the Functional Decomposition of Polynomials. Master's thesis, Department of Computer Science, University of Toronto. Technical Report 209/88 (1988). http://arxiv.org/abs/1004.5433

Gogia, S.K., Luthar, I.S.: Norms from certain extensions of $F_q(T)$. Acta Arith. **38**(4), 325–340 (1981). ISSN 0065-1036

Goss, D.: Basic Structures of Function Field Arithmetic. Springer, Heidelberg (1996). ISBN 3-540-61087-1

Gutierrez, J., Kozen, D.: Polynomial decomposition. In: Grabmeier, J., Kaltofen, E., Weispfenning, V. (eds.) Computer Algebra Handbook, Sect. 2.2.4, pp. 26–28 (2003). http://www.springer.com/978-3-540-65466-7

Gutierrez, J., Sevilla, D.: On Ritt's decomposition theorem in the case of finite fields. Finite Fields Their Appl. **12**(3), 403–412 (2006). http://dx.doi.org/10.1016/j.ffa. 2005.08.004. http://arxiv.org/abs/0803.3976

Hayes, D.R.: The distribution of irreducibles in GF$[q, x]$. Trans. Am. Math. Soc. **117**, 101–127 (1965). http://dx.doi.org/10.2307/1994199

Henderson, M., Matthews, R.: Composition behaviour of sub-linearised polynomials over a finite field. In: Finite Fields: Theory, Applications, and Algorithms, Waterloo, ON, 1997. Contemporary Mathematics, vol. 225, pp. 67–75. American Mathematical Society, Providence (1999)

Hou, X., Mullen, G.L.: Number of irreducible polynomials and pairs of relatively prime polynomials in several variables over finite fields. Finite Fields Their Appl. **15**(3), 304–331 (2009). http://dx.doi.org/10.1016/j.ffa.2008.12.004

Kozen, D., Landau, S.: Polynomial decomposition algorithms. J. Symbolic Comput. **7**, 445–456 (1989). http://dx.doi.org/10.1016/S0747-7171(89)80027-6 (1989). An earlier version was published as Technical Report 86-773, Cornell University, Department of Computer Science, Ithaca, New York (1986)

Kozen, D., Landau, S., Zippel, R.: Decomposition of algebraic functions. J. Symbolic Comput. **22**, 235–246 (1996)

Landau, S., Miller, G.L.: Solvability by radicals is in polynomial time. J. Comput. Syst. Sci. **30**, 179–208 (1985)

Levi, H.: Composite polynomials with coefficients in an arbitrary field of characteristic zero. Am. J. Math. **64**, 389–400 (1942)

Mullen, G.L., Panario, D.: Handbook of Finite Fields. Discrete Mathematics and Its Applications. CRC Press, Boca Raton (2013). ISBN 978-1-4398-7378-6 (Hardback). http://www.crcpress.com/product/isbn/9781439873786

Ore, O.: On a special class of polynomials. Trans. Am. Math. Soc. **35**, 559–584 (1933)

Schinzel, A.: Selected Topics on Polynomials. The University of Michigan Press, Ann Arbor (1982). ISBN 0-472-08026-1

Schinzel, A.: Polynomials with Special Regard to Reducibility. Cambridge University Press, Cambridge (2000). ISBN 0521662257

Tortrat, P.: Sur la composition des polynômes. Colloq. Math. **55**(2), 329–353 (1988)

Turnwald, G.: On Schur's conjecture. J. Australian Math. Soc. Ser. A **58**, 312–357 (1995). http://anziamj.austms.org.au/JAMSA/V58/Part3/Turnwald.html

Wan, D.: Zeta functions of algebraic cycles over finite fields. Manuscripta Math. **74**, 413–444 (1992)

Williams, K.S.: Polynomials with irreducible factors of specified degree. Can. Math. Bull. **12**, 221–223 (1969). ISSN 0008-4395

Zannier, U.: Ritt's Second Theorem in arbitrary characteristic. J. Reine Angew. Math. **445**, 175–203 (1993)

Zannier, U.: On composite lacunary polynomials and the proof of a conjecture of Schinzel. Invent. Math. **174**, 127–138 (2008). ISSN 0020-9910 (Print) 1432-1297 (Online). http://dx.doi.org/10.1007/s00222-008-0136-8. http://arxiv.org/abs/0705.0911v1

Ziegler, K.: Tame decompositions and collisions. In: Nabeshima, K. (ed.) Proceedings of the 2014 International Symposium on Symbolic and Algebraic Computation ISSAC '14, Kobe, Japan, pp. 421–428. ACM Press (2014). http://dx.doi.org/10.1145/2608628.2608653. http://arxiv.org/abs/1402.5945

Zieve, M.: Personal communication (2011)

Zieve, M.E., Müller, P.: On Ritt's Polynomial Decomposition Theorems, Submitted, 38 pp. (2008). http://arxiv.org/abs/0807.3578

Zippel, R.: Rational function decomposition. In: Watt, S.M. (ed.) Proceedings of the 1991 International Symposium on Symbolic and Algebraic Computation, ISSAC 1991, Bonn, Germany, pp. 1–6. ACM Press, Bonn (1991). ISBN 0-89791-437-6

Zsigmondy, K.: Über die Anzahl derjenigen ganzzahligen Functionen n-ten Grades von x, welche in Bezug auf einen gegebenen Primzahlmodul eine vorgeschriebene Anzahl von Wurzeln besitzen. Sitzungsber. Kaiserl. Akad. Wiss. Abt. II **103**, 135–144 (1894)

Orbit Closures of Linear Algebraic Groups

Willem A. de Graaf[✉]

Department of Mathematics, University of Trento, Trento, Italy
degraaf@science.unitn.it

Abstract. Let an algebraic group be given, acting on a vector space. We consider the problem of deciding whether a given element of the vector space lies in the closure of the orbit of another given element. We describe three methods for dealing with this problem that have appeared in the literature. We illustrate the methods by examples.

1 Introduction

Actions of linear algebraic groups appear in many contexts. In some of them the following problem is of interest: given an algebraic group $G \subset \mathrm{GL}(V)$ and v, $w \in V$, decide whether w lies in the (Zariski-) closure of the orbit $G \cdot v$. This problem arises for example in the theory of degenerations of algebraic structures (see [3]), and when one studies geometric properties of the orbits of an algebraic group ([21,22]), and in the theory of unipotent classes of semisimple Lie groups ([29]).

We note that since the closure of $G \cdot v$ is G-stable, the problem above is equivalent to deciding whether the orbit $G \cdot w$ lies in the closure of $G \cdot v$.

In this paper we review three approaches to this problem. The first one is a straightforward reduction to elimination using Gröbner bases (Sect. 3). The second method, due to Popov, is based on the effective Nullstellensatz (Sect. 4). Both these methods use an open subset of the group, isomorphic to an "easy" affine algebraic set. This is described in Sect. 2. The third method is specific for reductive groups that are constructed as so-called θ-groups (Sect. 5). Throughout we assume that the base field is algebraically closed and of characteristic 0, as many constructions that we use (e.g., the correspondence between a linear algebraic group and its Lie algebra) only work well in characteristic 0.

We illustrate the methods by examples. We also comment on the practical usefulness of them. Although it is possible to compute some instances of the problem using the first method, the use of Gröbner bases makes it very difficult to use it for larger examples. The second method elegantly reduces the problem to a system of linear equations. However, the number of unknowns in these equations is so large that it is questionable whether this method can be used for any nontrivial instance of the problem. The third method is rather efficient, and can be used to compute many nontrivial cases. However, it is only applicable to θ-groups.

The main ideas underlying the algorithms presented here, are all taken from sources in the literature. However, the examples that we computed have not

© Springer International Publishing Switzerland 2015
J. Gutierrez et al. (Eds.): Computer Algebra and Polynomials, LNCS 8942, pp. 76–93, 2015.
DOI: 10.1007/978-3-319-15081-9_4

appeared elsewhere. Also we remark that in some of the examples we report running times of the various programs used; these have been obtained on a 3.16 GHz machine with 32 GB of memory.

Notation and Terminology

Throughout we use standard terminology of algebraic geometry (see [27]) and linear algebraic groups (see [15]). In particular, by $\mathbb{A}^n$ we denote the n-dimensional affine space. By a closed set we mean the zero locus in some $\mathbb{A}^n$ of a set of polynomials in n indeterminates. By k we will denote an algebraically closed field of characteristic 0. A linear algebraic group is a subgroup of $\mathrm{GL}(n, k)$ given as the zero locus of a set of polynomials in the matrix entries. If G is an algebraic group, then $\mathfrak{g}$ will denote its Lie algebra.

2 An Embedding

Let $G \subset \mathrm{GL}(n, k)$ be an algebraic group. Here we describe a method to obtain a closed set X in an affine space, together with a regular map $\sigma : X \to G$ such that the image of σ contains a dense and open subset of G. The closed set X will be given by "easy" polynomial equations, so that it will be straightforward to compute in the coordinate ring $k[X]$ of X. The use of such an embedding for algorithmic purposes was first proposed in [26].

Lemma 1. *Let $G \subset \mathrm{GL}(n, k)$ be a connected algebraic group with Lie algebra $\mathfrak{g}$. Suppose that there are algebraic subalgebras $\mathfrak{g}_1, \mathfrak{g}_2$ of $\mathfrak{g}$ such that $\mathfrak{g}$ is the direct sum of vector spaces, $\mathfrak{g} = \mathfrak{g}_1 \oplus \mathfrak{g}_2$. Let G_1, G_2 be the connected subgroups of G with Lie algebras $\mathfrak{g}_1, \mathfrak{g}_2$ respectively. Furthermore, let X_1, X_2 be closed sets in two affine spaces, and let $\phi_i : X_i \to G_i$ be regular maps. Let $x_i \in X_i$ and suppose that the differential of ϕ_i at x_i, $(d\phi_i)_{x_i} : T_{x_i}(X_i) \to \mathfrak{g}_i$ is surjective, for $i = 1, 2$. Define $\phi : X_1 \times X_2 \to G$ by $\phi(y_1, y_2) = \phi_1(y_1)\phi_2(y_2)$, for $y_i \in X_i$. Then the differential of ϕ at (x_1, x_2) is surjective.*

Proof. Note that ϕ is the composition of the maps $X_1 \times X_2 \to G_1 \times G_2$, given by $(y_1, y_2) \mapsto (\phi_1(y_1), \phi_2(y_2))$ and $G_1 \times G_2 \to G$, given by $(g_1, g_2) \mapsto g_1 g_2$. The differential of the latter map is $(x, y) \mapsto x + y$. This implies the lemma. $\square$

2.1 Unipotent Case

Let $U \subset \mathrm{GL}(n, k)$ be a unipotent algebraic group with Lie algebra $\mathfrak{u} \subset \mathfrak{gl}(n, k)$. Let $u_1, \ldots, u_s$ be a basis of $\mathfrak{u}$. Let $\alpha_1, \ldots, \alpha_s \in k$, then $\exp(\sum_{i=1}^s \alpha_i u_i)$ lies in U. Moreover, the map $\mathbb{A}^s \to U$ given by $(\alpha_1, \ldots, \alpha_s) \mapsto \exp(\sum_{i=1}^s \alpha_i u_i)$ is bijective and regular (see [4], Chap. V, Sect. 3, No. 4, Proposition 14). Its differential (at the point $(0, \ldots, 0)$) is the map $(\alpha_1, \ldots, \alpha_s) \mapsto \sum_{i=1}^s \alpha_i u_i$.

2.2   Diagonalisable Case

For a subset $E \subset \mathbb{Z}^n$ set

$$T_E = \{\text{diag}(\alpha_1, \ldots, \alpha_n) \in k^n \mid \prod_{i=1}^{n} \alpha_i^{e_i} = 1 \text{ for all } e = (e_1, \ldots, e_n) \in E\}.$$

Then T_E is an algebraic subgroup of $\text{GL}(n, k)$, with Lie algebra

$$\mathfrak{t}_E = \{\text{diag}(\alpha_1, \ldots, \alpha_n) \in k^n \mid \sum_{i=1}^{n} e_i \alpha_i = 0 \text{ for all } e = (e_1, \ldots, e_n) \in E\}.$$

It is clear that $T_E = T_\Lambda$, where $\Lambda \subset \mathbb{Z}^n$ is the lattice generated by E.

Now let $T \subset \text{GL}(n, k)$ be an algebraic group consisting of diagonal matrices. Then there is an $E \subset \mathbb{Z}^n$ such that $T = T_E$ (see [30], Sect. 22.5). So, if $\mathfrak{t} \subset \mathfrak{gl}(n, k)$ is an algebraic Lie algebra consisting of diagonal matrices, then there is an $E \subset \mathbb{Z}^n$ with $\mathfrak{t} = \mathfrak{t}_E$. Let $\Lambda_\mathfrak{t} \subset \mathbb{Z}^n$ be the lattice defined by

$$\Lambda_\mathfrak{t} = \{(e_1, \ldots, e_n) \in \mathbb{Z}^n \mid \sum_{i=1}^{n} e_i \alpha_i = 0 \text{ for all } \text{diag}(\alpha_1, \ldots, \alpha_n) \in \mathfrak{t}\},$$

then $\mathfrak{t} = \mathfrak{t}_E$, where E is a basis of $\Lambda_\mathfrak{t}$. Moreover, $\mathfrak{t}$ is the Lie algebra of the group T_E.

Remark 2. Concerning the problem to compute a basis of $\Lambda_\mathfrak{t}$ we remark the following. Of course it suffices to have $\sum_{i=1}^{n} e_i \alpha_i = 0$, where $\text{diag}(\alpha_1, \ldots, \alpha_n)$ runs through a basis of $\mathfrak{t}$. If the α_i lie in a fixed number field K, then this leads to linear equations for the e_i with coefficients in $\mathbb{Q}$ (after expressing the α_i as linear combinations of a fixed basis of K over $\mathbb{Q}$). Then a basis of $\Lambda_\mathfrak{t}$ can be found by the *saturation* algorithm. The paper [11] contains a description of such an algorithm, based on the Smith normal form algorithm. MAGMA ([2]) contains an implementation of a different algorithm for computing saturations.

Let $E = \{e^1, \ldots, e^r\}$ be a basis of $\Lambda_\mathfrak{t}$. Let $L \subset \mathbb{Z}^n$ be the lattice consisting of all $d = (d_1, \ldots, d_n)$ such that $\sum_i d_i e_i^k = 0$ for $1 \leq k \leq r$. Let $\{d^1, \ldots, d^s\}$ be a basis of L. Define $\psi_j : k^* \to T_E$ by $\psi_j(t) = \text{diag}(t^{d_1^j}, \ldots, t^{d_n^j})$. Then the differential of ψ_j at 1 maps $\alpha \in k$ to $\alpha \text{diag}(d_1^j, \ldots, d_n^j)$. Now consider the map $\psi : (k^*)^s \to T_E$, given by $\psi(t_1, \ldots, t_s) = \psi_1(t_1) \cdots \psi_s(t_s)$. By Lemma 1 its differential at the point $(1, \ldots, 1)$ is surjective.

Finally, let $\widetilde{T} \subset \text{GL}(n, k)$ be a diagonalisable connected algebraic group, i.e., there is an $A \in \text{GL}(n, k)$ such that $T = A\widetilde{T}A^{-1}$ consists of diagonal matrices. By the construction above we get a regular map $\psi : (k^*)^s \to T$ with a surjective differential. So also the differential of $\tilde{\psi} : (k^*)^s \to \widetilde{T}$, $\tilde{\psi}(a) = A^{-1}\psi(a)A$ is surjective.

2.3   General Case

Let $G \subset \mathrm{GL}(n, k)$ be a connected algebraic group, with Lie algebra $\mathfrak{g} \subset \mathfrak{gl}(n, k)$. Let $\mathfrak{s}$ denote the solvable radical of $\mathfrak{g}$, and $\mathfrak{l}$ a Levi subalgebra of $\mathfrak{g}$, and $\mathfrak{n}$ the largest ideal of $\mathfrak{g}$ consisting of nilpotent elements. Then by [4], Chapter V, Sect. 4, Proposition 5 (see also [24]), $\mathfrak{s}$ has a commutative subalgebra $\mathfrak{d}$ consisting of semisimple elements, with the following properties

1. $\mathfrak{s} = \mathfrak{d} + \mathfrak{n}$ (semidirect sum),
2. $[\mathfrak{l}, \mathfrak{d}] = 0$,
3. $\mathfrak{n}$ is the set consisting of all nilpotent elements of $\mathfrak{s}$.

There exist algorithms to obtain bases of $\mathfrak{s}$ and $\mathfrak{l}$, cf. [10]. Moreover, in [11] algorithms are described for obtaining bases of $\mathfrak{d}$, $\mathfrak{n}$. Let $\mathfrak{h}$ be a Cartan subalgebra of $\mathfrak{l}$, and $\mathfrak{n}^+$, $\mathfrak{n}^-$ the sums of the positive, respectively negative, root spaces with respect to $\mathfrak{h}$ (relative to a choice of positive roots). Then

$$\mathfrak{g} = \mathfrak{n}^- \oplus \mathfrak{h} \oplus \mathfrak{n}^+ \oplus \mathfrak{d} \oplus \mathfrak{n}.$$

All these subalgebras are algebraic. Furthermore, they either consist of nilpotent elements, or are diagonalisable. So by applying the constructions of Sects. 2.1, 2.2, along with Lemma 1, we get

- an affine space $\mathbb{A}^{2r+s}$, with coordinate ring $k[x_1, \ldots, x_r, y_1, \ldots, y_r, z_1, \ldots, z_s]$,
- a closed set $X^{r,s}$ of $\mathbb{A}^{2r+s}$, given by the equations $x_i y_i = 1$ for $1 \leq i \leq r$,
- a regular map $\sigma : X^{r,s} \to G$, and a point x of $X^{r,s}$ such that the differential of σ at x maps $T_x(X^{r,s})$ surjectively onto $\mathfrak{g}$.

The last statement implies that the image of σ contains a nonempty open subset of G (since the image of σ contains an open subset of its closure (see [27], Sect. I.5.3, Theorem 6) which is G).

Remark 3. Suppose that G and hence $\mathfrak{g}$ are defined over the subfield K of k (e.g, over $\mathbb{Q}$). Then it can happen that the Cartan subalgebra $\mathfrak{h}$ is not split over K. In that case the image of σ will not be defined over K, but over a finite extension of K.

Example 4. In this example we give a basis of an algebraic and reductive Lie subalgebra of $\mathfrak{gl}(6, k)$. For this the 6×6-matrix with a 1 on position (i, j) and zeros elsewhere will be denoted $e_{i,j}$.

Let $\mathfrak{g} = \mathfrak{n}^- \oplus \mathfrak{t} \oplus \mathfrak{n}^+ \subset \mathfrak{gl}(6, k)$, where $\mathfrak{t}$ is diagonalisable (in fact, diagonal) and spanned by

$$-e_{3,3} - e_{4,4} + e_{5,5} + e_{6,6}, \quad -e_{1,1} - e_{2,2} + e_{3,3} + e_{4,4}, \quad 2e_{1,1} + e_{3,3} - e_{4,4}$$
$$+ e_{5,5} - e_{6,6}, \quad 2e_{2,2} - e_{3,3} + e_{4,4} - e_{5,5} + e_{6,6},$$

$\mathfrak{n}^+$ is spanned by

$$x_{\alpha_1} = -e_{5,3} - e_{6,4}, \quad x_{\alpha_2} = -e_{3,1} - e_{4,2}, \quad x_{\alpha_3} = -e_{5,1} - e_{6,2},$$

and $\mathfrak{n}^-$ is spanned by

$$x_{-\alpha_1} = -e_{3,5} - e_{4,6}, \quad x_{-\alpha_2} = -e_{1,3} - e_{2,4}, \quad x_{-\alpha_3} = -e_{1,5} - e_{2,6}.$$

This Lie algebra is reductive, with semisimple part of type A_2. The $x_{\pm\alpha_i}$ are the root vectors, with $\alpha_3 = \alpha_1 + \alpha_2$.

We apply the construction of Sect. 2.2 to $\mathfrak{t}$. Here we get that $\Lambda_{\mathfrak{t}}$ is spanned by

$$(1, -1, 0, 0, -1, 1), \quad (0, 0, 1, -1, -1, 1).$$

Furthermore, a basis of L (the lattice perpendicular to $\Lambda_{\mathfrak{t}}$) consists of

$$(1, 0, 0, -1, 0, -1), \quad (0, 1, 0, 1, 0, 1), \quad (0, 0, 1, 1, 0, 0), \quad (0, 0, 0, 0, 1, 1).$$

Consider the affine space $\mathbb{A}^8$ with coordinate ring $k[x_1, \ldots, x_4, y_1, \ldots, y_4]$. Let $X_{\mathfrak{t}}$ be the closed set in $\mathbb{A}^8$ given by $x_i y_i = 1$ for $1 \leq i \leq 4$. Then we get the map $\psi : X_{\mathfrak{t}} \to T$ (where $T \subset \mathrm{GL}(6, k)$ is the connected algebraic group with Lie algebra $\mathfrak{t}$), given by

$$\psi(t_1, \ldots, t_4, s_1, \ldots, s_4) = \mathrm{diag}(t_1, t_2, t_3, s_1 t_2 t_3, t_4, s_1 t_2 t_4).$$

Let N^+ (respectively N^-) be the connected algebraic subgroup of $\mathrm{GL}(6, k)$ with Lie algebra $\mathfrak{n}^+$ (respectively, $\mathfrak{n}^-$). Using the construction of Sect. 2.1 we get maps $\phi_+ : \mathbb{A}^3 \to N^+$, $\phi_- : \mathbb{A}^3 \to N^-$ given by

$$\phi_+(u_1, u_2, u_3) = \exp(u_1(-e_{5,3} - e_{6,4}) + u_2(-e_{3,1} - e_{4,2}) + u_3(-e_{5,1} - e_{6,2}))$$

$$= \begin{pmatrix} 1 & 0 & 0 & 0 & 0 & 0 \\ 0 & 1 & 0 & 0 & 0 & 0 \\ -u_2 & 0 & 1 & 0 & 0 & 0 \\ 0 & -u_2 & 0 & 1 & 0 & 0 \\ \frac{1}{2}u_1 u_2 - u_3 & 0 & -u_1 & 0 & 1 & 0 \\ 0 & \frac{1}{2}u_1 u_2 - u_3 & 0 & -u_1 & 0 & 1 \end{pmatrix},$$

and

$$\phi_-(v_1, v_2, v_3) = \exp(v_1(-e_{3,5} - e_{4,6}) + v_2(-e_{1,3} - e_{2,4}) + v_3(-e_{1,5} - e_{2,6}))$$

$$= \begin{pmatrix} 1 & 0 & -v_2 & 0 & \frac{1}{2}v_1 v_2 - v_3 & 0 \\ 0 & 1 & 0 & -v_2 & 0 & \frac{1}{2}v_1 v_2 - v_3 \\ 0 & 0 & 1 & 0 & -v_1 & 0 \\ 0 & 0 & 0 & 1 & 0 & -v_1 \\ 0 & 0 & 0 & 0 & 1 & 0 \\ 0 & 0 & 0 & 0 & 0 & 1 \end{pmatrix}.$$

Denote the coordinate ring of $k[\mathbb{A}^{14}]$ by $k[x_1, \ldots, x_4, y_1, \ldots, y_4, z_1, \ldots, z_6]$. Let $X^{4,6} \subset \mathbb{A}^{14}$ be given by the equations $x_i y_i = 1$, $i = 1, \ldots, 4$. Then we get the map $\sigma : X^{4,6} \to G$, given by

$$\sigma(t_1, \ldots, t_4, s_1, \ldots, s_4, u_1, u_2, u_3, v_1, v_2, v_3) =$$
$$\phi_-(v_1, v_2, v_3)\psi(t_1, \ldots, t_4, s_1, \ldots, s_4)\phi_+(u_1, u_2, u_3).$$

It has the property that its image contains a non-empty open subset of G.

3   Computing the Closure by Gröbner Elimination

Here we show how polynomial equations defining the closure of an orbit can be computed by a technique based on Gröbner bases. To the best of our knowledge, this approach was first described by Popov in [26].

Let $G \subset \mathrm{GL}(n, k)$ be a connected algebraic group with Lie algebra $\mathfrak{g} \subset \mathfrak{gl}(n, k)$. Let $X^{r,s} \subset \mathbb{A}^{2r+s}$ and $\sigma : X^{r,s} \to G$ be as in Sect. 2. Let $V = k^n$ be the space on which G acts, and let $v \in V$. Let $Y \subset V$ be the closure of $G \cdot v$. Define a map $\sigma_v : X^{r,s} \to Y$ by $\sigma_v(x) = \sigma(x) \cdot v$. Then σ_v is a regular map, and its image contains a nonempty open subset of Y. So the closure of the image of σ_v is Y.

Defining equations for the closure of the image of a regular map can be computed using elimination techniques based on Gröbner bases (see, for example, [5]). In the situation considered here this works as follows. Let $e_1, \ldots, e_n$ be a fixed basis of V. Write $\sigma_v(x) = \sum_{i=1}^{n} \sigma_v^i(x) e_i$. Then $\sigma_v^i : X^{r,s} \to k$ is a regular map, i.e., it is the restriction of a polynomial in $k[x_1, \ldots, x_r, y_1, \ldots, y_r, z_1, \ldots, z_s]$ to $X^{r,s}$; we denote this polynomial also by σ_v^i. Now let $T_1, \ldots, T_n$ be auxiliary indeterminates, and let R be the polynomial ring over k with the indeterminates x_i, y_i $(1 \le i \le r)$, z_j $(1 \le j \le s)$ and T_m $(1 \le m \le n)$. Let I be the ideal of R generated by $x_i y_i - 1$ for $1 \le i \le r$ and $T_m - \sigma_v^m$ for $1 \le m \le n$. Use a monomial order relative to which any monomial involving at least one of x_i, y_i or z_j is bigger than any monomial involving only the indeterminates T_m. Let $\mathcal{G}$ be a Gröbner basis of I with respect to such an order. Then $\mathcal{G} \cap k[T_1, \ldots, T_n]$ is a Gröbner basis of $J = I \cap k[T_1, \ldots, T_n]$. Moreover, the closed set in V defined by J is Y.

Example 5. Let the notation be as in Example 4. Let $v = e_5 + e_6$ (where e_i denotes the i-th elementary basis element of k^6). Then $\sigma(t_1, \ldots, t_4, s_1, \ldots, s_4, u_1, u_2, u_3, v_1, v_2, v_3) \cdot v$ is

$$(t_4(\tfrac{1}{2}v_1v_2 - v_3), s_1t_2t_4(\tfrac{1}{2}v_1v_2 - v_3), -t_4v_1, -s_1t_2t_4v_1, t_4, s_1t_2t_4).$$

Let I be the ideal constructed as above. MAGMA computes a Gröbner basis $\mathcal{G}$ of I, with respect to an elimination order, in 0.02 s. The intersection of $\mathcal{G}$ with $k[T_1, \ldots, T_6]$ is

$$\{T_2T_3 - T_1T_4, T_2T_5 - T_1T_6, T_4T_5 - T_3T_6\}.$$

So the closure Y of $G \cdot v$ is the set of $\sum_{i=1}^{6} \alpha_i e_i$ such that $\alpha_2\alpha_3 = \alpha_1\alpha_4$, and so on. We see that e_5 and e_6 lie in Y, so their orbits do as well. On the other hand, $e_4 + e_5$ does not lie in Y.

4   Popov's Algorithm

In this section we describe an algorithm due to V. L. Popov ([26]). It is a method for deciding whether a given w lies in the closure of $G \cdot v$.

4.1 Conical Groups

We say that an algebraic group $G \subset \mathrm{GL}(n, k)$ is *conical* if $\lambda I_n \in G$ for all nonzero $\lambda \in k$, where I_n is the $n \times n$-identity matrix. If G is conical, then also an orbit $G \cdot v$ is *conical*, i.e., $w \in G \cdot v$ if and only if $\lambda w \in G \cdot v$ for all nonzero $\lambda \in k$.

Now let $G \subset \mathrm{GL}(n, k)$ be an algebraic group. Set $V = k^n$, the natural G-module. Let $V_0 = k^{n+1}$ and write elements of V_0 as (α_0, v), where $\alpha_0 \in k$ and $v \in V$. A $g \in G$ acts on V_0 by $g \cdot (\alpha_0, v) = (\alpha_0, g \cdot v)$. By G_0 we denote the corresponding image of G in $\mathrm{GL}(V_0)$. Let I_{V_0} be the identity endomorphism of V_0 and let $D_0 = \{\lambda I_{V_0} \mid \lambda \in k, \lambda \neq 0\}$. Set $\widetilde{G} = G_0 D_0$. Then $\widetilde{G}$ is an algebraic subgroup of $\mathrm{GL}(V_0)$.

Lemma 6. *Let $v \in V$. A w in V lies in the closure of $G \cdot v$ if and only if $(1, w)$ lies in the closure of $\widetilde{G} \cdot (1, v)$.*

Proof. Let I be the vanishing ideal of $G \cdot v$ in $k[x_1, \ldots, x_n]$, and J the vanishing ideal of $\widetilde{G} \cdot (1, v)$ in $k[x_0, x_1, \ldots, x_n]$. Since the latter orbit is conical, J is spanned by homogeneous polynomials. By $k[x_1, \ldots, x_n]_{\leq d}$ we denote the space of polynomials of degree at most d. By $k[x_0, x_1, \ldots, x_n]_d$ we denote the space of polynomials of degree d. Consider the maps $\gamma : k[x_0, x_1, \ldots, x_n]_d \rightarrow k[x_1, \ldots, x_n]_{\leq d}$, $\delta : k[x_1, \ldots, x_n]_{\leq d} \rightarrow k[x_0, x_1, \ldots, x_n]_d$ given by $\gamma(f) = f(1, x_1, \ldots, x_n)$, $\delta(g) = x_0^d g(x_1/x_0, \ldots, x_n/x_0)$. These are inverses of each other. Set $I_{\leq d} = I \cap k[x_1, \ldots, x_n]_{\leq d}$ and $J_d = J \cap k[x_0, x_1, \ldots, x_n]_d$. It is straightforward to see that $\gamma(J_d) \subseteq I_{\leq d}$, $\delta(I_{\leq d}) \subseteq J_d$. So in fact these inclusions are equalities. Let w lie in the closure of $G \cdot v$, then $f(w) = 0$ for all $f \in I$. Let $h \in J_d$, then $h = \delta(f)$ for a certain $f \in I_{\leq d}$. This implies that $h(1, w) = 0$. We conclude that $(1, w)$ lies in the closure of $\widetilde{G} \cdot (1, v)$. The reverse implication is shown analogously. $\square$

We conclude that we may assume that the algebraic group, and hence the orbit, is conical.

4.2 The Degree of G

Let $X \subset \mathbb{A}^n$ be a closed set of dimension ℓ. Let $\Omega \subset \mathbb{A}^n$ be a plane of dimension $n - \ell$, "in general position". Then the number of points of $X \cap \Omega$ is called the *degree* of X. For a closed set $Y \subset \mathbb{P}^n$ the concept of degree is defined in the same way (see [14]). Now suppose that the vanishing ideal of X is spanned by homogeneous polynomials. Then we can also view X as a closed set in $\mathbb{P}^{n-1}$. Moreover, the notions of degree of X, seen as closed set in $\mathbb{A}^n$, and degree of X, seen as closed set in $\mathbb{P}^{n-1}$, coincide.

A closed set Y in projective space gives rise to a polynomial known as the *Hilbert polynomial* (see [14]). If the degree of Y is d and the dimension is ℓ, then the leading term of this polynomial is $\frac{d}{\ell!} x^\ell$.

Bayer and Stillman ([1]) have described algorithms for computing the Hilbert series, and Hilbert polynomial of a closed set in projective space. So using that it is possible to compute the degree of such a closed set, or of a conical closed set in affine space. Since, by the previous subsection, we may assume that the group that we are dealing with is conical, (this implies that it is an open set of a conical closed set) we can compute the degree of the group in this way.

Example 7. Polynomial equations for the group G, given its Lie algebra, can be computed using the methods of [11]. For the group of Example 4 this computation, performed in MAGMA, took $19669\,$s. The result is a set of homogeneous equations defining a closed set G' in $\mathrm{End}(V)$. Furthermore, G consists of all $g \in G'$ such that $\det(g) \neq 0$. The Hilbert polynomial of G' is then computed in $0.000\,$s. It is

$$\frac{1}{403200}X^{10} + \frac{83}{725760}X^9 + \frac{23}{10080}X^8 + \frac{625}{24192}X^7 + \frac{1171}{6400}X^6 + \frac{29111}{34560}X^5 +$$
$$\frac{102293}{40320}X^4 + \frac{177911}{36288}X^3 + \frac{291197}{50400}X^2 + \frac{9397}{2520}X + 1.$$

From the leading term we see that the dimension is 10 (we knew that already), and the degree is 9.

   We see that a disadvantage of this approach is the necessity to know defining polynomials of G. However, if they are available from the start, then this is a good way to compute the degree.

In the case where G is reductive there exists a different method based on a formula due to Kazarnovskiĭ ([20]). Here we do not describe this formula (for that see Kazarnovskiĭ's paper, or [7], Sect. 4.7, or [26]), but instead illustrate it for the group G of Example 4.

Example 8. Let the notation be as in Example 4. Then T is a maximal torus of G. The elements of T are

$$t = \mathrm{diag}(t_1, t_2, t_3, t_1^{-1}t_2t_3, t_4, t_1^{-1}t_2t_4).$$

The characters, $\chi : T \to k^*$ are given by $\chi(t) = t_1^{d_1} \cdots t_4^{d_4}$, where $d_i \in \mathbb{Z}$. We let X be the character group of T, written additively: so it is a free $\mathbb{Z}$-module with basis $\chi_1, \ldots, \chi_4$, where $\chi_i(t) = t_i$, and a character χ as above corresponds to $d_1\chi_1 + \cdots + d_4\chi_4$. We also represent χ by the row vector $(d_1, \ldots, d_4)$. Now the basis vectors $e_1, \ldots, e_6$ of $V = k^6$ are weight vectors of T. The corresponding weights are

$$(1,0,0,0),\ (0,1,0,0),\ (0,0,1,0),\ (-1,1,1,0),\ (0,0,0,1),\ (-1,1,0,1).$$

We let $\mathcal{P} \subset E = \mathbb{R}^4$ be the polytope that is the convex hull of these points along with $(0,0,0,0)$.

   The group G acts on $\mathfrak{g}$ via $g \cdot x = gxg^{-1}$. For this action the $x_{\pm\alpha_i}$ are weight vectors of T, and we denote the corresponding characters by $\pm\alpha_i$. Then

$$\alpha_1 = (0,0,-1,1),\ \alpha_2 = (-1,0,1,0),\ \alpha_3 = (-1,0,0,1).$$

Let $W = N_G(T)/T$ be the Weyl group of G. This group acts on X, and hence on E. We need a W-invariant inner product $(\,,\,)$ on E, which we get as follows.

We can also view the elements of X as elements of the dual space of $\mathfrak{t}$. Here a $\chi \in X$ corresponds its differential (at the identity) $d\chi \in \mathfrak{t}^*$: for χ as above and $u \in \mathfrak{t}$ we have $d\chi(u) = d_1 u_{11} + d_2 u_{22} + d_3 u_{33} + d_4 u_{55}$. As $W \cong N_G(\mathfrak{t})/Z_G(\mathfrak{t})$, it also acts on $\mathfrak{t}$. A W-invariant nondegenerate positive definite bilinear form on $\mathfrak{t}$ is given by $(x,y) = \mathrm{Tr}(xy)$. Using this bilinear form we identify $\mathfrak{t}$ and its dual space. By this identification we then also get a W-invariant bilinear form on the dual of $\mathfrak{t}$, and hence on E. In the basis of the χ_i this form is given by the matrix

$$\begin{pmatrix} 4 & 2 & 1 & 1 \\ 2 & 4 & -1 & -1 \\ 1 & -1 & 4 & 1 \\ 1 & -1 & 1 & 4 \end{pmatrix}.$$

Now we consider the functions $\alpha_i^\vee : E \to \mathbb{R}$ given by $\alpha_i^\vee(v) = \frac{2(v,\alpha_i)}{(\alpha_i,\alpha_i)}$. Let $v = \sum_i d_i \chi_i$. Then $\alpha_1^\vee(v) = -d_3 + d_4$, $\alpha_2^\vee(v) = -d_1 - d_2 + d_3$, $\alpha_3^\vee(v) = -d_1 - d_2 + d_4$. Let I be the value of the integral of the function $(\alpha_1^\vee)^2(\alpha_2^\vee)^2(\alpha_3^\vee)^2$ over $\mathcal{P}$. The package LATTE INTEGRALE ([6]) can perform this integration. After $0.01\,\mathrm{s}$ the program returned $I = \frac{1}{16800}$.

Now let $m_1, \ldots, m_r$ be the *exponents* of W (see [16], Sect. 3.16). Here the root system is of type A_2, and therefore $r = 2$, $m_1 = 1$, $m_2 = 2$. According to Kazarnovskiĭ's formula we have

$$\deg(G) = \frac{\dim(G)!}{|W|(m_1! m_2!)^2} I = 9.$$

4.3   The Algorithm

Here we let $G \subset \mathrm{GL}(n,k)$ be a conical algebraic group. By d we denote its degree. Let $\sigma : X^{r,s} \to G$ be an embedding as constructed in Sect. 2. Here $X^{r,s} \subset \mathbb{A}^{2r+s}$, and the coordinate ring of $X^{r,s}$ is $R_{r,s} = k[x_1, \ldots, x_r, y_1, \ldots, y_r, z_1, \ldots, z_s]/I_{r,s}$, where $I_{r,s}$ is generated by $x_i y_i - 1$, $1 \le i \le r$. Note that a basis of $R_{r,s}$ is given by the set of monomials $x_1^{k_1} \cdots x_r^{k_r} z_1^{m_1} \cdots z_s^{m_s}$, with $k_i \in \mathbb{Z}$ and $m_i \in \mathbb{N}$ (here we write x_i^{-1} in place of y_i). So it is straightforward to compute in the ring $R_{r,s}$.

Let $v \in k^n$, and let Y be the closure of the orbit $G \cdot v$. Then we consider the regular map $\psi : X^{r,s} \to Y$ given by $\psi(a) = \sigma(a) \cdot v$. Since the image of σ contains a dense and open subset of G, also the image of ψ is dense in Y. Let $k[T_1, \ldots, T_n]$ be the coordinate ring of k^n. We view the elements of $k[T_1, \ldots, T_n]$ as functions $k^n \to k$ (i.e., by fixing a basis of k^n), and let $\overline{T}_i$ be the restriction of $\overline{T}_i$ to Y. Let $\psi^* : k[Y] \to R_{r,s}$ be the comorphism of ψ (i.e., $\psi^*(f)(a) = f(\psi(a))$). Set $S_i = \psi^*(\overline{T}_i)$.

Theorem 9. *Let* $w \in k^n$ *and write* $w = (w_1, \ldots, w_n)$. *Let* $j_1, \ldots, j_m$ *be the indices such that* $\overline{T}_{j_k} \ne 0$. *Suppose that* $w_i = 0$ *if* $i \notin \{j_1, \ldots, j_m\}$. *Then* $w \notin Y$ *if and only if there are* $F_1, \ldots, F_m \in k[Z_1, \ldots, Z_m]$ *such that* $\deg(F_i) \le 2d - 2$ *and*

$$\sum_{k=1}^m (S_{j_k} - w_{j_k}) F_k(S_{j_1}, \ldots, S_{j_m}) = 1.$$

Proof. (Sketch) Note that $w \notin Y$ if and only if there is no point of Y where the functions $\overline{T}_{j_k} - w_{j_k}$ simultaneously vanish. By Hilbert's Nullstellensatz this is equivalent to the existence of $f_1, \ldots, f_m \in k[Y]$ such that $\sum_{i=1}^{m}(\overline{T}_{j_k} - w_{j_k})f_k = 1$. Now the effective Nullstellensatz (see [17]) strengthens this by also giving a bound on the degrees of the f_i. Here the bound turns out to be $\deg((\overline{T}_{j_k} - w_{j_k})f_k) \leq 2\deg(X) - 1$.

The degree of an $f \in k[Y]$ is the minimum degree of an $\hat{f} \in k[T_1, \ldots, T_n]$ whose restriction to Y equals f. Since Y is irreducible and conical we have that $\deg(fg) = \deg(f) + \deg(g)$ for all $f, g \in k[Y]$. Moreover, $\deg(\overline{T}_{j_k}) = 1$ for $1 \leq k \leq m$.

Also, since Y is (the closure of) the image of G under the map $\mathrm{End}(V) \to V$, $g \mapsto g \cdot v$, we have $\deg(Y) \leq \deg(G) = d$ (see [7], Proposition 4.7.10).

It follows that we get the bound $\deg(f_k) \leq 2d - 2$. Now mapping the equality $\sum_{i=1}^{m}(\overline{T}_{j_k} - w_{j_k})f_k = 1$ with ψ^* to $R_{r,s}$, we get the statement of the lemma. Note that ψ^* is injective, as the image of ψ contains a dense open subset of Y. $\qquad\qquad\qquad\qquad\qquad\qquad\qquad\qquad\qquad\qquad\qquad\qquad\qquad\square$

The idea of the algorithm is to write $F_1, \ldots, F_m \in k[Z_1, \ldots, Z_m]$ of degree $2d - 2$, with unknowns as coefficients. Then the condition in the theorem yields a set of linear equations for these unknowns, since we have a basis of $R_{r,s}$. Then $w \in Y$ if and only if these equations have no solution. So the number of unknowns in the equations that we get is

$$m\binom{m + 2d - 2}{m}.$$

The number of equations is equal to the number of different monomials in $R_{r,s}$ that appear when evaluating a monomial of $k[Z_1, \ldots, Z_m]$, of degree $\leq 2d - 2$, in $S_1, \ldots, S_m$.

Example 10. We let the notation be as in Example 4. As seen in Examples 7 and 8, here we have $d = 9$. With v as in Example 5 we get $m = 6$. So the number of unknowns in this case will be 447678.

Remark 11. It is also possible to base a different algorithm on Theorem 9. Let A be the subalgebra of $R_{r,s}$ generated by $S_{j_1}, \ldots, S_{j_m}$. By elimination based on Gröbner bases we can compute all polynomial relations among these generators. So we get an isomorphism $k[U_1, \ldots, U_m]/J \to A$. Now $w \notin X$ if and only if the ideal of $k[U_1, \ldots, U_m]$ generated by J and $U_k - w_{j_k}$ contains 1. This again can be checked by a Gröbner basis computation. However, note that computing the polynomial relations among the S_{j_k} leads to the same computation as computing defining polynomials for the closure of $G \cdot v$, cf. Example 5.

5 Closures of Nilpotent Orbits of θ-Groups

The algebraic groups known as θ-groups are reductive, and arise from gradings of (semi-) simple Lie algebras. They have been introduced and studied by Vinberg in the 70's ([31–33]). We divide this section in a number of subsections. In the first of these we sketch how θ-groups come about; for more detailed information we refer to the above mentioned papers of Vinberg. The second contains some general facts on nilpotent orbits of θ-groups. In the remaining sections we sketch an algorithm from [9] to decide whether a given nilpotent orbit is contained in the closure of another one.

5.1 θ-Groups

Let $\mathfrak{g}$ be a (semi-) simple Lie algebra over k. Let G be the connected component of the identity of the automorphism group of $\mathfrak{g}$. This group has Lie algebra $\mathrm{ad}\mathfrak{g}$, which we identify with $\mathfrak{g}$. By $\mathbb{Z}_m$ we denote the group $\mathbb{Z}/m\mathbb{Z}$, which, by convention, is equal to $\mathbb{Z}$ if $m = \infty$. We consider $\mathbb{Z}_m$-gradings of $\mathfrak{g}$:

$$\mathfrak{g} = \bigoplus_{i \in \mathbb{Z}_m} \mathfrak{g}_i.$$

This means that the $\mathfrak{g}_i$ are subspaces of $\mathfrak{g}$ such that $[\mathfrak{g}_i, \mathfrak{g}_j] \subset \mathfrak{g}_{i+j}$. In particular, $\mathfrak{g}_0$ is a subalgebra, and it can be shown that it is reductive. Let G_0 be the connected subgroup of G with Lie algebra $\mathfrak{g}_0$ (or, more precisely, $\mathrm{ad}\mathfrak{g}_0$). Since $[\mathfrak{g}_0, \mathfrak{g}_1] \subset \mathfrak{g}_1$ we also get that G_0 stabilizes $\mathfrak{g}_1$. The corresponding representation of G_0 is called a θ-representation, and G_0 is called a θ-group.

Example 12. In this example we consider the Lie algebra of type D_4. We start by giving a multiplication table of $\mathfrak{g}$. Let Φ be the root system of type D_4, with basis of simple roots $\Delta = \{\beta_1, \beta_2, \beta_3, \beta_4\}$. The simple roots correspond to the nodes of the Dynkin diagram in the following way

Let Q denote the $\mathbb{Z}$-span of Δ. We define a bilinear form $(\ ,\ )$ on Q by $(\beta_i, \beta_i) = 2$, $(\beta_i, \beta_j) = -1$ if $i \neq j$ and i, j are connected in the Dynkin diagram. Otherwise $(\beta_i, \beta_j) = 0$. Also we define a group homomorphism $\varepsilon : Q \times Q \to \{1, -1\}$ (where the latter is a multiplicative group), by $\varepsilon(\beta_i, \beta_j) = -1$ if $i = j$ or $i \neq j$ and $i < j$ and i, j are connected in the Dynkin diagram. Otherwise $\varepsilon(\beta_i, \beta_j) = 1$. Also we define $\tilde{\varepsilon} : \Phi \times \Phi \to \{1, -1,\}$ by $\tilde{\varepsilon}(\beta, \beta') = \varepsilon(\beta, \beta')$ if both β and β' are positive, or exactly one of them is positive and $\beta + \beta'$ is negative. Otherwise $\tilde{\varepsilon}(\beta, \beta') = -\varepsilon(\beta, \beta')$.

Let $\mathfrak{g}$ be the 28-dimensional vector space spanned by $h_1, \ldots, h_4$, along with x_β for $\beta \in \Phi$. On $\mathfrak{g}$ we define the Lie bracket by

$$[h_i, h_j] = 0$$
$$[h_i, x_\beta] = (\beta, \alpha_i)x_\beta$$
$$[x_\beta, x_{-\beta}] = h_\beta$$
$$[x_\beta, x_{\beta'}] = \tilde{\varepsilon}(\beta, \beta')x_{\beta+\beta'},$$

where we set $h_\beta = \sum_{i=1}^{4} m_i h_i$ for $\beta = \sum_{i=1}^{4} m_i \beta_i$, and $x_\gamma = 0$ if $\gamma \in Q \setminus \Phi$. With this Lie bracket, $\mathfrak{g}$ is a simple Lie algebra of type D_4 (cf. [18], Proposition 7.8, and also [10], Proposition 5.13.4). (We use $\tilde{\varepsilon}$ instead of ε because in the mentioned references the relation $[x_\beta, x_{-\beta}] = -h_\beta$ is used.)

Next we define a grading on $\mathfrak{g}$. The simple roots corresponding to the black nodes in the Dynkin diagram have degree 1, the others have degree 0. Also, the degree of a sum of simple roots will be the sum of the degrees. Finally, the degree of a negative root β is minus the degree of $-\beta$. Then we define a $\mathbb{Z}$-grading of $\mathfrak{g}$ by letting $\mathfrak{g}_0$ be equal to the span of the h_i along with all x_β such that the degree of β is 0. Furthermore, $\mathfrak{g}_i$ will be the span of all x_β such that the degree of β is i.

Then $\mathfrak{g}_0$ is the sum of a simple Lie algebra of type A_2, and a 2-dimensional torus. The space $\mathfrak{g}_1$ is spanned by $x_{\beta_3}, x_{\beta_4}, x_{\beta_2+\beta_3}, x_{\beta_2+\beta_4}, x_{\beta_1+\beta_2+\beta_3}, x_{\beta_1+\beta_2+\beta_4}$.

The image of $\mathfrak{g}_0$ in $\mathfrak{gl}(6, k)$ (corresponding to the action on $\mathfrak{g}_1$) is the Lie algebra considered in Example 4. The matrices given in that example are defined relative to the basis above.

5.2 Nilpotent Orbits of θ-Groups

Let the notation be as in the previous subsection. Let $e \in \mathfrak{g}_1$. Then the orbit $G_0 e$ is said to be *nilpotent* if the (Zariski-) closure of $G_0 e$ contains 0. We say that $e \in \mathfrak{g}_1$ is nilpotent if the orbit $G_0 e$ is nilpotent. It can be shown that this is equivalent to $\mathrm{ad} e$ being a nilpotent endomorphism of $\mathfrak{g}$.

Let $e \in \mathfrak{g}_1$ be nilpotent. Then there are $h \in \mathfrak{g}_0$, $f \in \mathfrak{g}_{-1}$ such that $[e, f] = h$, $[h, e] = 2e$, $[h, f] = -2f$. We say that e lies in the homogeneous $\mathfrak{sl}_2$-triple (h, e, f). The element h is called a *characteristic* of e.

Let $e_i \in \mathfrak{g}_1$, for $i = 1, 2$, be nilpotent elements lying in homogeneous $\mathfrak{sl}_2$-triples (h_i, e_i, f_i). Then the following are equivalent:

- e_1, e_2 are G_0-conjugate,
- $(h_1, e_1, f_1), (h_2, e_2, f_2)$ are G_0-conjugate,
- h_1, h_2 are G_0-conjugate.

This equivalence yields an algorithm to list the nilpotent orbits of G_0 (see [12]). Alternatively, Vinberg has devised a method, using so-called carrier algebras, to list the nilpotent orbits of a θ-group ([33], see also [12]). Also an algorithm by Popov ([25]) can be used to list the nilpotent orbits of a θ-group. Furthermore, Kawanake ([19]) considered the question of listing the nilpotent orbits of a θ-group also for base fields of characteristic $p > 0$.

We let $\mathfrak{h}_0$ be a fixed Cartan subalgebra of $\mathfrak{g}_0$. Then, after acting with an element of G_0, we may assume that $h_1, h_2 \in \mathfrak{h}_0$. Furthermore, $h_1, h_2 \in \mathfrak{h}_0$ are G_0-conjugate if and only if they are conjugate under the Weyl group $W_0 = N_{G_0}(\mathfrak{h}_0)/Z_{G_0}(\mathfrak{h}_0)$, which is a finite group, isomorphic to the Weyl group of the root system of $\mathfrak{g}_0$.

Example 13. Let the notation be as in Example 12. We let $\mathfrak{h}_0$ be the Cartan subalgebra of $\mathfrak{g}_0$ spanned by h_1, h_2, h_3, h_4 (this is a Cartan subalgebra of $\mathfrak{g}$ as well). Since we deal with a $\mathbb{Z}$-grading there are only nilpotent orbits. Using the implementation of the algorithms of [12], in the SLA package ([13]) of GAP4 ([8]) we immediately get the list of nilpotent orbits, $\mathcal{O}_1, \ldots, \mathcal{O}_4$, each given by a homogeneous $\mathfrak{sl}_2$-triple:

$$\mathcal{O}_1 : (h_1 + h_2 + h_3, x_{\beta_1+\beta_2+\beta_3}, x_{-\beta_1-\beta_2-\beta_3})$$
$$\mathcal{O}_2 : (h_1 + h_2 + h_4, x_{\beta_1+\beta_2+\beta_4}, x_{-\beta_1-\beta_2-\beta_4})$$
$$\mathcal{O}_3 : (2h_1 + 2h_2 + h_3 + h_4, x_{\beta_1+\beta_2+\beta_3} + x_{\beta_1+\beta_2+\beta_4}, x_{-\beta_1-\beta_2-\beta_3} + x_{-\beta_1-\beta_2-\beta_4})$$
$$\mathcal{O}_4 : (2h_1 + 4h_2 + 2h_3 + 2h_4, x_{\beta_2+\beta_4} + x_{\beta_1+\beta_2+\beta_3}, 2x_{-\beta_2-\beta_4} + 2x_{-\beta_1-\beta_2-\beta_3}).$$

In this example W_0 is the Weyl group of type A_2, generated by $s_i = s_{\beta_i}$, $i = 1, 2$. A reflection s_β acts on $\mathfrak{h}_0$ by $s_\beta(h) = h - (\beta, \beta_h)h_\beta$, where $\beta_h = \sum_{i=1}^{4} a_i \beta_i$ if $h = \sum_{i=1}^{4} a_i h_i$.

5.3 The Algorithm

We set $V = \mathfrak{g}_1$. Corresponding to a homogeneous $\mathfrak{sl}_2$-triple (h, e, f) we define the spaces

$$V_k(h) = \{v \in V \mid [h, v] = kv\} \text{ and } V_{\geq 2}(h) = \bigoplus_{k \geq 2} V_k(h).$$

Let (h', e', f'), (h, e, f) be two homogeneous $\mathfrak{sl}_2$-triples. The algorithm to decide whether $G_0 e'$ lies in the closure of $G_0 e$ is based on the following theorem. One of the main ingredients of the proof, which here is omitted, is the following result from [34], Theorem 5.6: the closure of $G_0 e$ is equal to $G_0(V_{\geq 2}(h))$.

Theorem 14. $G_0 e'$ *is contained in the closure of* $G_0 e$ *if and only if there is a* $w \in W_0$ *such that* $U_w = V_2(h') \cap V_{\geq 2}(wh)$ *contains a point of* $G_0 e'$. *Moreover, in that case the intersection of U and $G_0 e'$ is open and dense in* U_w.

Deciding whether a given $u \in U_w$ lies in $G_0 e'$ is straightforward: this is equivalent to the existence of $f'' \in \mathfrak{g}_{-1}$ such that (h', u, f'') is an $\mathfrak{sl}_2$-triple ([9], Lemma 2.6), and the latter can be established by solving a small system of linear equations.

So the previous theorem yields a straightforward algorithm:

1. For each $w \in W_0$ take a random element u of U_w, and check whether $u \in G_0 e'$
2. If this holds, then $G_0 e'$ lies in the closure of $G_0 e$.
3. If $u \notin G_0 e'$ then show that $U_w \cap G_0 e' = \emptyset$, and go to the next element of W_0.

The last step contains an algorithmic problem that we haven't solved yet. This will be the subject of the next subsection.

Example 15. Let the notation be as in Examples 12 and 13. Let $\rho : \mathfrak{g}_0 \to \mathfrak{gl}(\mathfrak{g}_1)$ be the representation given by the action of $\mathfrak{g}_0$ on $\mathfrak{g}_1$. We give a $\rho(x)$ by its matrix relative to the basis of $\mathfrak{g}_1$ given in Example 12. Then for the characteristics of the nilpotent orbits we have

$$\rho(h_1 + h_2 + h_3) = \mathrm{diag}(1, -1, 1, -1, 2, 0)$$
$$\rho(h_1 + h_2 + h_4) = \mathrm{diag}(-1, 1, -1, 1, 0, 2)$$
$$\rho(2h_1 + 2h_2 + h_3 + h_4) = \mathrm{diag}(0, 0, 0, 0, 2, 2)$$
$$\rho(2h_1 + 4h_2 + 2h_3 + 2h_4) = \mathrm{diag}(0, 0, 2, 2, 2, 2).$$

From this, the given representatives of the orbits in Example 13, and Theorem 14 we immediately get that $\mathcal{O}_1, \mathcal{O}_2 \subset \overline{\mathcal{O}_3}$ and $\mathcal{O}_3 \subset \overline{\mathcal{O}_4}$. The only possible inclusion relation remaining is the one between $\mathcal{O}_1$ and $\mathcal{O}_2$. They have the same dimension (3), so one cannot be contained in the closure of the other. It can also be seen using Theorem 14. The W_0-conjugates of $h_1 + h_2 + h_4$ are $h_2 + h_4$ and h_4. In all cases we get that $U_w = 0$.

5.4 Deciding Emptiness

Let (h, e, f) be a homogeneous $\mathfrak{sl}_2$-triple. Let $U \subset V_2(h)$ be a given subspace. Here we describe an algorithm for deciding whether $U \cap G_0 e = \emptyset$.

Consider the subgroup $Z_0(h) = \{g \in G_0 \mid g \cdot h = h\}$, with Lie algebra $\mathfrak{z}_0(h) = \{x \in \mathfrak{g}_0 \mid [x, h] = 0\}$. From the representation theory of $\mathfrak{sl}_2$ it follows that $\mathrm{ad}e : \mathfrak{z}_0(h) \to V_2(h)$ is surjective. But that is the differential of the map $Z_0(h) \to V_2(h)$, $g \mapsto g \cdot e$. It follows that $Z_0(h)e$ contains a nonempty open subset of $V_2(h)$. Moreover, $G_0 e \cap V_2(h) = Z_0(h)e$ (see the proof of [9], Lemma 2.6). It follows that a given $v \in V_2(h)$ lies in $G_0 e$ if and only if it lies in $Z_0(h)e$; but that happens precisely when $\dim Z_0(h)v = \dim V_2(h)$. Furthermore, $\dim Z_0(h)v = \dim[\mathfrak{z}_0(h), v]$.

Now let $v_1, \ldots, v_s, x_1, \ldots, x_n$ be bases of $V_2(h)$ and $\mathfrak{z}_0(h)$ respectively. Let $\psi_1, \ldots, \psi_s$ be the dual basis of $V_2(h)$ (i.e., $\psi_j(v_i) = \delta_{i,j}$). For a $v \in V$ we define the $n \times s$-matrix B_v by

$$B_v(i, j) = \psi_j([v, x_i]).$$

Let $u_1, \ldots, u_r$ be a basis of U, and consider the field of rational functions $F = k(U_1, \ldots, U_r)$. Let $u_0 = U_1 u_1 + \cdots + U_r u_r \in F \otimes U$, and set $B_U = B_{u_0}$. (The entries of this matrix are linear polynomials in the U_k.)

Lemma 16. $U \cap G_0 e \neq \emptyset$ if and only if the rank of B_U is s.

Proof. Let $\mathfrak{z}_0(h)_v = \{x \in \mathfrak{z}_0(h) \mid [x, v] = 0\}$. Let $x = \sum_i \alpha_i x_i$ and write $\alpha = (\alpha_1, \ldots, \alpha_n)$. Then $x \in \mathfrak{z}_0(h)_v$ if and only if $\psi_j([v, x]) = 0$ for all j if and only if $\alpha \cdot B_v = 0$. Hence $\dim[\mathfrak{z}_0(h), v] = \mathrm{rank}(B_v)$. So from what is said above, U contains elements of $G_0 e$ if and only if it contains elements u such that $\mathrm{rank}(B_u) = s$, if and only if $\mathrm{rank}(B_U) = s$. □

Example 17. Here is an example of a matrix that has been produced by a calculation of this type:

$$
\begin{array}{cccccccccccccc}
0 & U_1 & 0 & 0 & 0 & 0 & 0 & 0 & 0 & -U_6 & 0 & -U_8 & 0 & 0 \\
0 & 0 & U_2 & 0 & 0 & 0 & U_4 & 0 & -U_5 & 0 & 0 & 0 & -U_9 & 0 \\
0 & 0 & 0 & 0 & 0 & U_3 & 0 & 0 & 0 & -U_7 & -U_8 & 0 & 0 & 0 \\
0 & 0 & U_1 & 0 & 0 & 0 & 0 & -U_4 & U_6 & 0 & 0 & -U_9 & 0 & 0 \\
0 & 0 & 0 & U_2 & 0 & 0 & -U_3 & 0 & -U_7 & 0 & U_9 & 0 & 0 & 0 \\
0 & 0 & 0 & U_1 & 0 & 0 & 0 & U_3 & U_8 & U_9 & 0 & 0 & 0 & 0 \\
U_2 & 0 & 0 & 0 & 0 & 0 & 0 & 0 & 0 & 0 & -U_5 & 0 & -U_7 & 0 \\
0 & 0 & 0 & 0 & 0 & 0 & 0 & 0 & 0 & 0 & 0 & 0 & 0 & -U_8 \\
0 & 0 & 0 & 0 & U_4 & 0 & 0 & 0 & 0 & 0 & 0 & -U_5 & -U_6 & 0 \\
0 & 0 & 0 & 0 & 0 & 0 & 0 & 0 & 0 & 0 & 0 & 0 & 0 & -U_7 \\
0 & 0 & 0 & 0 & 0 & 0 & 0 & 0 & 0 & 0 & 0 & 0 & 0 & U_6 \\
0 & 0 & 0 & 0 & 0 & 0 & 0 & 0 & 0 & 0 & 0 & 0 & 0 & U_5 \\
0 & 0 & 0 & 0 & 0 & 0 & 0 & 0 & 0 & U_5 & U_6 & U_7 & U_8 & U_9 \\
2U_1 & U_2 & 0 & 0 & 0 & 0 & 0 & 0 & 0 & -U_5 & 0 & -U_7 & 0 & 0 \\
0 & -U_2 & 0 & 0 & U_3 & U_4 & 0 & 0 & 0 & -U_5 & 0 & -U_7 & 0 & 0 \\
-U_1 & U_2 & 0 & 0 & 0 & 0 & 0 & 0 & 0 & U_5 & -U_6 & U_7 & -U_8 & 0 \\
0 & -U_2 & 0 & 0 & 0 & -U_4 & 0 & 0 & 0 & -U_5 & 0 & 0 & U_8 & -U_9 \\
0 & 0 & 0 & 0 & -U_3 & U_4 & 0 & 0 & 0 & U_5 & U_6 & -U_7 & -U_8 & 0
\end{array}
$$

It has rank 13.

Lemma 16 reduces the problem of deciding whether $U \cap G_0 e = \emptyset$ to computing the rank of a matrix over a function field, whose entries are linear polynomials. This is known as Edmonds' problem (see [23]). There is a straightforward algorithm, namely to do a Gaussian elimination over a function field. That method suffers from coefficient explosion, but works reasonably well if the matrices are not too big. For the matrices arising from the examples considered in [9] (including the adjoint representation of E_8, and a $\mathbb{Z}/3\mathbb{Z}$-grading of E_8), MAGMA was able to compute the rank.

Remark 18. The straightforward algorithm, based on Lemma 16, is not always used in the algorithm of [9]. In some cases a more efficient variant can be used; we will not go into that here. Furthermore, in some cases there are easy criteria that show that there is no inclusion (for example, if the orbits have equal dimension). Also, in [9], in order to loop over the orbit $W_0 \cdot h$, a tree structure is used, due to Snow ([28]), where the nodes correspond to the elements of the orbit. A criterion is given that makes it possible, in many cases, to "prune" this tree, that is, if a space U_w, corresponding to a certain node of the tree, has no point of $G_0 e'$, and the criterion is fulfilled, then the same thing immediately follows for the entire subtree below, so that it can be skipped. This makes it possible to execute the algorithm also for cases where W_0 is very large, as for example, when considering the adjoint representation of E_8 (where $|W_0| = 696729600$).

6 A Somewhat Larger Example

Here we consider a θ-group arising from the Lie algebra of type E_6. The Dynkin diagram is

The corresponding Lie algebra $\mathfrak{g}$ is constructed exactly as in Example 12. The $\mathbb{Z}$-grading is also constructed in the same way: the simple root α_4 has degree 1, and the others have degree 0. In this case $\mathfrak{g}_0$ is the direct sum of a semisimple Lie algebra of type $2A_2 + A_1$ and a 1-dimensional torus. The space $\mathfrak{g}_1$ has dimension 18. Moreover, there are 17 orbits. Regarding the various methods described here to compute the orbit closures we remark:

- When computing the equations defining the closure of one particular orbit (Sect. 3), MAGMA ran for 7 days and 8 hours, and got out of memory.
- Computing the degree of G_0 with Kazarnovskiĭ's formula (as in Example 8) led to the number

$$\frac{20!}{72 \cdot (2!2!)^2} \frac{1}{169344000} = 12471030.$$

If this number is correct, then the number of unknowns in the equations produced by the method of Sect. 4 is truly astronomical.
- The method of Sect. 5, which has been implemented in the SLA package of GAP4, computed all the orbit closures in 0.9 s.

This illustrates what in our view is a general pattern. Algorithms for computing with linear algebraic groups tend to be efficient if they use the correspondence with Lie algebras, and even more so if the combinatorics of roots and Weyl groups can be employed. On the other hand, algorithms that mostly rely on the geometric properties of the group tend to more generally applicable, but also to have difficulties when used in practice.

References

1. Bayer, D., Stillman, M.: Computation of Hilbert functions. J. Symb. Comput. **14**(1), 31–50 (1992)
2. Bosma, W., Cannon, J., Playoust, C.: The Magma algebra system. I. The user language. J. Symb. Comput. **24**(3–4), 235–265 (1997). (Computational algebra and number theory. London (1993))
3. Burde, D., Steinhoff, C.: Classification of orbit closures of 4-dimensional complex Lie algebras. J. Algebra **214**(2), 729–739 (1999)
4. Chevalley, C.: Théorie des Groupes de Lie. Tome III. Théorèmes généraux sur les algèbres de Lie. Hermann, Paris (1955)
5. Cox, D., Little, J., O'Shea, D.: Ideals, Varieties and Algorithms: An Introduction to Computational Algebraic Geometry and Commutative Algebra. Springer, Heidelberg (1992)

6. De Loera, J.A., Dutra, B., Köppe, M., Moreinis, S., Pinto, G., Wu, J.: Software for exact integration of polynomials over polyhedra. Comput. Geom. **46**(3), 232–252 (2013)
7. Derksen, H., Kemper, G.: Computational Invariant Theory. Springer, Heidelberg (2002)
8. The GAP Group. GAP–Groups, Algorithms, and Programming, Version 4.5 (2012). http://www.gap-system.org
9. de Graaf, W.A., Vinberg, E.B., Yakimova, O.S.: An effective method to compute closure ordering for nilpotent orbits of θ-representations. J. Algebra **371**, 38–62 (2012)
10. de Graaf, W.A.: Lie Algebras: Theory and Algorithms, North-Holland Mathematical Library. Elsevier Science, New York (2000)
11. de Graaf, W.A.: Constructing algebraic groups from their Lie algebras. J. Symb. Comput. **44**, 1223–1233 (2009)
12. de Graaf, W.A.: Computing representatives of nilpotent orbits of θ-groups. J. Symb. Comput. **46**, 438–458 (2011)
13. de Graaf, W.A.: SLA—computing with simple Lie algebras. A GAP package, version 0.13 (2013). http://science.unitn.it/degraaf/sla.html
14. Harris, J.: Algebraic Geometry, Graduate Texts in Mathematics, vol. 133. Springer, New York (1995). (A first course, Corrected reprint of the 1992 original)
15. Humphreys, J.E.: Linear Algebraic Groups. Springer, Heidelberg (1975)
16. Humphreys, J.E.: Reflection Goups and Coxeter Groups. Cambridge University Press, Cambridge (1990)
17. Jelonek, Z.: On the effective Nullstellensatz. Invent. Math. **162**(1), 1–17 (2005)
18. Kac, V.G.: Infinite Dimensional Lie Algebras, 3rd edn. Cambridge University Press, Cambridge (1990)
19. Kawanaka, N.: Orbits and stabilizers of nilpotent elements of a graded semisimple Lie algebra. J. Fac. Sci. Univ. Tokyo Sect. IA Math. **34**(3), 573–597 (1987)
20. Kazarnovskiĭ, B.: Ya.: Newton polyhedra and Bezout's formula for matrix functions of finite-dimensional representations. Funktsional. Anal. i Prilozhen. **21**(4), 73–74 (1987)
21. Witold, K., Weyman, J.: Geometry of orbit closures for the representations associated to gradings of Lie algebras of types E_6, F_4 and G_2. arXiv:1201.1102 [math.RT]
22. Witold, K., Weyman, J.: Geometry of orbit closures for the representations associated to gradings of Lie algebras of types E_7.arXiv:1301.0720 [math.RT]
23. Lovász, L.: On Determinants, matchings, and Random Algorithms. In: Fundamentals of Computation Theory (Proc. Conf. Algebraic, Arith. and Categorical Methods in Comput. Theory, Berlin/Wendisch-Rietz, 1979) of Mathematics Research, vol. 2, pp. 565–574. Akademie, Berlin (1979)
24. Mostow, G.D.: Fully reducible subgroups of algebraic groups. Amer. J. Math. **78**, 200–221 (1956)
25. Popov, V.L.: The cone of Hilbert null forms. Tr. Mat. Inst. Steklova (Teor. Chisel, Algebra i Algebr. Geom.) **241**, 192–209 (2003). (English translation. In: Proc. Steklov Inst. Math. 241(1), 177–194 (2003))
26. Popov, V.L.: Two orbits: when is one in the closure of the other? Tr. Mat. Inst. Steklova (Mnogomernaya Algebraicheskaya Geometriya) **264**, 152–164 (2009). (English translation. In: Proc. Steklov Inst. Math. 264(1) 146–158 (2009))
27. Shafarevich, I.R.: Basic Algebraic Geometry 1. Springer, Heidelberg (1994)
28. Snow, D.M.: Weyl group orbits. ACM Trans. Math. Softw. **16**(1), 94–108 (1990)
29. Spaltenstein, N.: Classes unipotentes et sous-groupes de Borel. Lecture Notes in Mathematics, vol. 946. Springer, Berlin (1982)

30. Tauvel, P., Rupert, W.T.Yu.: Lie Algebras and Algebraic Groups. Springer, Heidelberg (2005)
31. Vinberg, È.B.: The classification of nilpotent elements of graded Lie algebras. Dokl. Akad. Nauk SSSR **225**(4), 745–748 (1975)
32. Vinberg, È.B.: The Weyl group of a graded Lie algebra. Izv. Akad. Nauk SSSR Ser. Mat. **40**(3), 488–526 (1976). (English translation: Math. USSR-Izv. 10, 463–495 (1976))
33. Vinberg, È.B.: Classification of homogeneous nilpotent elements of a semisimple graded Lie algebra. Trudy Sem. Vektor. Tenzor. Anal. **19**, 155–179 (1979). (English translation: Selecta Math. Sov. 6, 15–35 (1987))
34. Vinberg, È.B., Popov, V.L.: Invariant theory. In: Algebraic geometry, 4 (Russian), Itogi Nauki i Tekhniki, pp. 137–314. Akad. Nauk SSSR Vsesoyuz. Inst. Nauchn. i Tekhn. Inform. Moscow (1989). (English translation. In: Popov, V.L., Vinberg, È.B.: Invariant theory. In: Algebraic Geometry IV, Encyclopedia of Mathematical Sciences, vol. 55, pp. 123–284. Springer, Berlin (1994))

Symbolic Solutions of First-Order Algebraic ODEs

Georg Grasegger[1] and Franz Winkler[2]([⊠])

[1] Doctoral Program Computational Mathematics, Research Institute
for Symbolic Computation, Johannes Kepler University Linz, 4040 Linz, Austria
[2] Research Institute for Symbolic Computation,
Johannes Kepler University Linz, 4040 Linz, Austria
Franz.Winkler@risc.jku.at

Abstract. Algebraic ordinary differential equations are described by
polynomial relations between the unknown function and its derivatives.
There are no general solution methods available for such differential
equations. However, if the hypersurface determined by the defining polynomial of an algebraic ordinary differential equation admits a parametrization, then solutions can be computed and the solvability in certain
function classes may be decided. After an overview of methods developed in the last decade we present a new and rather general method for
solving algebraic ordinary differential equations.

Keywords: Ordinary differential equations · General solutions · Algebraic curves · Algebraic surfaces · Rational parametrizations · Radical
parametrizations

1 Introduction

Consider the field of rational functions $\mathbb{K}(x)$ for a field $\mathbb{K}$. Let $'$ be the uniquely
defined derivation on $\mathbb{K}(x)$ with constant field $\mathbb{K}$ and $x' = 1$. Then $\mathbb{K}(x)$ is
a differential field. By $\mathbb{K}(x)\{y\}$, we denote the ring of differential polynomials in y over $\mathbb{K}(x)$. Its elements are polynomials in y and the derivatives of y,
i.e. $\mathbb{K}(x)\{y\} = \mathbb{K}(x)[y, y', y'', \ldots]$. An algebraic ordinary differential equation
(AODE) is of the form
$$F(x, y, y', \ldots, y^{(n)}) = 0,$$
where $F \in \mathbb{K}(x)\{y\}$ and F is also polynomial in x. The AODE is called autonomous
if $F \in \mathbb{K}\{y\}$, i.e. if the coefficients of F do not depend on the variable of dif-
ferentiation x. For a given AODE we are interested in deciding whether it has
rational or radical solutions and, in the affirmative case, determining all of them.

G. Grasegger was supported by the Austrian Science Fund (FWF): W1214-N15,
project DK11.
F. Winkler was partially supported by the Spanish Ministerio de Economía y Competitividad under the project MTM2011-25816-C02-01.

© Springer International Publishing Switzerland 2015
J. Gutierrez et al. (Eds.): Computer Algebra and Polynomials, LNCS 8942, pp. 94–104, 2015.
DOI: 10.1007/978-3-319-15081-9_5

In order to define the notion of a general solution, we go a little more into detail. Let Σ be a prime differential ideal in $\mathbb{K}(x)\{y\}$. Then we call η a generic zero of Σ if for any differential polynomial P, we have $P(\eta) = 0 \Leftrightarrow P \in \Sigma$. Such an η exists in a suitable extension field.

Let F be an irreducible differential polynomial of order n. Then $\{F\}$, the radical differential ideal generated by F, can be decomposed into two parts. There is one component where the separant $\frac{\partial F}{\partial y^{(n)}}$ also vanishes. This component represents the singular solutions. The component we are interested in is the one where the separant does not vanish. It is a prime differential ideal $\Sigma_F := \{F\} : \langle \frac{\partial F}{\partial y^{(n)}} \rangle$ and represents the general component (see for instance Ritt [15]) comprising the regular solutions. A generic zero of Σ_F is called a general solution of $F = 0$. We call it a rational general solution if it is of the form $y = \frac{a_k x^k + \ldots + a_1 x + a_0}{b_m x^m + \ldots + b_1 x + b_0}$, where the a_i and b_i are algebraic over $\mathbb{K}$.

Here we consider only first order AODEs. For solving differential equations $G(x, y, y') = 0$ or $F(y, y') = 0$, we will look at the corresponding surface $G(x, y, z) = 0$ or curve $F(y, z) = 0$, respectively, where we replace y' by a transcendental variable z.

A plane algebraic curve $\mathcal{C} = \{(a, b) \in \mathbb{K}^2 \,|\, f(a, b) = 0\}$ over $\mathbb{K}$ is a one-dimensional algebraic variety, i.e. the zero set of a square-free bivariate polynomial $f \in \mathbb{K}[x, y]$. The polynomial f is called the defining polynomial of $\mathcal{C}$. An important aspect of algebraic curves is their parametrizability. Consider an irreducible plane algebraic curve defined by an irreducible polynomial f. A tuple of rational functions $\mathcal{P}(t) = (r(t), s(t))$ is called a rational parametrization of the curve if $f(r(t), s(t)) = 0$ and not both $r(t)$ and $s(t)$ are constant. A rational parametrization can be considered as a rational map $\mathcal{P}(t) : \mathbb{A} \to \mathcal{C}$. By abuse of notation we also call this map a (rational) parametrization. Later we will see other kinds of parametrizations. We call a parametrization $\mathcal{P}(t)$ proper if it is a birational map or, in other words, if for almost every point (x, y) on the curve we find exactly one t such that $\mathcal{P}(t) = (x, y)$. Parametrizations of higher dimensional algebraic varieties are defined in a similar way.

In this paper we use parametrizations of curves for solving AODEs. Hubert [8] already studies solutions of AODEs of the form $F(x, y, y') = 0$. She gives a method for finding a basis of the general solution of the equation by computing a Gröbner basis of the prime differential ideal of the general component. The solutions, however, are given implicitly. Later Feng and Gao [2,3] start using parametrizations for solving first order autonomous AODEs, i.e., AODEs of the form $F(y, y') = 0$. They provide an algorithm for actually solving such AODEs with coefficients in $\mathbb{Q}$ by using rational parametrizations of the algebraic curve $F(x, y) = 0$. The key fact, they are proving, is that any rational solution of the AODE gives a proper parametrization of the corresponding algebraic curve. For this they use a degree bound derived in Sendra and Winkler [19]. On the other hand, if a proper parametrization of the algebraic curve fulfills certain requirements, Feng and Gao can generate a rational solution of the AODE. All proper parametrizations of a plane algebraic curve are related by linear transformations. So one only needs to check whether the given parametrization can be linearly transformed into a parametrization whose second component is the derivative of the first. From the rational solution it is then possible to create a rational general solution by shifting the variable by a constant. This approach takes advantage of the well known theory of algebraic curves and rational parametrizations (see for instance [20,21]). For this reason, we speak of the algebro-geometric solution method for AODEs.

In [1] Aroca, Cano, Feng and Gao give a necessary and sufficient condition for an autonomous AODE to have an algebraic solution. They also provide a polynomial time algorithm to find such a solution if it exists. Their solution, however, is implicit, whereas we are interested in explicit solutions.

In Sect. 2 we give a brief overview of the algebro-geometric solution method for AODEs. In Sect. 3 we give an overview of our method for finding symbolic solutions to algebraic ordinary differential equations. For a more detailed version containing the proofs of all statements, the interested reader is referred to the report [4]. Here we restrict to the case of first-order autonomous AODEs, but try to extend the results to radical solutions using radical parametrizations (see Sect. 3.2). We do so by investigating a procedure based on a rather general form of parametrizations. In case the parametrization is rational, this general procedure simply contains the solution method for rational solutions as a special case (see Sect. 3.1). As shown in [1,2] it is enough to look for a single non-trivial solution, for if $y(x)$ is a solution, so is $y(x + c)$ for a constant c and the latter is also a general solution. In Sect. 3.3 we give examples of non-radical solutions that can be found by the given procedure. Finally in Sect. 3.4 we look for advantages of the procedure and compare it to existing algorithms.

2   Overview of the Algebro-Geometric Solution Method for AODEs

Consider an autonomous first-order AODE $F(y, y') = 0$. Suppose this AODE has a rational regular solution $y(x)$. Then $\mathcal{P}(t) = (y(t), y'(t))$ is a rational parametrization of the curve $\mathcal{C}$ defined by $F(y, z) = 0$ in the affine plane over the field $\mathbb{K}$. Indeed, $\mathcal{P}$ is a proper parametrization, i.e., a birational map from the affine line onto $\mathcal{C}$. So if $\mathcal{C}$ does not admit a rational parametrization, then the given AODE has no rational solution. But if $\mathcal{C}$ is rationally parametrizabel, then there are infinitely many proper parametrizations, all of which can be determined from a single one by linear transformations. So we can set up a general proper parametrization and determine the existence of a solution of the AODE by solving a system of algebraic equations in the unknown coefficients of the linear transformation. This was first realized by Feng and Gao in [2] and later elaborated in [3]. The decision algorithm is based on exact degree bounds for parametrizations, as derived in [19]. For a detailed discussion of parametrizations, we refer to [20].

In [11] and [13] Ngô and Winkler have generalized this solution method for autonomous first-order AODEs to general first-order AODEs. For an overview of this algebro-geometric method, we refer to [12].

Example 2.1. *Consider the non-autonomous first-order AODE*

$$F(x, y, y') = y'^2 + 3y' - 2y - 3x = 0.$$

The rational general solution of $F(x, y, y') = 0$ is $y = \frac{1}{2}((x+c)^2 + 3c)$, where c is an arbitrary constant. The separant of F is $S = 2y' + 3$. So the singular solution of F is $y = -\frac{3}{2}x - \frac{9}{8}$. But how do we get this rational general solution? Let us repeat briefly the computation described in [12].

The solution surface $z^2 + 3z - 2y - 3x = 0$ has the proper rational parametrization

$$\mathcal{P}(s, t) = \left(\frac{st + 2s + t^2}{s^2}, -\frac{3s + t^2}{s^2}, \frac{t}{s} \right).$$

The original differential condition $F = 0$ can be transformed into the so-called associated system, which is a system of differential conditions on the parameters:

$$s' = st, \qquad t' = s + t^2.$$

The rational solutions of the original AODE are in 1-1 correspondence to the rational solutions of the associated system. Observe that these conditions are autonomous, also first-order, and of degree 1 in the derivatives of the parameters. All this will be true in general. Now we can consider the irreducible invariant algebraic curves of the associated system:

$$G(s, t) = s, \qquad G(s, t) = t^2 + 2s, \qquad G(s, t) = s^2 + ct^2 + 2cs.$$

Invariant algebraic curves are candidates for generating rational solutions of the associated system. The third algebraic curve $s^2 + ct^2 + 2cs = 0$ depends on a transcendental parameter c. It can be parametrized by

$$\mathcal{Q}(x) = \left(-\frac{2c}{1 + cx^2}, -\frac{2cx}{1 + cx^2} \right).$$

The rational solution of the associated system is

$$s(x) = -\frac{2c}{1 + cx^2}, \qquad t(x) = -\frac{2cx}{1 + cx^2}.$$

In general we might have to apply a linear transformation in order to get a solution of the differential problem. Since $G(s, t)$ contains a transcendental constant, the above solution is a rational general solution of the associated system. This rational general solution of the associated system can be tranformed into a rational general solution of $F(x, y, y') = 0$; in this case

$$y = \frac{1}{2}x^2 + \frac{1}{c}x + \frac{1}{2c^2} + \frac{3}{2c},$$

which, after a change of parameter, can be written as $y = \frac{1}{2}((x+c)^2 + 3c)$.

Of course, not all AODEs are first-order. In [6,7] Huang, Ngô and Winkler have described how some of these methods can be generalized to AODEs of higher order. Also, one might try to find transformations of the ambient space which transform non-autonomous AODEs into autonomous ones. Indeed, the AODE considered in Example 2.1 can be transformed into the autonomous equation $y'^2 - 2y - \frac{9}{4} = 0$ in such a way that the rational solutions are in 1-1 correspondence. A complete characterization of affine transformations preserving rational solvability has been achieved in [10].

In the present paper we investigate a new way of generalizing this solution method. We consider different or more general classes in which we are looking for solutions.

3   A General Solution Procedure for First-Order Autonomous AODEs

In this section we introduce a procedure for finding solutions of first-order autonomous AODEs not only in the class of rational functions, but in more general classes of functions such as radical or even transcendental functions. For proofs and a detailed discussion of the procedure, we refer to Grasegger [4].

Let $F(y, y') = 0$ be an autonomous first-order AODE. We consider the corresponding algebraic curve $F(y, z) = 0$. Then obviously $\mathcal{P}_y := (y(t), y'(t))$ (for a solution y of the AODE) is a parametrization of F.

Now we take an arbitrary parametrization $\mathcal{P}(t) = (r(t), s(t))$ of F, i.e. functions r and s not both constant such that $F(r(t), s(t)) = 0$. We define $A_{\mathcal{P}}(t) := \frac{s(t)}{r'(t)}$. If it is clear which parametrization is considered, we simply write A. Assume the parametrization is of the form

$$\mathcal{P}_g = (r(t), s(t)) = (y(g(t)), y'(g(t))),$$

for unknown y and g. In this case, $A_{\mathcal{P}_g}$ turns out to be $1/g'(t)$. If we could find g (an integration problem), and its inverse g^{-1}, we also find y:

$$y(x) = r(g^{-1}(x)).$$

So we can determine a solution if we can solve the corresponding integration and function inversion problems.

Kamke [9] already mentions this procedure where he restricts to continuously differential functions r and s which satisfy $F(r(t), s(t)) = 0$. However, he gives no indication how to determine these functions.

In general g is not a bijective function. Hence, when we talk about an inverse function we actually mean one branch of a multivalued inverse. Each branch inverse will give us a solution of the differential equation.

We might add any constant c to the solution of the indefinite integral. Assume $g(t)$ is a solution of the integral and g^{-1} its inverse. Then $\bar{g}(t) = g(t) + c$ is also a solution and $\bar{g}^{-1}(t) = g^{-1}(t - c)$. We know that if $y(x)$ is a solution of the

AODE, so is $y(x + c)$. Hence, we may postpone the introduction of c to the end of the procedure.

The procedure finds a solution if we can compute the integral and the inverse function. On the other hand, it does not give us any clue on the existence of a solution in case either part does not work. Neither do we know whether we have found all solutions.

3.1 Rational Solutions

Feng and Gao [2] found an algorithm for computing all rational general solutions of an autonomous first order AODE. They use the fact that $(y(x), y'(x))$ is a proper rational parametrization. The main part of their algorithm says that there is a rational general solution if and only if for any proper rational parametrization $\mathcal{P}(t) = (r(t), s(t))$ we have that $A_{\mathcal{P}}(t) = q \in \mathbb{Q}$ or $A_{\mathcal{P}}(t) = a(t-b)^2$ with $a, b \in \mathbb{Q}$. The solutions therefore are $r(q(x + c))$ or $r(b - \frac{1}{a(x+c)})$ respectively.

It can be easily shown that our procedure coincides with the results from Feng and Gao. As mentioned above, our procedure does not give an answer to the question whether the AODE has a rational solution in case A is not of this special form. It might, however, find non-rational solutions for some AODEs. Nevertheless, Feng and Gao [2] already proved that there is a rational general solution if and only if A is of the special form mentioned above and all rational general solutions can be found by the algorithm.

3.2 Radical Solutions

Now we extend our class of possible parametrizations and also the class in which we are looking for solutions to functions including radical expressions.

The research area of radical parametrizations is rather new. Sendra and Sevilla [17] recently published a paper on parametrizations of curves using radical expressions. In this paper Sendra and Sevilla define the notion of radical parametrization and provide algorithms to find such parametrizations in certain cases which include, but are not restricted to, curves of genus less or equal 4. Every rational parametrization will be a radical one, but obviously not the other way around. Further considerations of radical parametrizations can be found in Schicho and Sevilla [16] and Harrison [5]. There is also a paper on radical parametrization of surfaces by Sendra and Sevilla [18]. Nevertheless, for the beginning we will restrict to the case of first-order autonomous equations and hence to algebraic curves.

Definition 3.1. *Let $\mathbb{K}$ be an algebraically closed field of characteristic zero. A field extension $\mathbb{K} \subseteq \mathbb{L}$ is called a* radical field extension *iff $\mathbb{L}$ is the splitting field of a polynomial of the form $x^k - a \in \mathbb{K}[x]$, where k is a positive integer and $a \neq 0$. A* tower of radical field extensions *of $\mathbb{K}$ is a finite sequence of fields*

$$\mathbb{K} = \mathbb{K}_0 \subseteq \mathbb{K}_1 \subseteq \mathbb{K}_2 \subseteq \ldots \subseteq \mathbb{K}_m$$

such that for all $i \in \{1, \ldots, m\}$, the extension $\mathbb{K}_{i-1} \subseteq \mathbb{K}_i$ is radical.

A field $\mathbb{E}$ is a radical extension field of $\mathbb{K}$ iff there is a tower of radical field extensions of $\mathbb{K}$ with $\mathbb{E}$ as its last element.

A polynomial $h(x) \in \mathbb{K}[x]$ is solvable by radicals over $\mathbb{K}$ iff there is a radical extension field of $\mathbb{K}$ containing the splitting field of h.

Let now $\mathcal{C}$ be an affine plane curve over $\mathbb{K}$ defined by an irreducible polynomial $f(x, y)$. According to [17], $\mathcal{C}$ is parametrizable by radicals iff there is a radical extension field $\mathbb{E}$ of $K(t)$ and a pair $(r(t), s(t)) \in \mathbb{E}^2 \backslash \mathbb{K}^2$ such that $f(r(t), s(t)) = 0$. Then the pair $(r(t), s(t))$ is called a radical parametrization of the curve $\mathcal{C}$.

We call a function $f(x)$ over $\mathbb{K}$ a radical function if there is a radical extension field of $\mathbb{K}(x)$ containing $f(x)$. Hence, a radical solution of an AODE is a solution that is a radical function. A radical general solution is a general solution which is radical.

Computing radical parametrizations as in [17] goes back to solving algebraic equations of degree less or equal 4. Depending on the degree we might therefore get more than one solution to such an equation. Each solution yields one branch of a parametrization. Therefore, we use the notation $a^{\frac{1}{n}}$ for any n-th root of a.

We will now see that the procedure mentioned above yields information about solvability in some cases.

Theorem 3.2. Let $\mathcal{P}(t) = (r(t), s(t))$ be a radical parametrization of the curve $F(y, z) = 0$. Assume $A_{\mathcal{P}}(t) = a(b + t)^n$ for some $n \in \mathbb{Q} \backslash \{1\}$.

Then $r(h(t))$, with $h(t) = -b + (-(n-1)a(t+c))^{\frac{1}{1-n}}$, is a radical general solution of the AODE $F(y, y') = 0$.

For the proof we refer to [4]. The algorithm of Feng and Gao is therefore a special case of this one with $n = 0$ or $n = 2$ and a rational parametrization. In exactly these two cases, g^{-1} (in the setting of Theorem 3.2) is a rational function. Furthermore, Feng and Gao [2] showed that all rational solutions can be found like this, assuming the usage of a rational parametrization. The existence of a rational parametrization is of course necessary to find a rational solution. However, in the procedure we might use a radical parametrization of the same curve which is not rational and we can still find a rational solution.

In Theorem 3.2, the case $n = 1$ is excluded because in this case the function g contains a logarithm and its inverse an exponential term.

Example 3.3. The equation $y^5 - y'^2 = 0$ gives rise to the radical parametrization $\left(\frac{1}{t}, -\frac{1}{t^{5/2}}\right)$ with corresponding $A(t) = \frac{1}{\sqrt{t}}$. We can compute $g(t) = \frac{2t^{3/2}}{3}$ and $g^{-1}(t) = \left(\frac{3}{2}\right)^{2/3} t^{2/3}$. Hence, $\frac{\left(\frac{2}{3}\right)^{2/3}}{(x+c)^{2/3}}$ is a solution of the AODE.

As a corollary of Theorem 3.2 we get the following statement for AODEs where the parametrization yields another form of $A(t)$.

Corollary 3.4. Assume we have a radical parametrization $\mathcal{P}(t) = (r(t), s(t))$ of an autonomous curve $F(y, z) = 0$ and assume $A(t) = \frac{a(b+t^k)^n}{kt^{k-1}}$ with $k \in \mathbb{Q}$. Then the AODE has a radical solution.

In contrast to the rational case there are more possible forms for A now. In the following we will see another rather simple form of A which might occur. Here we do not know immediately whether or not the procedure will lead to a solution.

Theorem 3.5. *Let $\mathcal{P}(t) = (r(t), s(t))$ be a radical parametrization of the curve $F(y, z) = 0$. Assume $A(t) = \frac{at^n}{b+t^m}$ for some $a, b \in \mathbb{Q}$ and $m, n \in \mathbb{Q}$ with $m \neq n-1$ and $n \neq 1$. Then the AODE $F(y, y') = 0$ has a radical solution if the function*

$$g(t) = \frac{1}{a}t^{1-n}\left(\frac{b}{1-n} + \frac{t^m}{1+m-n}\right) \tag{1}$$

has a radical inverse $h(x)$. A general solution of the AODE is then $r(h(x+c))$.

In case we have $n = 1$ or $m = n - 1$ the integral is a function containing a logarithm and the inverse function yields an expression containing the Lambert W-Function.

Example 3.6. *The equation $-y^5 - y' + y^8 y' = 0$ gives rise to the radical parametrization $\mathcal{P}(t) = \left(\frac{1}{t}, \frac{t^3}{1-t^8}\right)$ with corresponding $A(t) = \frac{t^5}{-1+t^8}$. Then equation (1) has a solution, e.g. $-\left(2t - \sqrt{-1+4t^2}\right)^{1/4}$. Hence, $-(2(x+c)-\sqrt{(-1+4(x+c)^2)}\,)^{-1/4}$ is a solution of the AODE.*

It remains to show when g as in (1) has an inverse expressible by radicals. Based on results of Ritt [14] on function decomposition, Grasegger shows in [4] that the following holds.

Theorem 3.7. *Assume $1-n = \frac{z_1}{d_1}$ and $m-n+1 = \frac{z_2}{d_2}$ with $z_1, z_2 \in \mathbb{Z}, d_1, d_2 \in \mathbb{N}$ such that $\gcd(z_1, d_1) = \gcd(z_2, d_2) = 1$. Let $\bar{n} = \frac{(1-n)d_1 d_2}{d}$, $\bar{m} = \frac{(m-n+1)d_1 d_2}{d}$ and $d = \gcd(z_1 d_2, z_2 d_1)$.*

The function $g(t) = \frac{1}{a}t^{1-n}\left(\frac{b}{1-n} + \frac{t^m}{1+m-n}\right)$ from Theorem 3.5 has an inverse expressible by radicals if

- $b = 0$, or
- $\pm(\bar{m}, \bar{n}) \in \mathbb{N}^2$ and $\max(|\bar{m}|, |\bar{n}|) \leq 4$, or
- $\pm(-\bar{m}, \bar{n}) \in \mathbb{N}^2$ and $|\bar{m}| + |\bar{n}| \leq 4$.

It has no inverse expressible by radicals in the case

- $\pm(\bar{m}, \bar{n}) \in \mathbb{N}^2$ and $\max(|\bar{m}|, |\bar{n}|) > 4$.

Hence, in some cases we are able to decide the solvability of an AODE with properties as in Theorem 3.5. Nevertheless, the procedure is not complete, since even Theorem 3.7 does not cover all possible cases for m and n.

3.3 Non-radical Solutions

The procedure is not restricted to the radical case but might also solve some AODEs with non-radical solutions.

Example 3.8. *Consider the equation $y^3 + y^2 + y'^2 = 0$. The corresponding curve has the parametrization $\mathcal{P}(t) = (-1 - t^2, t(-1 - t^2))$. We get $A(t) = \frac{1}{2}(1 + t^2)$ and hence, $g(t) = \int \frac{1}{A(t)} dt = 2\arctan(t)$. The inverse function is $g^{-1}(t) = \tan(\frac{t}{2})$ and hence, $y(x) = -1 - \tan(\frac{x+c}{2})^2$ is a solution.*

Beside trigonometric solutions we might also find exponential solutions.

Example 3.9. *Consider the AODE $y^2 + y'^2 + 2yy' + y = 0$. We get the rational parametrization $(-\frac{1}{(1+t)^2}, -\frac{t}{(1+t)^2})$. With $A(t) = -\frac{1}{2}t(1 + t)$, we compute $g(t) = -2\log(t) + 2\log(1 + t)$ and hence $g^{-1}(t) = \frac{1}{-1 + e^{t/2}}$, which leads to the solution $-e^{-(x+c)}(-1 + e^{(x+c)/2})^2$.*

We see that it is not even necessary to use radical parametrizations in order to find non-radical solutions.

3.4 Comparison

In many books on differential equations, we can find a method for transforming an autonomous ODE of any order $F(y, y', \ldots, y^{(n)}) = 0$ to an equation of lower order by substituting $u(y) = y'$ and solving $\int \frac{1}{u(y)} dy = x$ for y (see for instance [9, 22]). For the case of first order ODEs, this method yields a solution of the ODE. It turns out that in this case the method is a special case of our procedure, which uses a particular kind of parametrization $\mathcal{P}(t) = (t, s(t))$.

We will now give some arguments concerning the possibilities and benefits of the general procedure. Since in the procedure any radical parametrization can be used, we might take advantage of picking a good one as we will see in the following example.

Example 3.10. *We consider the AODE $y'^6 + 49yy'^2 - 7$ and find a parametrization of the form $(t, s(t))$:*

$$\left(t, \frac{\sqrt{\left(756 + 84\sqrt{28812t^3 + 81} \right)^{2/3} - 588t}}{\sqrt{6} \left(756 + 84\sqrt{28812t^3 + 81} \right)^{1/6}} \right).$$

Neither Mathematica 8 *nor* Maple 16 *can solve the corresponding integral explicitly and hence, the procedure stops. Neither of them is capable of solving the differential equation in explicit form by the built in functions for solving ODEs. Nevertheless, we can try our procedure using other parametrizations. An obvious one to try next is*

$$(r(t), s(t)) = \left(-\frac{-7 + t^6}{49t^2}, t \right).$$

For this parametrization both, the integral and the inverse function, are computable and the procedure yields the general solution

$$y(x) = -\frac{4\left(-7 + \frac{1}{64}\left(-147(c+x) - \sqrt{7}\sqrt{32 + 3087(c+x)^2}\right)^2\right)}{49\left(-147(c+x) - \sqrt{7}\sqrt{32 + 3087(c+x)^2}\right)^{2/3}}.$$

The procedure might find a radical solution of an AODE by using a rational parametrization as we have seen in Examples 3.6 and 3.10. As long as we are looking for rational solutions only, the corresponding curve has to have genus zero. Now we can also solve some examples where the genus of the corresponding curve is higher than zero and hence, there is no rational parametrization. The AODE in Example 3.11 below corresponds to a curve with genus 1.

Example 3.11. *Consider the AODE $-y^3 - 4y^5 + 4y^7 - 2y' - 8y^2y' + 8y^4y' + 8yy'^2 = 0$. We compute a parametrization and get*

$$\left(\frac{1}{t}, \frac{-4 + 4t^2 + t^4}{t\left(4t^2 - 4t^4 - t^6 - \sqrt{-16t^4 + 16t^8 + 8t^{10} + t^{12}}\right)}\right)$$

as one of the branches. The procedure yields the general solution

$$y(x) = -\frac{\sqrt{1 + c + x}}{\sqrt{1 + (c+x)^2}}.$$

Again Mathematica 8 *cannot compute a solution in reasonable time and* Maple 16 *only computes constant and implicit ones.*

4   Conclusion

We have presented a new general method for determining solutions of first-order autonomous algebraic ordinary differential equations. Our method relies crucially on the availability of a parametrization of the solution surface of the given AODE. In case the parametrization is rational, and we are considering rational solution functions, our new method simply specializes to the well known algebro-geometric method. But we may also consider non-rational parametrizations and non-rational classes of solution functions, thereby drastically enlarging the applicability of our method. Currently we are able to determine solutions for various classes of solution functions, in particular radical functions. But we are still lacking a complete decision algorithm.

References

1. Aroca, J.M., Cano, J., Feng, R., Gao, X.-S.: Algebraic general solutions of algebraic ordinary differential equations. In: Kauers, M. (ed.) ISSAC'05. Proceedings of the 30th International Symposium on Symbolic and Algebraic Computation, Beijing, China, pp. 29–36. ACM Press, New York (2005)

2. Feng, R., Gao, X.-S.: Rational general solutions of algebraic ordinary differential equations. In: Gutierrez, J. (ed.) Proceedings of the 2004 International Symposium on Symbolic and Algebraic Computation (ISSAC), pp. 155–162. ACM Press, New York (2004)

3. Feng, R., Gao, X.-S.: A polynomial time algorithm for finding rational general solutions of first order autonomous ODEs. J. Symb. Comput. **41**(7), 739–762 (2006)

4. Grasegger, G.: A procedure for solving autonomous AODEs. Technical report 2013-05, Doctoral Program "Computational Mathematics". Johannes Kepler University Linz, Austria (2013)

5. Harrison, M.: Explicit solution by radicals, gonal maps and plane models of algebraic curves of genus 5 or 6. J. Symb. Comput. **51**, 3–21 (2013)

6. Huang, Y., Ngô, L.X.C., Winkler, F.: Rational general solutions of trivariate rational differential systems. Math. Comput. Sci. **6**(4), 361–374 (2012)

7. Huang, Y., Ngô, L.X.C., Winkler, F.: Rational general solutions of higher order algebraic ODEs. J. Syst. Sci. Complex. **26**(2), 261–280 (2013)

8. Hubert, E.: The general solution of an ordinary differential equation. In: Lakshman, Y.N. (ed.) Proceedings of the 1996 International Symposium on Symbolic and Algebraic Computation (ISSAC), pp. 189–195. ACM Press, New York (1996)

9. Kamke, E.: Differentialgleichungen: Lösungsmethoden und Lösungen. B. G. Teubner, Stuttgart (1997)

10. Ngô, L.X.C., Sendra, J.R., Winkler, F.: Classification of algebraic ODEs with respect to rational solvability. In: Computational Algebraic and Analytic Geometry. Contemporary Mathematics, vol. 572, pp. 193–210. American Mathematical Society, Providence, RI (2012)

11. Ngô, L.X.C., Winkler, F.: Rational general solutions of first order non-autonomous parametrizable ODEs. J. Symb. Comput. **45**(12), 1426–1441 (2010)

12. Ngô, L.X.C., Winkler, F.: Rational general solutions of parametrizable AODEs. Publicationes Mathematicae Debrecen **79**(3–4), 573–587 (2011)

13. Ngô, L.X.C., Winkler, F.: Rational general solutions of planar rational systems of autonomous ODEs. J. Symb. Comput. **46**(10), 1173–1186 (2011)

14. Ritt, J.F.: On algebraic functions, which can be expressed in terms of radicals. Trans. Am. Math. Soc. **24**, 21–30 (1924)

15. Ritt, J.F.: Differential Algebra. Colloquium publications, vol. 33. American Mathematical Society, New York (1950)

16. Schicho, J., Sevilla, D.: Effective radical parametrization of trigonal curves. Computational Algebraic and Analytic Geometry. volume 572 of Contemporary Mathematics, pp. 221–231. American Mathematical Society, Providence, RI (2012)

17. Sendra, J.R., Sevilla, D.: Radical parametrizations of algebraic curves by adjoint curves. J. Symb. Comput. **46**(9), 1030–1038 (2011)

18. Sendra, J.R., Sevilla, D.: First steps towards radical parametrization of algebraic surfaces. Comput. Aided Geom. Des. **30**(4), 374–388 (2013)

19. Sendra, J.R., Winkler, F.: Tracing index of rational curve parametrizations. Comput. Aided Geom. Des. **18**(8), 771–795 (2001)

20. Sendra, J.R., Winkler, F., Pérez-Díaz, S.: Rational Algebraic Curves, A Computer Algebra Approach. In: Algorithms and Computation in Mathematics, vol. 22. Springer, Heidelberg (2008)

21. Walker, R.J.: Algebraic Curves. Springer, Heidelberg (1978). Reprint of the 1st ed. 1950 by Princeton University Press

22. Zwillinger, D.: Handbook of Differential Equations, 3rd edn. Academic Press, San Diego (1998)

Ore Polynomials in Sage

Manuel Kauers[1]([⊠]), Maximilian Jaroschek[2], and Fredrik Johansson[3]

[1] Research Institute for Symbolic Computation (RISC),
Johannes Kepler University (JKU), 4040 Linz, Austria
mkauers@risc.jku.at
[2] Max-Planck-Institut für Informatik, Campus E1 4, 66123 Saarbrücken, Germany
mjarosch@risc.jku.at
[3] INRIA Bordeaux-Sud-Ouest & IMB,
200, avenue de la Vieille Tour, 33405 Talence cedex, France
fjohanss@risc.jku.at

Abstract. We present a Sage implementation of Ore algebras. The main features for the most common instances include basic arithmetic and actions; GCRD and LCLM; D-finite closure properties; natural transformations between related algebras; guessing; desingularization; solvers for polynomials, rational functions and (generalized) power series. This paper is a tutorial on how to use the package.

1 Introduction

In computer algebra, objects are often described implicitly through equations they satisfy. For example, the exponential function $\exp(x)$ is uniquely specified by the linear differential equation $f'(x) - f(x) = 0$ and the initial value $f(0) = 1$. Likewise, the sequence F_n of Fibonacci numbers is uniquely determined by the linear recurrence $F_{n+2} - F_{n+1} - F_n = 0$ and the two initial values $F_0 = 0$, $F_1 = 1$. Especially for representing functions or sequences that cannot be expressed in "closed form", the differential or difference equations they may satisfy provide an attractive way to store them on the computer. The question is then how to calculate with objects which are given in this form.

Algorithms for Ore algebras provide a systematic answer to this question [3,5]. Invented in the first half of the 20th century [15] with the objective of providing a unified theory for various kinds of linear operators, they have been used for many years in computer algebra systems, for example in the Maple packages OreTools [1], gfun [16] or Mgfun [4], or in the Mathematica packages by Mallinger [14] and Koutschan [11,12].

The purpose of this paper is to introduce an implementation of a collection of algorithms related to Ore algebras for the computer algebra system Sage [17]. It is addressed to first-time users who are already familiar with Sage, and with the theory of Ore algebras and its use for doing symbolic computation related to special functions. Readers unfamiliar with Sage are referred to [17], and readers unfamiliar with Ore algebras may wish to consult the recent tutorial [10] and the references given there for an introduction to the subject.

All three authors were supported by the Austrian FWF grant Y464-N18.

© Springer International Publishing Switzerland 2015
J. Gutierrez et al. (Eds.): Computer Algebra and Polynomials, LNCS 8942, pp. 105–125, 2015.
DOI: 10.1007/978-3-319-15081-9_6

At the time of writing, the package we describe here is still under construction and has not yet been incorporated into the official Sage distribution. Readers who want to try it out are invited to download the current version from

http://www.risc.jku.at/research/combinat/software/ore_algebra

and are encouraged to send us bug reports or other comments. We hope that the community will find the code useful.

The following instructions show how to load the code and then create an Ore algebra A of linear differential operators and an Ore algebra B of recurrence operators. Observe the correct application of the respective commutation rules in both cases.

```
sage: from ore_algebra import *
```

```
sage: R.<x> = PolynomialRing(ZZ); A.<Dx> = OreAlgebra(R)
```

```
sage: A
```

$$\mathbf{Z}[x]\langle \mathrm{Dx}\rangle$$

```
sage: A.random_element()
```

$$(2x - 2)\,\mathrm{Dx}^2 + \left(-6x^2 - 2x - 1\right)\mathrm{Dx} - 9x^2 - 21$$

```
sage: Dx*x
```

$$x\mathrm{Dx} + 1$$

```
sage: B.<Sx> = OreAlgebra(R)
```

```
sage: B
```

$$\mathbf{Z}[x]\langle \mathrm{Sx}\rangle$$

```
sage: Sx*x
```

$$(x + 1)\,\mathrm{Sx}$$

More details on the construction of Ore algebras are given in the following section. The construction and manipulation of elements of Ore algebras is discussed in Sect. 3.

The package also supports Ore algebras with several generators already. However, so far we offer hardly more functionality than addition and multiplication for these. Much more functionality is available for algebras with one generator. Some of it is described in Sect. 4. We plan to add more code for the multivariate case in the future.

2   Ore Algebras

An Ore algebra is determined by a base ring and a finite number of generators. In the examples above, the base ring was $\mathbf{Z}[x]$, and the generators were Dx and Sx, respectively. If no other information is provided in the arguments, the OreAlgebra constructor chooses the nature of the generators according to their name: a generator called Dt represents the standard derivation d/dt acting on the generator t of the base ring, a generator called Sn represents the standard shift operator sending the generator n of the base ring to $n + 1$.

For this way of generating algebras, generator names must be composed of one of the following single-letter prefixes followed by the name of a generator of the base ring.

Prefix	Name	Commutation rule
D	Standard derivation d/dx	$\mathrm{D}x x = x\,\mathrm{D}x + 1$
S	Standard shift $x \rightsquigarrow x + 1$	$\mathrm{S}x x = (x + 1)\,\mathrm{S}x$
T or Θ	Eulerian derivation $x\,d/dx$	$\mathrm{T}x x = x\,\mathrm{T}x + x$
F or Δ	Forward difference Δ_x	$\mathrm{F}x x = (x + 1)\mathrm{F}x + 1$
Q	q-shift $x \rightsquigarrow q\,x$	$\mathrm{Q}x x = q\,x\,\mathrm{Q}x$
J	q-derivation ("Jackson derivation")	$\mathrm{J}x x = q\,x\,\mathrm{J}x + 1$
C	commutative generator	$\mathrm{C}x x = x\,\mathrm{C}x$

For the q-shift and the q-derivation, the base ring must contain an element q. The element playing the role of q can be specified as an optional argument.

```
sage: R.<x> = PolynomialRing(ZZ['q'])
```

```
sage: A.<Qx> = OreAlgebra(R)
```

```
sage: Qx*x
```

$$qx\,\mathrm{Qx}$$

```
sage: A.<Qx> = OreAlgebra(R, q=2)
```

```
sage: Qx*x
```

$$2x\,\mathrm{Qx}$$

In general, the commutation rules of a generator X of an Ore algebra A with base ring R are governed by two maps, $\sigma\colon R \to R$ and $\delta\colon R \to R$, where σ is a ring endomorphism (i.e., $\sigma(a + b) = \sigma(a) + \sigma(b)$ and $\sigma(ab) = \sigma(a)\sigma(b)$ for all $a, b \in R$) and δ is a skew-derivation for σ (i.e., $\delta(a + b) = \delta(a) + \delta(b)$ and

$\delta(ab) = \delta(a)b + \sigma(a)\delta(b)$ for all $a, b \in R$). With two such maps being given, the generator X satisfies the commutation rule $Xa = \sigma(a)X + \delta(a)$ for every $a \in R$. If there is more than one generator, then each of them has its own pair of maps σ, δ. Different generators commute with each other; noncommutativity only takes place between generators and base ring elements.

It is possible to create an Ore algebra with user specified commutation rules. In this form, each generator must be declared by a tuple (X, σ, δ), where X is the name of the generator (a string), and σ and δ are dictionaries which contain the images of the generators of the base ring under the respective map. Here is how to specify an algebra of difference operators in this way:

```
sage: R.<x> = ZZ['x']
```

```
sage: A = OreAlgebra(R, ('X', {x:x+1}, {x:1}))
```

```
sage: X = A.gen()
```

```
sage: X*x
```

$$(x + 1) X + 1$$

As another example, here is how to define an algebra of differential operators whose base ring is a differential field $K = \mathbb{Q}(x, y, z)$ where y represents $\exp(x)$ and z represents $\log(x)$:

```
sage: K = ZZ['x','y','z'].fraction_field()
```

```
sage: x,y,z = K.gens()
```

```
sage: A = OreAlgebra(K, ('D', {}, {x:1, y:y, z:1/x}))
```

```
sage: D = A.gen()
```

```
sage: D*x, D*y, D*z
```

$$\left(xD + 1, yD + y, zD + \frac{1}{x} \right)$$

In the dictionary specifying σ, omitted generators are understood to be mapped to themselves, so that $\{\}$ in the definition of A in the example above is equivalent to $\{x:x,y:y,z:z\}$. In the dictionaries specifying δ, omitted generators are understood to be mapped to zero.

For Ore algebras with several generators, it is possible to mix specifications of generators via triples (X, σ, δ) with generators using the naming convention shortcuts as explained before. Continuing the previous example, here is a way to define an algebra A over K with two generators, a D that behaves like before, and in addition an Sx which acts like the standard shift on x and leaves the other generators fixed.

```
sage: A = OreAlgebra(K, ('D', {}, {x:1, y:y, z:1/x}), 'Sx')
```

```
sage: D, Sx = A.gens()
```

```
sage: D*x, Sx*x
```

$$(xD + 1, (x + 1)\mathrm{Sx})$$

```
sage: D*y, Sx*y
```

$$(yD + y, y\mathrm{Sx})$$

```
sage: D*z, Sx*z
```

$$\left(zD + \frac{1}{x}, z\mathrm{Sx}\right)$$

In theory, any integral domain can serve as base ring of an Ore algebra. Not so in our implementation. Here, base rings must themselves be polynomial rings (univariate or multivariate), or fraction fields of polynomial rings. Their base rings in turn may be either $\mathbb{Z}$, $\mathbb{Q}$, a prime field $GF(p)$, or a number field $\mathbb{Q}(\alpha)$, or — recursively — some ring which itself would be suitable as base ring of an Ore algebra.

```
sage: ZZ['x'].fraction_field()['y','z'] ### OK
```

$$\mathrm{Frac}(\mathbf{Z}[x])[y, z]$$

```
sage: GF(1091)['x','y','z']['u'] ### OK
```

$$\mathbf{F}_{1091}[x, y, z][u]$$

```
sage: ( ZZ['x','y','z'].quotient_ring(x^2+y^2+z^2-1) )['u'] ### not OK
```

$$\mathbf{Z}[x, y, z]/\left(x^2 + y^2 + z^2 - 1\right)\mathbf{Z}[x, y, z][u]$$

```
sage: GF(9, 'a')['x'] ### not OK
```

$$\mathbf{F}_{3^2}[x]$$

Note that the maps σ and δ must leave all the elements of the base ring's base ring fixed. They may only have nontrivial images for the top level generators.

The constituents of an Ore algebra A can be accessed through the methods summarized in the following table. Further methods can be found in the documentation.

Method name	Short description
`associated_commutative_algebra()`	returns a polynomial ring with the same base ring as A and whose generators are named like the generators of A
`base_ring()`	Returns the base ring of A
`delta(i)`	Returns a callable object representing the delta map associated to the ith generator (default: $i = 0$)
`gen(i)`	Returns the ith generator (default: $i = 0$)
`sigma(i)`	Returns a callable object representing the sigma map associated to the ith generator (default: $i = 0$)
`var(i)`	Returns the name of the ith generator (default: $i = 0$)

Examples:

```
sage: R.<x> = ZZ['x']; A.<Dx> = OreAlgebra(R)

sage: A
```

$$\mathbf{Z}[x]\langle \mathrm{Dx}\rangle$$

```
sage: A.associated_commutative_algebra()
```

$$\mathbf{Z}[x][\mathrm{Dx}]$$

```
sage: A.base_ring()
```

$$\mathbf{Z}[x]$$

```
sage: A.gen()
```

$$\mathrm{Dx}$$

```
sage: s = A.sigma(); d = A.delta();

sage: s(x^5), d(x^5)
```

$$\left(x^5, 5x^4\right)$$

3   Ore Polynomials

Ore polynomials are elements of Ore algebras, i.e., Sage objects whose parent is an Ore algebra object as described in the previous section. They can be constructed by addition and multiplication from generators and elements of the base ring.

```
sage: R.<x> = ZZ['x']; A.<Dx> = OreAlgebra(R)
```

```
sage: (5*x^2+3*x-7)*Dx^2 + (3*x^2+8*x-1)*Dx + (9*x^2-3*x+8)
```

$$\left(5x^2 + 3x - 7\right) \mathrm{Dx}^2 + \left(3x^2 + 8x - 1\right) \mathrm{Dx} + 9x^2 - 3x + 8$$

Alternatively, an Ore polynomial can be constructed from any piece of data that is also accepted by the constructor of the associated commutative algebra. The associated commutative algebra of an Ore algebra is the commutative polynomial ring with the same base ring as the Ore algebra and with generators that are named like the generators of the Ore algebra. In particular, it is possible to create an Ore polynomial from the corresponding commutative polynomial, from a coefficient list, or even from a string representation.

```
sage: R.<x> = ZZ['x']; A.<Dx> = OreAlgebra(R)
```

```
sage: Ac = A.associated_commutative_algebra()
```

```
sage: Ac
```

$$\mathbf{Z}[x][\mathrm{Dx}]$$

```
sage: A(Ac.random_element())
```

$$(x - 3) \mathrm{Dx}^2 + \left(x^2 + x + 445\right) \mathrm{Dx} - x^2 + x + 1$$

```
sage: A([5*x,7*x-3,3*x+1])
```

$$(3x + 1) \mathrm{Dx}^2 + (7x - 3) \mathrm{Dx} + 5x$$

```
sage: A("(5*x^2+3*x-7)*Dx^2 + (3*x^2+8*x-1)*Dx + (9*x^2-3*x+8)")
```

$$\left(5x^2 + 3x - 7\right) \mathrm{Dx}^2 + \left(3x^2 + 8x - 1\right) \mathrm{Dx} + 9x^2 - 3x + 8$$

Ore polynomials can also be created from Ore polynomials that belong to other algebras, provided that such a conversion is meaningful.

```
sage: R.<x> = ZZ['x']; A.<Dx> = OreAlgebra(R)

sage: L = (5*x^2+3*x-7)*Dx^2 + (3*x^2+8*x-1)*Dx + (9*x^2-3*x+8)

sage: L.parent()
```

$$\mathbf{Z}[x]\langle Dx \rangle$$

```
sage: B = OreAlgebra(QQ['x'], 'Dx')

sage: L = B(L)

sage: L.parent()
```

$$\mathbf{Q}[x]\langle Dx \rangle$$

In accordance with the Sage coercion model, such conversions take place automatically (if possible) when operators from different algebras are added or multiplied. Note that the result of such an operation need not belong to either of the parents of the operands but may instead have a suitable "common extension" as parent.

```
sage: A = OreAlgebra(ZZ['t']['x'], 'Dx')

sage: B = OreAlgebra(QQ['x'].fraction_field(), 'Dx')

sage: L = A.random_element() + B.random_element()

sage: L.parent()
```

$$\mathrm{Frac}(\mathbf{Q}[t][x])\langle Dx \rangle$$

4   Selected Methods

Besides basic arithmetic for Ore operators, the package provides a wide range of methods to create, manipulate and solve several different kinds of operators. Some of these methods are accessible in any Ore algebra while others are tied specifically to, e.g., recurrence operators or differential operators.

In this section, we give an overview of the functionality provided by the package. Because of space limitation, only some of the available methods can be discussed here. For further information, we refer to the documentation.

4.1   Methods for General Algebras

A univariate Ore algebra over a field is a left Euclidean domain, which means that it is possible to perform left division with remainder. Building upon this, the greatest common right divisor (GCRD) and the least common left multiple (LCLM) of two Ore polynomials can be computed. The package provides a number of methods to carry out these tasks.

Method name	Short description
`A.quo_rem(B)`	Returns the left quotient and the left remainder of A and B
`A.gcrd(B)`	Returns the greatest common right divisor of A and B
`A.xgcrd(B)`	Returns the greatest common right divisor of A and B and the according Bézout coefficients
`A.lclm(B)`	Returns the least common left multiple of A and B
`A.xlclm(B)`	Returns the least common left multiple L of A and B and the left quotients of L and A and of L and B
`A.resultant(B)`	Returns the resultant of A and B (see [13] for its definition and properties)

All these methods are also available for Ore operators living in univariate Ore algebras over a base ring R which does not necessarily have to be a field. The operators are then implicitly assumed to live in the respective Ore algebra over the quotient field K of R. The output will be the GCRD (LCLM, quotient, remainder) in the Ore algebra over K but not over R, in which these objects might not exist or might not be computable.

```
sage: A = OreAlgebra(ZZ['n'], 'Sn')

sage: G = A.random_element(2)

sage: L1, L2 = A.random_element(7), A.random_element(5)

sage: while L1.gcrd(L2) != 1: L2 = A.random_element(5)

sage: L1, L2 = L1*G, L2*G

sage: L1.gcrd(L2) == G.normalize()

                           True

sage: L3, S, T = L1.xgcrd(L2)

sage: S*L1 + T*L2 == L3

                           True

sage: LCLM = L1.lclm(L2)

sage: LCLM % L1 == LCLM % L2 == 0

                           True

sage: LCLM.order() == L1.order() + L2.order() - G.order()

                           True
```

The GCRD is only unique up to multiplication (from the left) by an element from the base ring. The method **normalize** called in line 6 of the listing above multiplies a given operator from the left by some element from the base ring such as to produce a canonical representative from the class of all the operators that can be obtained from each other by left multiplication of a base ring element. This facilitates the comparison of output.

The efficiency of computing the GCRD depends on the size of the coefficients of intermediate results, and there are different strategies to control this growth via so-called polynomial remainder sequences (PRS). The default is the improved PRS described in [7], which will usually be the fastest choice. Other strategies can be selected by the option **prs**.

```
sage: A = OreAlgebra(ZZ['n'], 'Sn')

sage: L1, L2 = A.random_element(3), A.random_element(2)

sage: algs = ["improved", "classic", "monic", "subresultant"]

sage: [L1.gcrd(L2, prs=a) for a in algs]
```

$$[1, 1, 1, 1]$$

If L_1, L_2 are operators, then the solutions of their GCRD are precisely the common solutions of L_1 and L_2. The LCLM, on the other hand, is the minimal order operator whose solution space contains all the solutions of L_1 and all the solutions of L_2. Because of this property, GCRD and LCLM are useful tools for constructing operators with prescribed solutions. For example, here is how to construct a differential operator which has the solutions x^5 and $\exp(x)$, starting from the obvious operators annihilating x^5 and $\exp(x)$, respectively.

```
sage: R.<x> = ZZ[]; A.<Dx> = OreAlgebra(R)

sage: L = (Dx - 1).lclm(x*Dx - 5)

sage: L
```

$$\left(x^2 - 5x\right) Dx^2 + \left(-x^2 + 20\right) Dx + 5x - 20$$

```
sage: L(x^5)
```

$$0$$

```
sage: L(exp(x)).full_simplify()
```

$$0$$

Observe how in the last two lines we apply the operator L to other objects. Such applications are not defined for every algebra and in general have to be specified by the user through an optional argument:

```
sage: A.<Qqn> = OreAlgebra(ZZ['q']['qn'])
```

```
sage: var('q', 'n', 'x')
```

$$(q, n, x)$$

```
sage: (Qqn^2+Qqn+1)(q^n, action=lambda expr: expr.substitute(n=n+1))
```

$$q^n + q^{(n+2)} + q^{(n+1)}$$

```
sage: (Qqn^2+Qqn+1)(x, action=lambda expr: expr.substitute(x=q*x))
```

$$q^2 x + qx + x$$

Thanks to the LCLM operation discussed above, we have the property that when f and g are two objects which are annihilated by some operators L_1, L_2 belonging to some Ore algebra A then this algebra contains also an operator which annihilates their sum $f + g$. In other words, the class of solutions of operators of A is closed under addition. It turns out that similar closure properties hold for other operations. The following table lists some of the corresponding methods. Methods for more special closure properties will appear further below.

Method name	Short description
`lclm()`	Computes an annihilating operator for $f + g$ from annihilating operators for f and g
`symmetric_product()`	Computes an annihilating operator for fg from annihilating operators for f and g
`symmetric_power()`	Computes an annihilating operator for f^n from an annihilating operator for f and a given positive integer n
`annihilator_of_associate()`	Computes an annihilating operator for $M(f)$ from an annihilating operator for f and a given operator M
`annihilator_of_polynomial()`	Computes an annihilating operator for the object $p(f, \partial f, \partial^2 f, \dots)$ from an annihilating operator for f and a given multivariate polynomial p

As an example application, let us prove Cassini's identity for Fibonacci numbers:

$$F_{n+1}^2 - F_n F_{n+2} = (-1)^n.$$

The idea is to derive, using commands from the table above, a recurrence satisfied by the left hand side, and then show that this recurrence is also valid for the right hand side.

```
sage: A.<Sn> = OreAlgebra(ZZ['n'])
```

```
sage: fib = Sn^2 - Sn - 1
```

```
sage: R.<x0,x1,x2> = ZZ['n']['x0','x1','x2']
```

```
sage: fib.annihilator_of_polynomial(x1^2 - x0*x2)
```

$$Sn + 1$$

As this operator obviously annihilates $(-1)^n$, the proof is complete after checking that the identity holds for $n = 0$. Another way of carrying out the same computation using the other commands would be as follows.

```
sage: A.<Sn> = OreAlgebra(ZZ['n'])
```

```
sage: fib = Sn^2 - Sn - 1
```

```
sage: L1 = fib.annihilator_of_associate(Sn).symmetric_power(2)
```

```
sage: L2 = fib.annihilator_of_associate(Sn^2).symmetric_product(fib)
```

```
sage: L1.lclm(L2)
```

$$Sn^3 - 2Sn^2 - 2Sn + 1$$

Observe that the resulting operator again annihilates $(-1)^n$, but its order is higher than the operator obtained before, so we need to check more initial values to complete the proof. For larger computations, the command `annihilator_of_polynomial` would also consume less computation time than the step-by-step approach.

4.2   Methods for Special Algebras

For the elements of some of the most important algebras, additional methods have been implemented. The following table lists some of the additional methods available for differential operators, i.e., elements of an Ore algebra of the form $R[x]\langle Dx \rangle$ or $K(x)\langle Dx \rangle$.

Method name	Short description
to_S()	Converts to a recurrence operator for the Taylor series solutions at the origin
to_F()	Converts to a difference operator for the Taylor series solutions at the origin
to_T()	Rewrites in terms of the Euler derivative
annihilator_of_integral()	Converts an annihilator for $f(x)$ to one for $\int f(x)dx$
annihilator_of_composition()	Converts an annihilator for $f(x)$ to one for $f(a(x))$ where $a(x)$ is algebraic over the base ring
desingularize()	Computes a left multiple of this operator with polynomial coefficients and lowest possible leading coefficient degree
associate_solutions(p)	Applied to an operator P, this computes, if possible, an operator M and a rational function m such that $DM = p + mP$ (see [2] for further information)
polynomial_solutions()	Computes the polynomial solutions of this operator
rational_solutions()	Computes the rational function solutions of this operator
power_series_solutions()	Computes power series solutions of this operator
generalized_series_solutions()	Computes generalized series solutions of this operator

As an example application, we compute an annihilator for the error function $\frac{2}{\sqrt{\pi}} \int_0^x \exp(-t^2)dt$, starting from the differential equation for $\exp(x)$, and produce the recurrence for the Taylor series coefficients at the origin. Finally, we compute the series solutions of the differential operator at infinity.

```
sage: R.<x> = ZZ['x']; A.<Dx> = OreAlgebra(R, 'Dx')
```

```
sage: (Dx - 1).annihilator_of_composition(-x^2)
```

$$Dx + 2x$$

```
sage: L = (Dx + 2*x).annihilator_of_integral()
```

```
sage: L
```

$$Dx^2 + 2xDx$$

```
sage: L.to_S(OreAlgebra(ZZ['n'], 'Sn'))
```

$$\left(n^2 + 3n + 2\right) Sn^2 + 2n$$

```
sage: L.power_series_solutions(10)
```

$$\left[x - \frac{1}{3}x^3 + \frac{1}{10}x^5 - \frac{1}{42}x^7 + \frac{1}{216}x^9 + O(x^{10}), 1 + O(x^{10})\right]$$

```
sage: L.annihilator_of_composition(1/x).generalized_series_solutions()
```

$$\left[\exp\left(-x^{-2}\right) \cdot x \cdot \left(1 - \frac{1}{2}x^2 + \frac{3}{4}x^4 + O(x^5)\right), 1 + O(x^5)\right]$$

The last output implies that the operator annihilating $\int_0^x \exp(-t^2)dt$ also admits a solution which behaves for $x \to \infty$ like $\frac{1}{x}\exp(-x^2)$.

The next example illustrates the methods for finding rational and polynomial solutions of an operator L. These methods accept as an optional parameter an inhomogeneous part consisting of a list (or tuple) of base ring elements, $(f_1, \ldots, f_r)$. They return as output a list of tuples $(g, c_1, \ldots, c_r)$ with $L(g) = c_1 f_1 + \cdots + c_r f_r$ where g is a polynomial or rational function and $c_1, \ldots, c_r$ are constants, i.e., elements of the base ring's base ring. The tuples form a vector space basis of the solution space.

In the example session below, we start from two polynomials p, q, then compute an operator L having p and q as solutions, and then recover p and q from L. Note that for consistency also the solutions of homogeneous equations are returned as tuples. At the end we give an example for solving an inhomogeneous equation.

```
sage: R.<x> = ZZ[]
```

```
sage: p = x^2 + 3*x + 8; q = x^3 - 7*x + 5
```

```
sage: A.<Dx> = OreAlgebra(R)
```

```
sage: L = (p*Dx - p.derivative()).lclm(q*Dx - q.derivative())
```

```
sage: L
```

$$\left(x^4 + 6x^3 + 31x^2 - 10x - 71\right)\mathrm{Dx}^2 + \left(-4x^3 - 18x^2 - 62x + 10\right)\mathrm{Dx} + 6x^2 + 18x + 14$$

```
sage: L.polynomial_solutions()
```

$$\left[\left(-3x^3 + 2x^2 + 27x + 1\right), \left(-8x^3 + 5x^2 + 71x\right)\right]$$

```
sage: M = (2*x+3)*Dx^2 + (4*x+5)*Dx + (6*x+7)
```

```
sage: sol = M.polynomial_solutions([1,x,x^2,x^3])
```

```
sage: sol
```

$$\left[(1, 7, 6, 0, 0), (x, 5, 11, 6, 0), \left(x^2, 6, 14, 15, 6\right)\right]$$

```
sage: map(lambda s: M(s[0]) == s[1]+s[2]*x+s[3]*x^2+s[4]*x^3, sol)
```

$$[\text{True}, \text{True}, \text{True}]$$

The functions `polynomial_solutions` and `rational_solutions` are not only defined for differential operators but also for recurrence operators, i.e., elements of an Ore algebra of the form $R[x]\langle Sx \rangle$ or $K(x)\langle Sx \rangle$. Some other methods defined for recurrence operators are listed in the following table.

Method name	Short description
`to_D()`	converts annihilator for the coefficients in a power series to a differential operator for the sum
`to_F()`	Converts shift operator to a difference operator
`to_T()`	Converts to a differential operator in terms of the Euler derivative
`annihilator_of_sum()`	Converts an annihilator for $f(n)$ to one for the sum $\sum_{k=0}^{n} f(k)$
`annihilator_of_composition()`	Converts an annihilator for $f(n)$ to one for $f(\lfloor un + v \rfloor)$ where $u, v \in$
`annihilator_of_interlacing()`	Interlaces two or more sequences
`desingularize()`	Computes a left multiple of this operator with polynomial coefficients and lowest possible leading coefficient degree
`associate_solutions(p)`	Applied to an operator P, this computes, if possible, an operator M and a rational function m such that $(S - 1)M = p + mP$ (see [2] for further information)
`polynomial_solutions()`	Computes the polynomial solutions of this operator
`rational_solutions()`	Computes the rational function solutions of this operator
`generalized_series_solutions()`	Computes asymptotic expansions of sequences annihilated by the operator
`to_list()`	Computes terms of a sequence annihilated by the operator

As an example application, we compute an annihilator for the sequence $c(n) = \sum_{k=0}^{n} 1/k!$:

```
sage: R.<n> = ZZ[]; A.<Sn> = OreAlgebra(R)

sage: inverse_factorials = (n + 1) * Sn - 1

sage: partial_sums = inverse_factorials.annihilator_of_sum()

sage: partial_sums
```

$$(n + 2)\, \mathrm{Sn}^2 + (-n - 3)\, \mathrm{Sn} + 1$$

The to_list method returns the first few values of a sequence, given the initial values:

```
sage: L = partial_sums.to_list([1, 2], 8)
```

```
sage: L
```

$$\left[1, 2, \frac{5}{2}, \frac{8}{3}, \frac{65}{24}, \frac{163}{60}, \frac{1957}{720}, \frac{685}{252}\right]$$

```
sage: N(L[7])
```

$$2.71825396825397$$

We compute the asymptotic expansion of the sequence of terms to estimate how many terms we need to approximate e to a given number of digits:

```
sage: digits = 10^5
```

```
sage: asymp = inverse_factorials.generalized_series_solutions(3)
```

```
sage: target = lambda x: log(abs(asymp[0](RR(x))), 10) + digits
```

```
sage: num_terms = ceil(find_root(target, 1, 10^6))
```

```
sage: num_terms
```

$$25207$$

In some cases, for example when the base ring is $\mathbb{Z}$ or $\mathbb{Z}[x]$, isolated values of a sequence can be computed asymptotically faster for large n than by listing all values, using the binary splitting technique. The forward_matrix_bsplit method, called with argument n, returns a matrix P and a polynomial Q such that P/Q multiplied by a column vector of initial values $c_0, c_1, \ldots$ yields $c_n, c_{n+1}, \ldots$. This way, computing 10^5 digits of e takes a fraction of a second:

```
sage: e_approx = N(e, 400000)
```

```
sage: P, Q = partial_sums.forward_matrix_bsplit(num_terms)
```

```
sage: u = Matrix([[e_approx], [e_approx]]) - P * Matrix([[1], [2]]) / Q
```

```
sage: u.change_ring(RealField(20))
```

$$\begin{pmatrix} 1.3053 \times 10^{-100009} \\ 5.1780 \times 10^{-100014} \end{pmatrix}$$

4.3   Guessing

Guessing is, in some sense, the reverse operation of to_list for recurrence operators, or of power_series_solutions for differential operators. It is one of the most popular features of packages like gfun, and there are even some special purpose packages dedicated to this technique [6,9]. The basic idea is simple. Given a finite array of numbers, thought of as the first terms of an infinite sequence, we want to know whether this sequence satisfies a recurrence. The algorithm behind a guessing engine searches for small equations matching the given data. Generically, no such equations exist, so if some are found, it is fair to "guess" that they are in fact valid equations for the whole infinite sequence.

We provide a guessing function which takes as input a list of terms and an Ore algebra, and returns as output an operator which matches the given data and which, in some measure, would be unlikely to exist for random data.

```
sage: data = [ 0, 1, 1, 2, 3, 5, 8, 13, 21, 34, 55 ]

sage: L = guess(data, OreAlgebra(ZZ['n'], 'Sn'))

sage: L
```

$$-\mathrm{Sn}^2 + \mathrm{Sn} + 1$$

```
sage: L(data)
```

$$[0,0,0,0,0,0,0,0,0]$$

```
sage: M = guess(data, OreAlgebra(ZZ['x'], 'Dx'))

sage: M
```

$$\left(-x^3 - x^2 + x\right)\mathrm{Dx} - x^2 - 1$$

```
sage: M(x/(1-x-x^2))
```

$$0$$

If an algebra of differential operators is supplied as second argument, the data is understood as the first few coefficients of a power series. The output operator is expected to have this power series as solution.

It can happen that the procedure is unable to find an operator matching the given data. In this case, an exception is raised. There are two possible explanations for such an event. Either the sequence in question does not satisfy any equations, or it does but the equations are so big that more data is needed to detect them.

Several options are available for customizing the search for relations. In order to explain them, we first need to give some details on the underlying algorithms. For simplicity of language, we restrict here to the case of recurrence operators. The situation for differential operators is very similar.

For the most typical situations, there are two important hyperbolas. One describes the region in the (r, d)-plane consisting of all points for which there exists an operator of order r and degree d truly satisfied by the sequence in question. (See [8] for an explanation why the boundary of this region is usually a hyperbola.) The second describes the region of all points (r, d) for which an operator of order r and degree d can be detected when N terms are provided as input. This region is determined by the requirement $(r + 1)(d + 2) < N$.

The method tests a sequence of points (r_1, d_1), (r_2, d_2), ... right below this second hyperbola. Success at a point (r_i, d_i) means that some evidence for an operator of order $\leq r_i$ and degree $\leq d_i$ has been found. This operator however is not explicitly computed. Instead, the method uses the partial information found about this operator to calculate an operator which with high probability is the minimal order operator satisfied by the sequence in question. This operator is usually more interesting than the one at (r_i, d_i), and its computation is usually more efficient.

Using the option `path`, the user can specify a list of points (r_i, d_i) which should be used instead of the standard path. By setting the options `min_degree`, `max_degree`, `min_order`, `max_order`, all points (r, d) of the path are discarded for which r or d is not within the specified bounds. These options can be used to accelerate the search in situations where the user has some knowledge (or intuition) about the size of the expected equations.

The figure on the right illustrates the typical situation for guessing problems that are not too small and not too artificial. The gray region indicates the area which is not accessible with the given amount of data. Only the points (r, d) below it can be tested for an operator of order r and degree d that fits to the given data. Let's assume that operators exist on and above the solid black hyperbola. The user will usually not know this curve in advance but may have some expectations about it and can restrict the search accordingly, for example to the dashed area

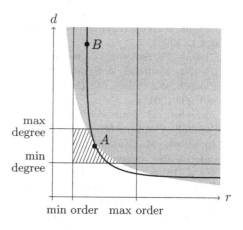

shown in the figure. The method will detect the existence of an operator, say at point A, and construct from the information gained at this point an operator of minimal possible order, which may correspond to point B. This operator is returned as output. Note that the degree of the output may exceed the value of `max_degree`, and its order may be smaller than `min_order`:

```
sage: data = [(n+1)^10*2^n + 3^n for n in xrange(200)]

sage: L = guess(data, OreAlgebra(ZZ['n'],'Sn'), min_order=3, max_degree=5)

sage: L.order(), L.degree()
```

$$(2, 10)$$

In order to test a specific point (r, d), the data array must contain at least $(r+1)(d+2)$ terms. If it has more terms, the guess becomes more trustworthy, but also the computation time increases. By setting the option ensure to a positive integer e, the user can request that only such points (r, d) should be tested for which the data array contains at least e more terms than needed. This increases the reliability. By setting the option cut to a positive integer c, the user requests that for testing a point (r, d), the method should take into account at most c more terms than needed. If the data array contains more terms, superfluous ones are ignored in the interest of better performance. We must always have $0 \le e \le c \le \infty$. The default setting is $e = 0$, $c = \infty$.

References

1. Abramov, S.A., Le, H.Q., Li, Z.: OreTools: a computer algebra library for univariate Ore polynomial rings. Technical report CS-2003-12, University of Waterloo (2003)
2. Sergei, S.A., van Hoeij, M.: Integration of solutions of linear functional equations. Integr. Transforms Spec. Funct. **9**, 3–12 (1999)
3. Bronstein, M., Petkovšek, M.: An introduction to pseudo-linear algebra. Theor. Comput. Sci. **157**(1), 3–33 (1996)
4. Chyzak, F.: Fonctions holonomes en calcul formel. Ph.D. thesis, INRIA Rocquencourt (1998)
5. Chyzak, F., Salvy, B.: Non-commutative elimination in Ore algebras proves multivariate identities. J. Symbolic Comput. **26**, 187–227 (1998)
6. Hebisch, W., Rubey, M.: Extended rate, more GFUN. J. Symbolic Comput. **46**(8), 889–903 (2011)
7. Jaroschek, M.: Improved polynomial remainder sequences for Ore polynomials. J. Symbolic Comput. **58**, 64–76 (2013)
8. Jaroschek, M., Kauers, M., Chen, S., Singer, M.F.: Desingularization explains order-degree curves for Ore operators. In: Kauers, M. (ed.) Proceedings of ISSAC'13, pp. 157–164 (2013)
9. Kauers, M.: Guessing handbook. Technical report 09–07, RISC-Linz (2009)
10. Kauers, M.: The holonomic toolkit. In: Blümlein, J., Schneider, C. (eds.) Computer Algebra in Quantum Field Theory. Texts & Monographs in Symbolic Computation, pp. 119–144. Springer, Vienna (2013)
11. Koutschan, C.: Advanced applications of the holonomic systems approach. Ph.D. thesis, RISC-Linz, Johannes Kepler Universität Linz (2009)
12. Koutschan, C.: HolonomicFunctions (User's Guide). Technical report 10–01, RISC Report Series, University of Linz, Austria, January 2010
13. Li, Z.: A Subresultant theory for linear differential, linear difference, and Ore polynomials, with applications. Ph.D. thesis, RISC-Linz (1996)

14. Mallinger, C.: Algorithmic manipulations and transformations of univariate holonomic functions and sequences. Master's thesis, J. Kepler University, Linz, August 1996
15. Ore, Ø.: Theory of non-commutative polynomials. Ann. Math. **34**, 480–508 (1933)
16. Salvy, B., Zimmermann, P.: Gfun: a Maple package for the manipulation of generating and holonomic functions in one variable. ACM Trans. Mathe. Softw. **20**(2), 163–177 (1994)
17. Stein, W.A., et al.: Sage Mathematics Software (Version 5.9). The Sage Development Team (2013). http://www.sagemath.org

Giac and GeoGebra – Improved Gröbner Basis Computations

Zoltán Kovács[1]($\boxtimes$) and Bernard Parisse[2]

[1] Johannes Kepler University, Altenberger Strasse 54, 4040 Linz, Austria
zoltan@geogebra.org
[2] Institut Fourier, UMR 5582 du CNRS, Université de Grenoble, 100 Rue des Maths,
BP 53, 38041 Grenoble Cedex 9, France
bernard.parisse@ujf-grenoble.fr

Abstract. GeoGebra is open source mathematics education software being used in thousands of schools worldwide. It already supports equation system solving, locus equation computation and automatic geometry theorem proving by using an embedded or outsourced CAS. GeoGebra recently changed its embedded CAS from Reduce to Giac because it fits better into the educational use. Also careful benchmarking of open source Gröbner basis implementations showed that Giac is fast in algebraic computations, too, therefore it allows heavy Gröbner basis calculations even in a web browser via Javascript.

Gröbner basis on $\mathbb{Q}$ for revlex ordering implementation in Giac is a modular algorithm (E. Arnold). Each $\mathbb{Z}/p\mathbb{Z}$ computation is done via the Buchberger algorithm using F4 linear algebra technics and "remake" speedups, they might be run in parallel for large examples. The output can be probabilistic or certified (which is much slower). Experimentation shows that the probabilistic version is faster than other open-source implementations, and about 3 times slower than the Magma implementation on one processor, it also requires less memory for big examples like Cyclic9.

Keywords: Gröbner basis · Computer algebra · Computer aided mathematics education

1 Introduction: Heavy Computations in the Classroom

Mathematics education has always been influenced by culture and traditions, nevertheless development of technology also played an important role in changing the approach of the teacher and the subject of teaching. The availability of personal computers and the computer algebra systems (CAS) being widespread made possible to verify the results of manual solving of an equation, what is more, to solve equations automatically and concentrate on higher level problems in the classroom.

Solving an equation system in the high school is one of the most natural mathematical problems, required not only by pure mathematics but physics and

J. Gutierrez et al. (Eds.): Computer Algebra and Polynomials, LNCS 8942, pp. 126–138, 2015.
DOI: 10.1007/978-3-319-15081-9_7

chemistry as well. Despite of its importance the way of solving an equation system is not trivial. In general, there is no algorithmic technique known which covers all possible equation systems and returns all solutions of a problem in finite time. As Wikipedia explains,

> In general, given a class of equations, there may be no systematic method (algorithm) that is guaranteed to work. This may be due to a lack of mathematical knowledge; some problems were only solved after centuries of effort. But this also reflects that, in general, no such method can exist: some problems are known to be unsolvable by an algorithm, such as Hilbert's tenth problem, which was proved unsolvable in 1970 [20].

Even after some restrictions (for example assuming that the equations are algebraic and the solutions are real numbers) there is no guarantee that the system will be solvable quickly enough, however there are efficient methods already which can compute a large enough set of problems of "typical uses".

On the other hand, classroom use of modern equation solver algorithms (in the background, i.e. invisibly for the students) cannot be restricted to direct equation solving only. There are other fields of mathematics which seem distant or unrelated with computer algebra, but still use heavy algebraic computations. With no doubt two such fields in analytical geometry are automated theorem proving (for computing proofs for Euclidean theorems in elementary geometry, see Fig. 1) and locus computation (for example to introduce the notion of parabola analytically [14]).

In most classroom situations there is no importance in the applied algorithm when an algebraic equation system must be solved in the background. Some advanced problems which are normally present in the high school curriculum, however, may lead to slow computations if the applied algorithm is not fast enough.

In this paper first we show some possible classroom situations from the present challenges of mathematics education by utilizing the dynamic geometry software GeoGebra. In Sect. 2 we demonstrate the bottleneck of GeoGebra's formerly used CAS (Reduce [9]) in its web based version, and show an alternative CAS (Giac [18]) for the same modern classroom. Then we focus on benchmarking Giac's and other CAS's especially in computing solutions of algebraic equation systems, i.e. the Gröbner basis of a set of polynomials. In Sect. 3 we provide the main concepts of Giac's Gröbner basis algorithm with detailed benchmarks.

2   Computer Algebra in the Classroom

The introduction of smartphones and tablets with broadband Internet connection allows students to access educational materials from almost everywhere. Ancsin et al. [1] argues that today's classroom computers are preferably no longer workstation PCs but tablets. Thus one of the most important question for a software developer working for mathematics education is "will my program

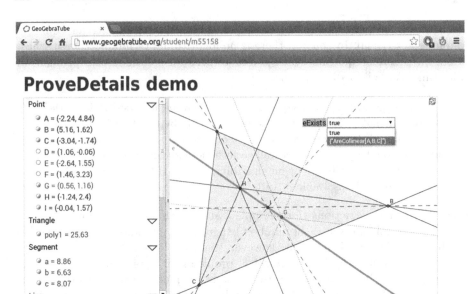

Fig. 1. Giac computes sufficient condition for the Euler's line theorem in JavaScript. Here the elimination ideal of $2v_7 - v_3, 2v_8 - v_4, 2v_9 - v_5 - v_3, 2v_{10} - v_6 - v_4, -v_{11}v_{10} + v_{12}v_9, -v_{11}v_8 + v_{12}v_7 + v_{11}v_6 - v_7v_6 - v_{12}v_5 + v_8v_5, -v_{14} - v_5 + v_3, -v_{13} + v_6 - v_4, -v_{16} + v_6 + v_3, -v_{15} + v_5 - v_4, v_{17}v_{14} - v_{18}v_{13}, v_{17}v_{16} - v_{18}v_{15} - v_{17}v_6 + v_{15}v_6 + v_{18}v_5 - v_{16}v_5, 2v_{19} - v_5 - v_3, 2v_{20} - v_6 - v_4, v_{22} - v_{20} - v_{19} + v_3, v_{21} + v_{20} - v_{19} - v_4, 2v_{23} - v_3, 2v_{24} - v_4, v_{26} - v_{24} - v_{23} + v_3, v_{25} + v_{24} - v_{23} - v_4, -v_{27}v_{22} + v_{28}v_{21} + v_{27}v_{20} - v_{21}v_{20} - v_{28}v_{19} + v_{22}v_{19}, -v_{27}v_{26} + v_{28}v_{25} + v_{27}v_{24} - v_{25}v_{24} - v_{28}v_{23} + v_{26}v_{23}, -1 + v_{29}v_{27}v_{18} - v_{29}v_{28}v_{17} - v_{29}v_{27}v_{12} + v_{29}v_{17}v_{12} + v_{29}v_{28}v_{11} - v_{29}v_{18}v_{11}$ is computed with respect to revlex ordering for variables $v_7, v_8, v_9, \ldots, v_{29}$. The result ideal contains $v_3v_6 - v_4v_5$ which yields the geometrical meaning "if triangle ABC is non-degenerate, then its orthocenter H, centroid I and circumcenter G are collinear".

work on a tablet"? Technically speaking, modern developments should move to the direction of the HTML5 standard with JavaScript (JS) empowered.

Online access of web server based computer algebra systems became very popular in the academic world during the last years, including many students and teachers of a number of universities world-wide. For example, *Sage*[1] has already been famous not only for being freely available to download, but for its free demonstration server http://cloud.sagemath.com. Similar approaches are the *SymPy Live/Gamma*[2] projects and the *IPython Notebook*[3]. On one hand, these systems are free of charge and thus they can be well used in education, and they are empowered by HTML5 and JS on the client side. On the other hand,

---

[1] http://sagemath.org.

[2] http://live.sympy.org, http://www.sympygamma.com.

[3] http://ipython.org/notebook.html.

when no or only slow Internet connection is available, none of these systems can be used conveniently because the server side computations are not available any longer. This is why it seems to be a more fruitful approach to develop an *offline* system which can be run locally on the user's machine inside a web browser, especially in those classrooms where no Internet connection is permitted.

GeoGebra developers, reported in [16], started to focus on implementing a full featured offline CAS using the HTML5 technology. The first visible result in May 2012 was the embedded system GGBReduce which offered many modules of the Reduce CAS, using the Google Web Toolkit for compiling the JLisp Lisp implementation into JS. This work is a official part of Reduce under the name JSLisp now [12].

The first tests back in May 2011 were very promising: the http://www.geogebra.org/mpreduce/mpreduce.html web page with the base system loaded below 1 s and used 1.7 MB of JS code. Unfortunately, after adding some extra modules and the Lisp heap, the initialization time of the Reduce system increased drastically: see http://dev.geogebra.org/qa/?f=_CAS_mixture.ggb& c=w-42-head for an example of GeoGebraWeb 4.2 (2.2 MB JS), the notification message "CAS initializing" disappears only after 10 s or even more. By contrast, http://dev.geogebra.org/qa/?f=_CAS_mixture.ggb&c=w-44-head loads below 5 s, using the Giac CAS in GeoGebraWeb 4.4 (7.5 MB JS).

Test case http://dev.geogebra.org/qa/?f=_CAS-commands.ggb&... shows some typical classroom computations related to analyze a rational function as an exercise. It includes factorizing a polynomial, turning a rational function into partial fractions, computing the limit of a rational function at infinity, and computing the asymptotes. Also conversion of symbolic to numeric, computation of derivative and a solution of an equation in one variable are included. Finally, GeoGebraWeb 4.4 displays the graph of the function and its derivative. The entire process completes in 7 s. Each update takes 5 s.

These benchmarks were collected, however, on a modern PC. On low cost machines the statistics will be worse, expecting further work on speedup the computations preferably by using native code in the browser if possible.

2.1   JavaScript: The Assembly Language for the Web

Embedding a third party CAS usually requires a remarkable work in both the main software and the embedded part. Figure 2 shows former efforts to include a CAS in GeoGebra. JSCL, Jasymca [4], Jama [10] and Yacas/Mathpiper [13] were used as native Java software, but Reduce is written in Lisp and Giac in C++. (Not shown in the figure, but also Maxima was planned as an extra CAS shipped with GeoGebra separately—this plan was finally cancelled.)

The main criteria for using a CAS was to be able to run in both the "desktop" and "web" environments. The desktop environment, technically a Java Virtual Machine, has used to be the primary user interface for many years since the very beginning of the GeoGebra project, including platforms Microsoft Windows, Apple's Mac OS X, and Linux. The web environment is the new direction,

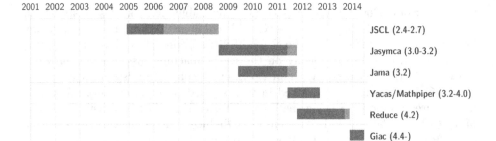

Fig. 2. Computer algebra systems used in GeoGebra from version 2.4 to 4.4. Orange bars show that the corresponding version of GeoGebra was no longer developed by the programmers, but still used by the community (Color figure online).

being capable of supporting platforms Windows 8, iPad, Android, and Google's Chromebook.

Our approach was to support both environments with the same codebase, i.e. to make it possible to use the same source code for all platforms. This has been succeeded by using the tools explained in Fig. 3. The speedup in the initialization time is obvious: Giac is much faster in its startup, even if there is a slowness factor between 1 and 10 compared to the Java Native Interface (JNI) version, depending on the type of the computation.

GeoGebra since version 4.2 not only supports typical CAS operations but dynamic geometry computations as well. An example is the **LocusEquation** command which computes the equation of the locus (if its construction steps can be described algebraically). The forthcoming version 5.0 will support **Envelope** equations and automated geometry proofs in elementary geometry by using the **Prove** command. It was essential to make benchmarks to test the underlying algebra commands in Giac before really using them officially in 4.4.

"JavaScript is Assembly Language for the Web", states Scott Hanselman from Microsoft, citing Erik Meijer, former head of Cloud Programmability Team at Microsoft [8]. The reasoning is as follows:

- JavaScript is ubiquitous.
- It's fast and getting faster.
- Javascript is as low-level as a web programming language goes.
- You can craft it manually or you can target it by compiling from another language.

On the other hand, despite being universal and fast, algebraic computations programmed in a non-scripting (i.e. compiled) language (e.g. C or Java) are still much faster than it is expectable for being run in a web browser using the normal browser standards. After the JavaScript engine race 2008–2011 [22], there is a second front of bleeding edge research to develop another standard language of the web (Google's Dart [23], for example), and a renewed focus on C and C++ to compile them into browser independent or native bytecode

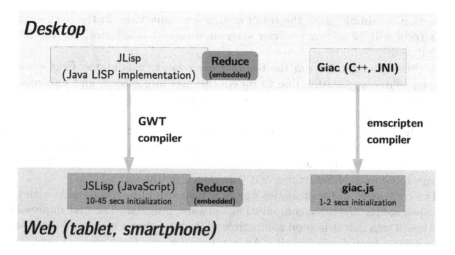

Fig. 3. Embedded computer algebra systems Reduce and Giac in GeoGebra 4.2 and 4.4. Reduce itself is embedded into JLisp and JSLisp.

(e.g. Google's Portable and Native Client [21])[4]. Despite these new experimental ways, JavaScript is still the de facto standard of portability and speed for modern computers, and probably the best approach to generate as fast JavaScript code for computer algebra algorithms as possible.[5]

To have an exact speed comparison of JNI and JS versions of Giac a modern headless scriptable browser will be helpful. The new version of *PhantomJS* [6] has already technical support to run Giac especially for benchmarking purposes: its stable version is expected to be publicly available soon.

2.2 Benchmarks from Automated Theorem Proving

The GeoGebra Team already developed a benchmarking system for testing various external computer algebra systems with different stress cases. In this

---

[4] Giac has already been successfully compiled into .pexe and .nexe applications at http://ggb1.idm.jku.at/~kovzol/data/giac. The native executables for 32/64 bits Intel and ARM achitecture binaries are between 9.7 and 11.9 MB, the portable executable is 6.3 MB, however, the .pexe → .nexe compilation takes too long, at least 1 min on a recent hardware. The runtime speed is comparable with the JNI version, i.e. much faster than the JavaScript version.

[5] Axel Rauschmayer, author of the forthcoming O'Reilly book *Speaking JavaScript*, predicts JavaScript to run near-native performance in 2014 (see http://www.2ality. com/2014/01/web-platform-2014.html for details). The speed is currently about 70 % of compiled C++ code by using `asm.js`. This will, however, doubtfully speed up the JS port of Giac like in a native client since it heavily uses the anonymous union of C/C++ to pack data (unsupported in JS).

[6] https://github.com/ariya/phantomjs/wiki/PhantomJS-2 contains a step-by-step guide to compile the development version of PhantomJS version 2.

subsection we simply show the result of this benchmarking. In the next section other tests will be shown between systems designed much more for algebraic computations.

In this subsection we run the tests on *open source* candidates. GeoGebra is an open source application due to its educational use: schools and universities may prefer using software free of charge than paying for licenses.

The test cases are chosen from a set of simple theorems in elementary geometry. All tests are Gröbner basis computations in a polynomial ring over multiple variables. The equation systems (i.e. the ideals to compute the Gröbner basis for) can be checked in details at [15], the final summary (generated with default settings in Giac as of September 2013) is shown in Table 1.

The conclusion of the statistics was that Giac would be comparable with the best open source algebraic computation software, Singular. The first impression of its speed was that it is good competitor of Reduce, thus a very good candidate to be a long term basis for all CAS computations in GeoGebra. (Later Giac was extended with an even faster algorithm for computing Gröbner basis, as described in Sect. 3.)

3   Gröbner Basis Algorithm in Giac[7]

Starting with version 1.1.0-26, Giac [18] has a competitive implementation of Gröbner basis for reverse lexicographic ordering, that will be described in this section.

3.1   Sketch of E. Arnold Modular Algorithm

Let $f_1, \ldots, f_m$ be polynomials in $\mathbb{Q}[x_1, \ldots, x_n]$, $I = \langle f_1, \ldots, f_m \rangle$ be the ideal generated by $f_1, \ldots, f_m$. Without loss of generality, we may assume that the f_i have coefficients in $\mathbb{Z}$ by multiplying by the least common multiple of the denominators of the coefficients of f_i. We may also assume that the f_i are primitive by dividing by their content.

Let $<$ be a total monomial ordering (for example `revlex` the total degree reverse lexicographic ordering). We want to compute the Gröbner basis G of I over $\mathbb{Q}$ (and more precisely the inter-reduced Gröbner basis, sorted with respect to $<$). Now consider the ideal I_p generated by the same f_i but with coefficients in $\mathbb{Z}/p\mathbb{Z}$ for a prime p. Let G_p be the Gröbner basis of I_p (also assumed to be inter-reduced, sorted with respect to $<$, and with all leading coefficients equal to 1).

Assume we compute G by the Buchberger algorithm [3] with Gebauer and Möller criterion [7], and we reduce in $\mathbb{Z}$ (by multiplying the s-poly to be reduced by appropriate leading coefficients), if no leading coefficient in the polynomials are divisible by p, we will get by the same process but computing modulo p the G_p Gröbner basis. Therefore the computation can be done in parallel in

---

7 The content of this section is released under the Public Domain.

Table 1. Outputs of Gröbner basis benchmarking in seconds on an Intel Xeon CPU E3-1220 V2 @ 3.10 GHz running Ubuntu Linux 11.10 in VirtualBox 4.2.10 on an Ubuntu 12.04.1 host. Timeout is 60 s, timed out tests are shown with empty cells. Average* shows the average by computing 60 s computation time for tests being timed out. Singular solved the first four tests below 0.01—the benchmarking system was unable to measure timing under this precision. The CoCoA column shows the results of Giac via CoCoALib.

Test	Maxima	JAS 2	Reduce	Singular	CoCoA	Giac
Thales	0.2	0.46	0.11	0.00	0.08	0.03
Heights	0.29	0.51	0.11	0.00	0.29	0.03
Medians	0.4	0.65	0.12	0.00	0.14	0.09
Bisectors	0.42	0.5	0.1	0.00	0.09	0.01
Euler's line		1.66	0.2	0.01	0.14	0.01
Nine points circle	1.19	1.5	0.11	0.01	0.13	0.01
Angle bisector	36.08	1.74	0.75	0.01	0.31	0.04
Simson's line						
Pappus		3.37		0.5	9.28	4.9
Simson (reduced)		5.77	6.07	0.07	0.87	0.15
Pappus (reduced)		2.33	2.18	0.02	0.34	0.4
Average	*6.43*	*1.85*	*1.08*	*0.06*	*1.17*	*0.57*
Average*	*56.43*	*7.85*	*14.41*	*6.06*	*7.17*	*6.56*

$\mathbb{Z}$ and in $\mathbb{Z}/p\mathbb{Z}$ except for a finite set of *unlucky* primes (since the number of intermediate polynomials generated in the algorithm is finite). If we are choosing our primes sufficiently large (e.g. about 30 bits), the probability to fall on an unlucky prime is very small (less than the number of generated polynomials divided by about 2^{30}, even for really large examples like Cyclic9 where there are a few 10^4 polynomials involved, it would be about 1e-5).

The Chinese remaindering modular algorithm works as follows: compute G_p for several primes, for all primes that have the same leading monomials in G_p, reconstruct $G_{\prod p_j}$ by Chinese remaindering, then reconstruct a candidate Gröbner basis G_c in $\mathbb{Q}$ by rational (Farey) reconstruction. Once it stabilizes, do the checking step described below, and return G_c on success.

Checking steps: check that the original f_i polynomials reduce to 0 with respect to G_c (fast check) and check that G_c is a Gröbner basis (slow check).

Theorem 1 *(Arnold). If the checking steps succeed, then G_c is the Gröbner basis of I.*

This is a consequence of ideal inclusions (first check) and dimensions (second check), for a complete proof, see [2]. The proof does not require that we reconstruct from Gröbner basis for all primes p, it is sufficient to have one Gröbner basis for one of the primes. This can be used to speedup computation like in F4remake (Joux-Vitse, see [11]).

3.2 Computation Modulo a Prime

The Buchberger algorithm with F4[5,6]-like linear algebra is implemented modulo primes smaller than 2^{31} using total degree as selection criterion for critical pairs.

1. Initialize the basis to the empty list, and a list of critical pairs to empty.
2. Add one by one all the f_i to the basis and update the list of critical pairs with Gebauer and Möller criterion, by calling the *gbasis update procedure* (described below at step 9).
3. Begin of a new iteration:
   All pairs of minimal total degree are collected to be reduced simultaneously, they are removed from the list of critical pairs.
4. The symbolic preprocessing step begins by creating a list of monomials, gluing together all monomials of the corresponding s-polys (this is done with a heap data structure).
5. The list of monomials is "reduced" by division with respect to the current basis, using heap division (like Monagan-Pearce [17]) without taking care of the real value of coefficients. This gives a list of all possible remainder monomials and a list of all possible quotient monomials and a list of all quotient times corresponding basis element monomial products. This last list together with the remainder monomial list is the list of all possible monomials that may be generated reducing the list of critical pairs of maximal total degree, it is ordered with respect to <. We record these lists for further prime runs (speeds up step 4 and 5) during the first prime computation.
6. The list of quotient monomials is multiplied by the corresponding elements of the current basis, this time doing the coefficient arithmetic. The result is recorded in a sparse matrix, each row has a pointer to a list of coefficients (the list of coefficients is in general shared by many rows, the rows have the same reductor with a different monomial shift), and a list of monomial indices (where the index is relative to the ordered list of possible monomials). We sort the matrix by decreasing order of leading monomial.
7. Each s-polynomial is written as a dense vector with respect to the list of all possible monomials, and reduced with respect to the sparse matrix, by decreasing order with respect to <. (To avoid reducing modulo p each time, we are using a dense vector of 128 bits integers on 64 bits architectures, and we reduce mod p only at the end of the reduction. If we work on 24 bit signed integers, we can use a dense vector of 63 bits signed integer and reduce the vector if the number of rows is greater than 2^{15}).
8. Then inter-reduction happens on all the dense vectors representing the reduced s-polynomials, this is dense row reduction to echelon form (0 columns are removed first). Care must be taken at this step to keep row ordering for further prime runs.
9. gbasis update procedure:
   We record zero reducing pairs during the first prime iteration, this information will be used during later iterations with other primes to avoid computing and reducing useless critical pairs (if a pair does not reduce to 0 on $\mathbb{Q}$,

it has in general a large number of monomials therefore the probability that it reduces to 0 on the first prime run is very small). Each non zero row will bring a new entry in the current basis. New critical pairs are created with this new entry (discarding useless pairs by applying Gebauer-Möller criterion). An old entry in the basis may be removed if its leading monomial has all partial degrees greater or equal to the leading monomial corresponding degree of the new entry. Old entries may also be reduced with respect to the new entries at this step or at the end of the main loop.

10. If there are new critical pairs remaining start a new iteration at step 3. Otherwise the current basis is the Gröbner basis modulo p.

3.3   Probabilistic and Deterministic Check, Benchmarks

We first perform the fast check that the original f_i polynomials reduce to 0 modulo G_c. Then the user has a choice between a probabilistic fast check (useful for conjectures) and a deterministic slower certification for a computer assisted proof.

Probabilistic checking algorithm: instead of checking that s-polys of critical pairs of G_c reduce to 0, we will check that the s-polys reduce to 0 modulo several primes that do not divide the leading coefficients of G_c and stop as soon as the inverse of the product of these primes is less than a fixed $\varepsilon > 0$ (the check is done only after the reconstructed basis stabilizes, with our examples the first check was always successful).

Deterministic checking algorithm: check that all s-polys reduce to 0 over $\mathbb{Q}$. The fastest way to check seems to make the reduction using integer computations. We have also tried reconstruction of the quotients over $\mathbb{Z}/p\mathbb{Z}$ for sufficiently many primes: once the reconstructed quotients stabilize, we can check the 0-reduction identity on $\mathbb{Z}$, and this can be done without computing the products quotients by elements of G_c if we have enough primes (with appropriate bounds on the coefficients of G_c and the lcm of the denominators of the reconstructed quotients).

Benchmarks Comparison of Giac (1.1.0-26) with Singular 3.1 (from Sage 5.10) on Mac OS X.6, Dual Core i5 2.3 Ghz, RAM 2 × 2Go:

– The benchmarks are the classical Cyclic and Katsura benchmarks, and a more random example, described in the Giac syntax below:

```
alea6 := [5*x^2*t+37*y*t*u+32*y*t*v+21*t*v+55*u*v,
39*x*y*v+23*y^2*u+57*y*z*u+56*y*u^2+10*z^2+52*t*u*v,
33*x^2*t+51*x^2+42*x*t*v+51*y^2*u+32*y*t^2+v^3,
44*x*t^2+42*y*t+47*y*u^2+12*z*t+2*z*u*v+43*t*u^2,
49*x^2*z+11*x*y*z+39*x*t*u+44*x*t*u+54*x*t+45*y^2*u,
48*x*z*t+2*z^2*t+59*z^2*v+17*z+36*t^3+45*u] ;
l:=[x,y,z,t,u,v] ;
```

```
p1:=prevprime(2^24); p2:=prevprime(2^29);
time(G1:=gbasis(alea6 % p1,1,revlex));
time(G2:=gbasis(alea6 % p2,1,revlex));
threads:=2; // <= to the number of CPU
// debug_infolevel(1); // uncomment for intermed. steps
proba_epsilon:=1e-7; // probabilistic algorithm.
time(H0:=gbasis(alea6,indets(cyclic5),revlex));
proba_epsilon:=0; // deterministic
time(H1:=gbasis(alea6,indets(cyclic5),revlex));
```

- Mod timings were computed modulo `nextprime(2^24)` and modulo 107374-1827 (`nextprime(2^30)`).
- Probabilistic check on $\mathbb{Q}$ depends linearly on log of precision, two timings are reported, one with error probability less than `1e-7`, and the second one for `1e-16`.
- Check on $\mathbb{Q}$ in Giac can be done with integer or modular computations hence two times are reported.

Table 2. Benchmarks between Giac and Singular. » means timeout (3/4 h or more) or memory exhausted (Katsura12 modular `1e-16` check with Giac) or test not done because it would obviously timeout (e.g. Cyclic8 or 9 on $\mathbb{Q}$ with Singular).

	Giac mod p 24, 31 bits	Giac run2	Singular mod p	Giac $\mathbb{Q}$ prob. 1e-7, 1e-16	Giac $\mathbb{Q}$ certified	Singular $\mathbb{Q}$
Cyclic7	0.5, 0.58	0.1	2.0	3.5, 4.2	20, 29.3	»
Cyclic8	7.2, 8.9	1.8	52.5	103, 107	258, 679	»
Cyclic9	633, 1340	200	?	1 day	»	»
Kat8	0.063, 0.074	0.009	0.2	0.33, 0.53	6.55, 4.35	4.9
Kat9	0.29, 0.39	0.05	1.37	2.1, 4.2	54, 36	41
Kat10	1.53, 2.27	0.3	11.65	14, 33.9	441, 335	480
Kat11	10.4, 13.8	2.8	86.8	170, 361	4610, 3530	6035
Kat12	76, 103	27	885	1950, 4000	25000, »	»
Alea6	0.83, 1.08	.26	4.18	187, 194	718	>1h

This leads to the following observations (see Table 2):

- Computation modulo p for 24 to 31 bits is faster that Singular, but seems also faster than Magma (and Maple). For smaller primes, Magma is 2 to 3 times faster.
- The probabilistic algorithm on $\mathbb{Q}$ is much faster than Singular on these examples (this probably means that Singular does not implement a modular algorithm). Compared to Maple 16, it is reported to be faster for Katsura10, and as fast for Cyclic8. Compared to Magma, it is about 3 to 4 times slower.
- If [19] is up to date (except about Giac), Giac is the third software and first open-source software to solve Cyclic9 on $\mathbb{Q}$ (the link is rather old, but we

believe it is still correct). It requires 378 primes of size 29 bits, takes about 1 day, requires 3 GB of memory on 1 processor, while with 6 processors it takes 6 h (requires 6 GB). The answer has integer coefficients of about 1600 digits (and not 800 as stated in J.-C. Faugère F4 article), for a little more than 1 million monomials, that's about 1.4 GB of RAM.

- The deterministic modular algorithm is much faster than Singular for Cyclic examples, and as fast for Katsura examples.
- For the random last example, the speed is comparable between Magma and Giac. This is where there are less pairs reducing to 0 ("F4remake" is not as efficient as for Cyclic or Katsura) and larger coefficients. This could suggest that advanced algorithms like F4/F5/etc. are probably not much more efficient than Buchberger algorithm for these kind of inputs without symmetries.
- Certification is the most time-consuming part of the process (except for Cyclic8). Integer certification is significantly faster than modular certification for Cyclic examples, and almost as fast for Katsura.
- We would like to stress that a computer assisted **mathematical proof** can not be performed with a closed-source software without a certification step. Therefore the relevant timings for comparison with closed-source pieces of software is the probabilistic check.

Acknowledgments. The first author thanks Michael Borcherds and Zbyněk Konečný for the test cases in benchmarking the embeddable computer algebra systems. The second author wishes to thank Vanessa Vitse for insightful discussions and Frédéric Han for testing.

References

1. Ancsin, G., Hohenwarter, M., Kovács, Z.: GeoGebra goes mobile. Electron. J. Math. Technol. **5**(2), 160–168 (2011)
2. Arnold, E.A.: Modular algorithms for computing Gröbner bases. J. Symbolic Comput. **35**(4), 403–419 (2003)
3. Buchberger, B.: Gröbner bases: an algorithmic method in polynomial ideal theory. In: Bose, N.K. (ed.) Multidimensional Systems Theory, pp. 184–232. Reidel Publishing Company, Dodrecht (1985)
4. Dersch, H.: Jasymca 2.0 – symbolic calculator for Java, Mar 2009. http://webuser. hs-furtwangen.de/~dersch/jasymca2/Jasymca2en/Jasymca2en.html
5. Faugère, J.-C.: A new efficient algorithm for computing Gröbner bases (F4). J. Pure Appl. Algebra **139**(1–3), 61–88 (1999)
6. Faugère, J.-C.: A new efficient algorithm for computing Gröbner bases without reduction to zero (F5). In: Proceedings of the 2002 International Symposium on Symbolic and Algebraic Computation, ISSAC 2002, pp. 75–83. ACM, New York (2002)
7. Gebauer, R., Möller, H.M.: On an installation of Buchberger's algorithm. J. Symbolic Comput. **6**(2–3), 275–286 (1988)
8. Hanselman, S.: JavaScript is assembly language for the web: semantic markup is dead! Clean vs. machine-coded HTML (2011). http://goo.gl/YKiO6B. Accessed 8 Jan 2014

9. Hearn, A.C.: REDUCE User's Manual Version 3.8, Feb 2004. http://reduce-algebra.com/docs/reduce.pdf
10. Hicklin, J., Moler, C., Webb, P., Boisvert, R.F., Miller, B., Pozo, R., Remington, K.: JAMA: JAva MAtrix package, Nov 2012. http://math.nist.gov/javanumerics/jama/
11. Joux, A., Vitse, V.: A variant of the F4 algorithm. In: Kiayias, A. (ed.) CT-RSA 2011. LNCS, vol. 6558, pp. 356–375. Springer, Heidelberg (2011)
12. Kosan, T.: JSLisp. http://sourceforge.net/p/reduce-algebra/code/HEAD/tree/trunk/jslisp
13. Kosan, T.: MathPiper, Jan 2011. http://www.mathpiper.org
14. Kovács, Z.: Definition of a parabola as a locus. GeoGebraTube material (2012). http://www.geogebratube.org/student/m23662
15. Kovács, Z.: GeoGebra developers' Trac wiki: theorem proving planning (2012). http://dev.geogebra.org/trac/wiki/TheoremProvingPlanning. Accessed 8 Jan 2014
16. Kovács, Z.: GeoGebraWeb offers CAS functionality, May 2012. http://blog.geogebra.org/2012/05/geogebraweb-cas/
17. Monagan, M., Pearce, R.: Sparse polynomial division using a heap. J. Symbolic Comput. **46**(7), 807–822 (2011)
18. Parisse, B., Graeve, R.D.: Giac/Xcas computer algebra system (2013). http://www-fourier.ujf-grenoble.fr/~parisse/giac_fr.html
19. Steel, A.: Gröbner basis timings page (2004). http://magma.maths.usyd.edu.au/~allan/gb/
20. Wikipedia. Equation solving — Wikipedia, the free encyclopedia (2013). http://en.wikipedia.org/w/index.php?title=Equation_solving&oldid=580349875. Accessed 14 Jan 2014
21. Wikipedia. Google Native Client — Wikipedia, the free encyclopedia (2013). http://en.wikipedia.org/w/index.php?title=Google_Native_Client&oldid=588335015. Accessed 8 Jan 2014
22. Wikipedia. JavaScript engine — Wikipedia, the free encyclopedia (2013). http://en.wikipedia.org/w/index.php?title=JavaScript_engine&oldid=586802475. Accessed 8 Jan 2014
23. Wikipedia. Dart (programming language) — Wikipedia, the free encyclopedia (2014). http://en.wikipedia.org/w/index.php?title=Dart_(programming_language)&oldid=589448479 . Accessed 8 Jan 2014

Polar Varieties Revisited

Ragni Piene[⊠]

Department of Mathematics, University of Oslo,
P.O.Box 1053, Blindern, 0316 Oslo, Norway
ragnip@math.uio.no
http://www.mn.uio.no/math/english/people/aca/ragnip/index.html

Abstract. We recall the definition of classical polar varieties, as well as those of affine and projective reciprocal polar varieties. The latter are defined with respect to a non-degenerate quadric, which gives us a notion of orthogonality. In particular we relate the reciprocal polar varieties to the "Euclidean geometry" in projective space. The Euclidean distance degree and the degree of the focal loci can be expressed in terms of the ranks, i.e., the degrees of the classical polar varieties, and hence these characters can be found also for singular varieties, when one can express the ranks in terms of the singularities.

Keywords: Schubert varieties · Polar varieties · Reciprocal polar varieties · Euclidean normal bundle · Euclidean distance degree · Focal locus

1 Introduction

The theory of polars and polar varieties has played an important role in the quest for understanding and classifying projective varieties. Their use in the definition of projective invariants is the very basis for the geometric approach to the theory of characteristic classes, such as Todd classes and Chern classes. In particular this approach gives a way of defining Chern classes for *singular* projective varieties (see e.g. [15,17]). The *local* polar varieties were used by Lê–Teissier, Merle and others in the study of singularities.

More recently, polar varieties have been applied to study the topology of real affine varieties and to find real solutions of polynomial equations by Bank et al., Safey El Din–Schost, and Mork–Piene [1–4,13,14,18], to complexity questions by Bürgisser–Lotz [5], to foliations by Soares [19] and others, to focal loci and caustics of reflection by Catanese and Trifogli [6,22] and Josse–Pène [8], to Euclidean distance degree by Draisma et al. [7].

In this note I will explore the relation of polar and reciprocal polar varieties of possibly singular varieties to the Euclidean normal bundle, the Euclidean distance degree, and the focal loci. For simplicity, we work over an algebraically closed field of characteristic 0.

J. Gutierrez et al. (Eds.): Computer Algebra and Polynomials, LNCS 8942, pp. 139–150, 2015.
DOI: 10.1007/978-3-319-15081-9_8

2   Classical Polar Varieties

Let $\mathbb{G}(m, n)$ denote the Grassmann variety of m-dimensional linear subspaces of $\mathbb{P}^n$. Let $L_k \subset \mathbb{P}^n$ be a linear subspace of dimension $n - m + k - 2$. Consider the special Schubert variety

$$\Sigma(L_k) := \{W \in \mathbb{G}(m, n) \,|\, \dim W \cap L_k \geq k - 1\}.$$

It has a natural structure as a determinantal scheme and is of codimension k in $\mathbb{G}(m, n)$.

As is well known, if L'_k is another linear subspace of dimension $n - m + k - 2$, then $\Sigma(L'_k)$ and $\Sigma(L_k)$ are projectively equivalent (in particular, their rational equivalence classes are equal). So we will often just write Σ_k for such a variety.

Example 1. The case $m = 1$, $n = 3$. For $\mathbb{G}(1, 3) = \{$lines in $\mathbb{P}^3\}$, the special Schubert varieties are

$$\Sigma_1 = \{\text{lines meeting a given line}\}$$
$$\Sigma_2 = \{\text{lines in a given plane}\}.$$

Example 2. The case $m = 2$, $n = 5$. For $\mathbb{G}(2, 5) = \{$planes in $\mathbb{P}^5\}$, the special Schubert varieties are

$$\Sigma_1 = \{\text{planes meeting a given plane}\}$$
$$\Sigma_2 = \{\text{planes intersecting a given 3-space in a line}\}$$
$$\Sigma_3 = \{\text{planes contained in a given hyperplane}\}$$

More general Schubert varieties are defined similarly, by giving conditions with respect to flags of linear subspaces (see e.g. [11]). For example, in Example 1, we could consider

$$\Sigma_{0,2} = \{\text{lines in a given plane through a given point in the plane}\}.$$

Let $X \subset \mathbb{P}^n$ be a (possibly singular) variety of dimension m. The *Gauss map* $\gamma \colon X \dashrightarrow \mathbb{G}(m, n)$ is the rational map that sends a nonsingular point $P \in X_{\mathrm{sm}}$ to the (projective) tangent space $T_P X$, considered as a point in the Grassmann variety. More precisely, if $V = H^0(\mathbb{P}^n, \mathcal{O}_{\mathbb{P}^n}(1))$, then γ is given on $X_{\mathrm{sm}} := X \setminus \mathrm{Sing}\, X$ by the restriction of the quotient

$$V_X \to \mathcal{P}^1_X(\mathcal{L}),$$

where $\mathcal{P}^1_X(\mathcal{L})$ denotes the bundle of 1st principal parts of the line bundle $\mathcal{L} := \mathcal{O}_{\mathbb{P}^n}(1)|_X$. Note that restricted to X_{sm} the sheaf of principal parts is locally free with rank $m + 1$, and that the fibers (over X_{sm}) of $\mathbb{P}(\mathcal{P}^1_X(\mathcal{L})) \subset X \times \mathbb{P}^n$ over X (with respect to projection on the first factor) define the (projective) tangent spaces of X.

The *polar varieties* of $X \subset \mathbb{P}^n$ are the closures of the inverse images

$$M_k := \overline{\gamma|_{X_{\mathrm{sm}}}^{-1} \Sigma_k}$$

of the special Schubert varieties via the Gauss map [15, p. 252].

In the situation of Example 1, let $X \subset \mathbb{P}^3$ be a curve. Then M_1 is the set of nonsingular points $P \in X$ such that the tangent line $T_P X$ meets a given line, i.e., it is the ramification points of the linear projection map $X \to \mathbb{P}^1$.

In Example 2, let $X \subset \mathbb{P}^5$ be a surface. Then M_1 is the ramification locus of the projection map $X \to \mathbb{P}^2$ with center a plane, and M_2 consists of points $P \in X_{\mathrm{sm}}$ such that the tangent plane $T_P X$ intersects a given 3-space in a line. One could also consider more general polar varieties, corresponding to general Schubert varieties, like the set of points P such that $T_P X$ meets a given line; this is the ramification locus of the linear projection of X to $\mathbb{P}^3$ with the line as center. Note that the cardinality of this 0-dimensional variety is equal to the degree of the tangent developable of X.

3   Polar Classes and Chern Classes

It was noted classically that the polar varieties are invariant under linear projections and sections. Therefore the *polar classes*, i.e., their rational equivalence classes, are *projective invariants* of the variety. Already Noether, Segre and Zeuthen observed that certain integral combinations of the polar classes of surfaces are *intrinsic* invariants, i.e., they depend only on the surface, not on the given projective embedding. This was pursued by Todd and Severi, who used the polar classes to define what are now called the Chern classes. The formula is the following:

$$c_k(X) = \sum_{i=0}^{k} (-1)^{k-i} \binom{m+1-k+i}{i} [M_i] h^i,$$

where h denotes the class of a hyperplane section. Since this expression makes sense also for singular varieties, it gives a definition of Chern classes for singular projective varieties, called the *Chern–Mather* classes (see e.g. [17]). The formula can be inverted to express the polar classes in terms of Chern classes. When the variety X is nonsingular, this is just the expression coming from the fact that in this case,

$$[M_k] = c_k(\mathcal{P}_X^1(\mathcal{L})).$$

Using the canonical exact sequence

$$0 \to \Omega_X^1 \otimes \mathcal{L} \to \mathcal{P}_X^1(\mathcal{L}) \to \mathcal{L} \to 0,$$

we compute, with $c_i(X) = c_i(\Omega_X^{1\vee}) = (-1)^i c_i(\Omega_X^1)$,

$$[M_k] = \sum_{i=0}^{k} (-1)^{k-i} \binom{m+1-k+i}{i} c_{k-i}(X) h^i.$$

When X is singular, we replace X by its Nash transform $\overline{X}$, i.e., $\nu\colon \overline{X} \to X$ is the smallest proper modification of X such that γ extends to a morphism $\overline{\gamma}\colon \overline{X} \to \mathbb{G}(m, n)$. If we denote by $V_{\overline{X}} \to \mathcal{P}$ the locally free quotient corresponding to $\overline{\gamma}$, then it follows from the definition that we have

$$[M_k] = \overline{\gamma}_* c_k(\mathcal{P}).$$

In many cases, this allows us to find formulas for the degrees of the polar classes in terms of the singularities of X, see [15–17].

4   Reciprocal Polar Varieties

In [3,4] Bank et al. introduced what they called *dual* polar varieties. These varieties were further studied in [13,14] under the name of *reciprocal* polar varieties. They are defined with respect to a non degenerate quadric in the ambient projective space, and sometimes with respect to the choice of a hyperplane at infinity.

Let $Q \subset \mathbb{P}^n$ be a non degenerate quadric. Then Q induces a *polarity*, classically called a *reciprocation*, between points and hyperplanes in $\mathbb{P}^n$. The *polar hyperplane* $P^\perp$ of P is the linear span of the points on Q such that the tangent hyperplane to Q at that point contains P:

$$P = (b_0 : \cdots : b_n) \mapsto P^\perp \colon \sum_{i=0}^{n} b_i \frac{\partial q}{\partial X_i} = 0.$$

The map $P \mapsto P^\perp$ is the isomorphism $\mathbb{P}^n \to (\mathbb{P}^n)^\vee$ given by the symmetric bilinear form associated with the quadratic polynomial q defining Q. The point $P^\perp$ in the dual projective space represents a hyperplane in $\mathbb{P}^n$. If $L \subset \mathbb{P}^n$ is a linear space of dimension r, then $L^\perp \subset (\mathbb{P}^n)^\vee$ gives a $(n - r - 1)$-dimensional linear subspace $(L^\perp)^\vee \subset \mathbb{P}^n$, which we will, by abuse of notation, also denote by $L^\perp$. Note that if H is a hyperplane, its *pole* $H^\perp$ is the intersection of the tangent hyperplanes to Q at the points of intersection with H. The map $H \mapsto H^\perp$ is the inverse of the above map $P \mapsto P^\perp$.

If the quadric is $q = \sum_{i=0}^{n} X_i^2$, then the polar of

$$P = (b_0 : \cdots : b_n)$$

is the hyperplane

$$P^\perp : b_0 X_0 + \ldots + b_n X_n = 0.$$

Let $X \subset \mathbb{P}^n$ be a (possibly singular) variety of dimension m. Let $K_i \subset \mathbb{P}^n$ be a linear subspace of dimension $i + n - m - 1$. Recall [14, Def.2.1,p. 104], [3, p. 527] the definition of the i-th reciprocal polar variety of X (with respect to K_i): $W_{K_i}^\perp(X)$, $1 \le i \le m$, is the Zariski closure of the set

$$\{P \in X_{\mathrm{sm}} \setminus K_i^\perp \,|\, T_P X \not\pitchfork \langle P, K_i^\perp \rangle^\perp\},$$

where $T_P X$ denotes the tangent space to X at the point P and the notation $M \not\pitchfork N$ means that the two linear spaces M, N are not transversal, i.e., that their linear span $\langle M, N \rangle$ is not equal to the whole ambient space $\mathbb{P}^n$. For general K_i, the ith reciprocal variety has codimension i.

The condition $T_P X \not\pitchfork \langle P, K_i^\perp \rangle^\perp$ is equivalent to the condition $\dim(T_P X \cap \langle P, K_i^\perp \rangle^\perp) \geq i - 1$, and $T_P X \cap \langle P, K_i^\perp \rangle^\perp$ is equal to $T_P X \cap P^\perp \cap K_i$, or to $\langle T_P X^\perp, P \rangle^\perp \cap K_i$, or to $\langle \langle T_P X^\perp, P \rangle, K_i^\perp \rangle^\perp$. That the dimension of the latter space is greater or equal to $i - 1$ is equivalent to $\dim \langle \langle T_P X^\perp, P \rangle, K_i^\perp \rangle \leq n - i$, or to $\dim(\langle T_P X^\perp, P \rangle \cap K_i^\perp) \geq 0$. Thus we obtain the following description of the ith polar variety:

$$W_{K_i}^\perp(X) = \overline{\{P \in X_{\mathrm{sm}} \setminus K_i^\perp \mid \dim(\langle T_P X^\perp, P \rangle \cap K_i^\perp) \geq 0\}}.$$

In the case $i = m$ we have that K_m is a hyperplane, so $K_m^\perp$ is a point. Assuming this is not a point on X, we see that the mth reciprocal polar variety of X with respect to K_m is the (finite) set of nonsingular points $P \in X$

$$W_{K_m}^\perp(X) = \{P \in X_{sm} \mid K_m^\perp \in \langle T_P X^\perp, P \rangle\}.$$

Let $H_\infty = Z(x_0) \subset \mathbb{P}^n$ denote "the hyperplane at infinity", set $\mathbb{A}^n = \mathbb{P}^n \setminus H_\infty$, and consider the affine part $Y := X \cap \mathbb{A}^n$ of X. Let $L = K_i \subseteq H_\infty$ be a linear space of dimension $i + n - m - 1$. Define the *affine reciprocal polar variety* to be the affine part of the reciprocal polar variety:

$$W_L^\perp(Y) := W_L^\perp(X) \cap \mathbb{A}^n.$$

The linear variety $\langle P, L^\perp \rangle^\perp$ is contained in the hyperplane H_∞, so we can consider the affine cone $I_{P,L^\perp}$ of $\langle P, L^\perp \rangle^\perp$ as a linear variety in the affine space $\mathbb{A}^n$. Then the affine reciprocal polar variety can be written as

$$W_L^\perp(Y) = \overline{\{P \in Y_{sm} \setminus L^\perp \mid t_P Y \not\pitchfork I_{P,L^\perp}\}},$$

where $t_P Y$ denotes the affine tangent space at P, translated to the origin.

Assume $X = Z(F_1, \ldots, F_r)$ (for some $r \geq n - m$). Consider the case $L = K_m = H_\infty$. Then we see that $W_{H_\infty}^\perp(Y)$ is the closure of the set of smooth points of Y where the $(n - m + 1)$-minors of the matrix

$$\begin{pmatrix} \frac{\partial q}{\partial x_1} & \cdots & \frac{\partial q}{\partial x_n} \\ \frac{\partial F_1}{\partial x_1} & \cdots & \frac{\partial F_1}{\partial x_n} \\ \vdots & \vdots & \vdots \\ \frac{\partial F_r}{\partial x_1} & \cdots & \frac{\partial F_r}{\partial x_n} \end{pmatrix}.$$

vanish. This generalizes [14, Proposition 3.2.5].

Example 3. Assume q is given by $q(x_0, x_1, \ldots, x_n) = x_0^2 + \sum_{i=1}^n (x_i - a_i x_0)^2$ for some $a_1, \ldots, a_n$ such that $\sum_{i=1}^n a_i \neq 1$. Then q restricts to (essentially) the square of the Euclidean distance function on $\mathbb{A}^n = \mathbb{P}^n \setminus H_\infty$, namely to $\sum_{i=1}^n (x_i - a_i)^2 + 1$. The affine reciprocal polar variety is given (on smooth points) by the vanishing of the $(n - m + 1)$-minors of the matrix

$$
\begin{pmatrix}
x_1 - a_1 & \cdots & x_n - a_n \\
\frac{\partial F_1}{\partial x_1} & \cdots & \frac{\partial F_1}{\partial x_n} \\
\vdots & \vdots & \vdots \\
\frac{\partial F_r}{\partial x_1} & \cdots & \frac{\partial F_r}{\partial x_n}
\end{pmatrix}.
$$

just as in [7, (2.1)].

5   Euclidean Normal Bundles

Consider a variety $X \subset \mathbb{P}^n$ and let $Q \subset \mathbb{P}^n$ be a non-degenerate quadric. As we saw in the previous section, the quadric induces a polarity on $\mathbb{P}^n$, which can be viewed as an orthogonality, like what one has in a Euclidean space. In [6] this was used to define *Euclidean normal spaces* at each point $P \in X_{\mathrm{sm}}$. Actually they considered a non-degenerate quadric in a hyperplane at ∞, essentially as we saw in the case of affine reciprocal polar varieties. Here we shall consider the orthogonality on all of $\mathbb{P}^n$, and we define the normal space at a smooth point P as follows:

$$
N_P X := \langle T_P X^\perp, P \rangle.
$$

We shall now see that, by passing to the Nash modification of X, the Euclidean normal spaces are the fibers of a projective bundle. The Nash modification $\nu \colon \overline{X} \to X$ is the "smallest" proper, birational map such that the pullback of the cotangent sheaf Ω_X^1 of X admits a locally free quotient of rank m. This is equivalent to $\nu^* \mathcal{P}_X^1(1)$ admitting a locally free quotient of rank $m + 1$. Denote this quotient by $\mathcal{P}$, and let $\mathcal{K}$ denote the kernel of the surjection $\mathcal{O}_{\overline{X}}^{n+1} \to \mathcal{P}$. Thus $\mathcal{K}$ is a modification of the conormal sheaf $\mathcal{N}_{X/\mathbb{P}^n}$ of X in $\mathbb{P}^n$ twisted by $\mathcal{O}_X(1)$.

The quadric Q gives an isomorphism $\mathcal{O}_X^{n+1} \cong (\mathcal{O}_X^{n+1})^\vee$, hence we get a quotient $\mathcal{O}_{\overline{X}}^{n+1} \to \mathcal{K}^\vee$, whose (projective) fibers are the spaces $T_P X^\perp$. Adding the point map $\mathcal{O}_X^{n+1} \to \mathcal{O}_X(1)$, we get a surjection

$$
\mathcal{O}_{\overline{X}}^{n+1} \to \mathcal{E} := \mathcal{K}^\vee \oplus \mathcal{O}_{\overline{X}}(1),
$$

whose (projective) fibers are the Euclidean normal spaces $N_P X$. Indeed, $\mathbb{P}(\mathcal{E}) \subset \overline{X} \times \mathbb{P}^n$, and the fibers of the structure map $\mathbb{P}(\mathcal{E}) \to \overline{X}$ above smooth points of X are the spaces $N_P X \subset \mathbb{P}^n$ defined above. We call $\mathcal{E}$ the *Euclidean normal bundle* of X with respect to Q (cf. [6] and [7]).

Let $p \colon \mathbb{P}(\mathcal{E}) \to \overline{X}$ denote the structure map, and let $q \colon \mathbb{P}(\mathcal{E}) \to \mathbb{P}^n$ denote the projection on the second factor. The map q is called the *endpoint map* (for an explanation of the name, see [6]).

Let $B \in \mathbb{P}^n$ be a (general) point. Then

$$p(q^{-1}(B)) = \{P \in X \mid B \in \langle T_P X^\perp, P \rangle\}.$$

Letting $L := B^\perp$, so that $B = L^\perp$, we see that

$$p(q^{-1}(B)) = \{P \in X \mid L^\perp \in \langle T_P X^\perp, P \rangle\} = W_L^\perp(X)$$

is a reciprocal polar variety. In particular, the degree of q is just the degree of the reciprocal polar variety.

Example 4. Assume $X \subset \mathbb{P}^2$ is a (general) plane curve of degree d. The *reciprocal polar variety* is the intersection of the curve with its reciprocal polar, which has degree d, so q has degree d^2.

In [7] the degree of the endpoint map $q \colon \mathbb{P}(\mathcal{E}) \to \mathbb{P}^n$ was called the *Euclidean distance degree* of X:

$$\text{ED deg } X = p_* c_1(\mathcal{O}_{\mathbb{P}(\mathcal{E})}(1))^n = s_m(\mathcal{E}),$$

where $m = \dim X$ and s_m denotes the mth Segre class. The reason for the name is the relationship to computing critical points for the distance function in the Euclidean setting. We refer to [7] for many more details. In the case of curves and surfaces (and in a slightly different setting), the degree of q is called the normal class in [9].

Since $\mathcal{E} = \mathcal{K}^\vee \oplus \mathcal{O}_{\overline{X}}(1)$, we get

$$s(\mathcal{E}) = s(\mathcal{K}^\vee)s(\mathcal{O}_{\overline{X}}(1)) = c(\mathcal{P})c(\mathcal{O}_{\overline{X}}(-1))^{-1}.$$

Therefore,

$$s_m(\mathcal{E}) = \sum_{i=0}^{m} c_i(\mathcal{P})c_1(\mathcal{O}_{\overline{X}}(1))^{m-i}.$$

Since $c_i(\mathcal{P})c_1(\mathcal{O}_{\overline{X}}(1))^{m-i}$ is the degree μ_i of the ith polar variety $[M_i]$ of X [15, p. 256], we conclude (cf. [7, Theorem 5.4]):

$$\text{ED deg } X = \sum_{i=0}^{m} \mu_i.$$

The μ_i are called the *ranks* (or classes) of X. Note that μ_0 is the degree of X and μ_{n-1} is the degree of the dual variety $X^\vee$ (provided the dimension of $X^\vee$ is $n-1$). It is known (see [15, Proposition 3.6,p. 266], [23, 3.3], and [10, (4),p. 189]) that the ith rank of X is equal to the $(n-1-i)$th rank of the dual variety $X^\vee$ of X. As observed in [7, Theorem 5.4], it follows that the Euclidean distance degree of X is equal to that of $X^\vee$. Moreover, whenever we have formulas for the ranks μ_i, we thus get formulas for $\text{ED deg } X$.

Example 5. If $X \subset \mathbb{P}^n$ is a smooth hypersurface of degree $\mu_0 = d$, then $\mu_i = d(d-1)^i$, hence in this case (cf. [7, (7.1)])

$$\text{ED deg}\, X = d\sum_{i=0}^{n-1}(d-1)^i = \frac{d((d-1)^n - 1)}{d-2}.$$

If X has only isolated singularities, then only μ_{n-1} is affected, and we get (from Teissier's formula [20, Corollary 1.5,p. 320], and the Plücker formula for hypersurfaces with isolated singularities ([21, II.3,p. 46] and [12, Corollary 4.2.1,p. 60])

$$\text{ED deg}\, X = \frac{d((d-1)^n - 1)}{d-2} - \sum_{P \in \text{Sing}(X)} (\mu_P^{(n)} + \mu_P^{(n-1)}),$$

where $\mu_P^{(n)}$ is the Milnor number and $\mu_P^{(n-1)}$ is the sectional Milnor number of X at P.

Example 6. Assume $X \subset \mathbb{P}^3$ is a generic projection of a smooth surface of degree $\mu_0 = d$, so that X has *ordinary* singularities: a double curve of degree ϵ, t triple points, and ν_2 pinch points. Then (using the formulas for μ_1 and μ_2 given in [16, p. 18])

$$\text{ED deg}\, X = \mu_0 + \mu_1 + \mu_2 = d^3 - d^2 + d - (3d-2)\epsilon - 3t - 2\nu_2.$$

Further examples can be deduced from results in [15, 16], and [17].

6   Focal Loci

The *focal locus* (see e.g. [6] for an explanation of the name, or [7], where it is denoted the *ED discriminant*) is the branch locus Σ_X of the endpoint map $q\colon \mathbb{P}(\mathcal{E}) \to \mathbb{P}^n$. More precisely, let R_X denote the ramification locus of q; by definition, R_X is the subscheme of $\mathbb{P}(\mathcal{E})$ given on the smooth locus $\mathbb{P}(\mathcal{E})_{\text{sm}}$ by the 0th Fitting ideal $F^0(\Omega^1_{\mathbb{P}(\mathcal{E})/\mathbb{P}^n}|_{\mathbb{P}(\mathcal{E})_{\text{sm}}})$. The focal locus Σ_X is the closure of the image $q(R_X)$.

Recall that we have, on the Nash modification $\nu\colon \overline{X} \to X$, the exact sequence

$$0 \to \mathcal{K} \to \mathcal{O}_{\overline{X}}^{n+1} \to \mathcal{P} \to 0,$$

where $\mathcal{K}$ and $\mathcal{P}$ are the Nash bundles of the sheaves $\mathcal{N}_{X/\mathbb{P}^n}(1)$ and $\mathcal{P}_X^1(1)$ respectively, and that $\mathcal{E} = \mathcal{K}^\vee \oplus \mathcal{O}_{\overline{X}}(1)$.

Let $Z \to \overline{X}$ be a resolution of singularities, and, by abuse of notation, denote by $\mathcal{K}, \mathcal{P}, \mathcal{E}$ also their pullbacks to Z. The class $[R_X]$ of the closure of the ramification locus of $q\colon \mathbb{P}(\mathcal{E}) \to \mathbb{P}^n$ is given by

$$[R_X] = c_1(\Omega^1_{\mathbb{P}(\mathcal{E})}) - q^*c_1(\Omega^1_{\mathbb{P}^n}) = c_1(\Omega^1_{\mathbb{P}(\mathcal{E})}) + (n+1)c_1(\mathcal{O}_{\mathbb{P}(\mathcal{E})}(1)).$$

Using the exact sequences

$$0 \to p^*\Omega^1_Z \to \Omega^1_{\mathbb{P}(\mathcal{E})} \to \Omega^1_{\mathbb{P}(\mathcal{E})/Z} \to 0$$

and

$$0 \to \Omega^1_{\mathbb{P}(\mathcal{E})/Z} \to p^*\mathcal{E} \otimes \mathcal{O}_{\mathbb{P}(\mathcal{E})}(-1) \to \mathcal{O}_{\mathbb{P}(\mathcal{E})} \to 0$$

we find

$$[R_X] = p^*\big(c_1(\Omega^1_Z) + c_1(\mathcal{P}) + c_1(\mathcal{O}_Z(1))\big) + mc_1(\mathcal{O}_{\mathbb{P}(\mathcal{E})}(1)).$$

Therefore the degree of R_X with respect to the map q is given by

$$\deg R_X = \big(c_1(\Omega^1_Z) + c_1(\mathcal{P}) + c_1(\mathcal{O}_Z(1))\big) p_* c_1(\mathcal{O}_{\mathbb{P}(\mathcal{E})}(1))^{n-1} + mp_* c_1(\mathcal{O}_{\mathbb{P}(\mathcal{E})}(1))^n.$$

In terms of Segre classes of $\mathcal{E}$ this gives

$$\deg R_X = \big(c_1(\Omega^1_Z) + c_1(\mathcal{P}) + c_1(\mathcal{O}_Z(1))\big) s_{m-1}(\mathcal{E}) + m s_m(\mathcal{E}),$$

which gives, since $c_1(\Omega^1_Z) = c_1(\mathcal{P}^1_Z(1)) - (m+1)c_1(\mathcal{O}_Z(1))$,

$$\deg R_X = \big(c_1(\mathcal{P}^1_Z(1)) + c_1(\mathcal{P}) - mc_1(\mathcal{O}_Z(1))\big) s_{m-1}(\mathcal{E}) + m s_m(\mathcal{E}).$$

Now $s_m(\mathcal{E}) = \sum_{i=0}^m \mu_i$ and $s_{m-1}(\mathcal{E}) c_1(\mathcal{O}_Z(1)) = \sum_{i=0}^{m-1} \mu_i$, hence

$$\deg R_X = \big(c_1(\mathcal{P}^1_Z(1)) + c_1(\mathcal{P})\big) s_{m-1}(\mathcal{E}) - m \sum_{i=0}^{m-1} \mu_i + m \sum_{i=0}^m \mu_i,$$

hence

$$\deg R_X = \big(c_1(\mathcal{P}^1_Z(1)) + c_1(\mathcal{P})\big) s_{m-1}(\mathcal{E}) + m\mu_m.$$

In the special case when $X \subset \mathbb{P}^n$ is a hypersurface ($m = n-1$), we know by [15, Corollary 3.4] that

$$c_1(\mathcal{P}^1_Z(1)) = c_1(\mathcal{P}) + c_1(\mathcal{R}^{-1}_{Z/X}),$$

where $\mathcal{R}_{Z/X} = F^0(\Omega^1_{Z/X})$ is the (invertible) ramification ideal of $Z \to X$. Hence we get

$$\deg R_X = \big(2c_1(\mathcal{P}) + c_1(\mathcal{R}^{-1}_{Z/X})\big) s_{n-2}(\mathcal{E}) + (n-1)\mu_{n-1},$$

or

$$\deg R_X = \big(2c_1(\mathcal{P}) + c_1(\mathcal{R}^{-1}_{Z/X})\big) \sum_{i=0}^{n-2} c_i(\mathcal{P}) c_1(\mathcal{O}_Z(1))^{n-2-i} + (n-1)\mu_{n-1}.$$

Example 7. Let $X \subset \mathbb{P}^2$ be a plane curve of degree μ_0 and class $\mu_1 = c_1(\mathcal{P})$. Then

$$\deg R_X = 2c_1(\mathcal{P}) + \kappa + \mu_1 = 3\mu_1 + \kappa,$$

where κ is the "total number of cusps" of X. Note that, again by [15, Corollary 3.4], $3\mu_1 + \kappa = 3\mu_0 + \iota$, where ι is the "total number of inflection points" of X. But the degree $\mu_0(X)$ of X is equal to the class $\mu_1(X^\vee)$ of the dual curve $X^\vee$, and $\iota(X)$ of X is $\kappa(X^\vee)$ of $X^\vee$. This shows that the degree of the focal locus, or ED discriminant, of the dual curve is equal to that of X:

$$\deg R_{X^\vee} = 3\mu_1(X^\vee) + \kappa(X^\vee) = 3\mu_0 + 3\iota = \deg R_X.$$

The focal locus of a plane curve is also known as the *evolute* or the *caustic by reflection*. So, provided the maps $R_X \to \Sigma_X$ and $R_{X^\vee} \to \Sigma_{X^\vee}$ are birational, we have shown that the degree of the evolute of X is equal to the degree of the evolute of the dual curve $X^\vee$. For more on evolutes, see [8]. In the case that X is a "Plücker curve" of degree $d = \mu_0$ and having only δ nodes and κ ordinary cusps as singularities, and ι ordinary inflection points, then the classical formula, due to Salmon, is

$$\deg R_X = 3d(d-1) - 6\delta - 8\kappa.$$

Since in this case $\mu_1 = d(d-1) - 2\delta - 3\kappa$, this checks with our formula. Moreover, since the number of inflection points is $\iota = 3d(d-2) - 6\delta - 8\kappa$, $\deg R_{X^\vee} = 3d + \iota = 3d(d-1) - 6\delta - 8\kappa = \deg R_X$, as it should.

If X is smooth, we have $Z = \overline{X} = X$, $\mathcal{E} = \mathcal{N}_{X/\mathbb{P}^n}(1)^\vee \oplus \mathcal{O}_X(1)$, and $s(\mathcal{N}_{X/\mathbb{P}^n}(1)^\vee) = c(\mathcal{P}^1_X(1))$. Hence we can compute the class of R_X in terms of the Chern classes of X and $\mathcal{O}_X(1)$. We get

$$\deg R_X = 2\Big(c_1(\Omega^1_X) \sum_{i=0}^{m-1} c_i(\mathcal{P}^1_X(1))c_1(\mathcal{O}_X(1))^{m-1-i} + (m+1)\sum_{i=0}^{m-1}\mu_i\Big) + m\mu_m.$$

Since $c_i(\mathcal{P}^1_X(1)) = \sum_{j=0}^{i}\binom{m+1-i+j}{j}c_{i-j}(\Omega^1_X)c_1(\mathcal{O}_X(1))^j$, and since the μ_i's can be expressed in terms of the Chern numbers $c_{m-j}(\Omega^1_X)c_1(\mathcal{O}_X(1))^j$, we see that also $\deg R_X$ can be expressed in terms of these Chern numbers and the Chern numbers $c_1(\Omega^1_X)c_{m-1-j}(\Omega^1_X)c_1(\mathcal{O}_X(1))^j$.

Example 8. Assume $X \subset \mathbb{P}^n$ is a smooth curve of degree d. Then

$$\deg R_X = 2(2g-2) + 4\mu_0 + \mu_1 = 2(2g-2) + 4d + 2d + 2g - 2 = 6(d+g-1),$$

as in [7, Ex.7.11].

Example 9. Let $X \subset \mathbb{P}^n$ be a smooth surface of degree $\mu_0 = d$. Then, as in [6, Section 5] we get:

$$\deg R_X = 2(15d + 9c_1(\Omega^1_X)c_1(\mathcal{O}_X(1)) + c_1(\Omega^1_X)^2 + c_2(\Omega^1_X)).$$

Example 10. Let $X \subset \mathbb{P}^n$ be a general hypersurface ($m = n-1$) of degree μ_0. It is known that in this case $R_X \to \Sigma_X$ is birational [22, Theorem 2]. Since $c_1(\Omega^1_X) = (\mu_0 - n - 1)c_1(\mathcal{O}_X(1))$ we get

$$\deg \Sigma_X = \deg R_X = (2\mu_0 - n - 1)s_{n-2}(\mathcal{E})c_1(\mathcal{O}_X(1)) + (n-1)s_{n-1}(\mathcal{E}).$$

Hence

$$\deg \Sigma_X = (2\mu_0 - n - 1) \sum_{i=0}^{n-2} \mu_i + (n-1) \sum_{i=0}^{n-1} \mu_i = (n-1)\mu_{n-1} + 2(\mu_0 - 1) \sum_{i=0}^{n-2} \mu_i.$$

For a smooth hypersurface of degree d in $\mathbb{P}^n$, we have $\mu_i = d(d-1)^i$. Hence

$$\deg \Sigma_X = (n-1)d(d-1)^{n-1} + 2d(d-1)((d-1)^{n-1} - 1)(d-2)^{-1},$$

which checks with the formula found in [22, Theorem 2].

References

1. Bank, B., Giusti, M., Heintz, J., Mbakop, G.M.: Polar varieties, real equation solving, and data structures: the hypersurface case. J. Complex. **13**, 5–27 (1997)
2. Bank, B., Giusti, M., Heintz, J., Mbakop, G.M.: Polar varieties and efficient real elimination. Math. Z. **238**, 115–144 (2001)
3. Bank, B., Giusti, M., Heintz, J., Pardo, L.M.: Generalized polar varieties and an efficient real elimination procedure. Kybernetika **40**, 519–550 (2004)
4. Bank, B., Giusti, M., Heintz, J., Pardo, L.M.: Generalized polar varieties: geometry and algorithms. J. Complex. **21**, 377–412 (2005)
5. Bürgisser, P., Lotz, M.: The complexity of computing the Hilbert polynomial of smooth equidimensional complex projective varieties. Found. Comput. Math. **7**, 51–86 (2007)
6. Catanese, F., Trifogli, C.: Focal loci of algebraic varieties I. Comm. Algebra **28**, 6017–6057 (2000)
7. Draisma, J., Horobeţ, E., Ottaviani, G., Sturmfels, B., Thomas, R.R.: The Euclidean distance degree of an algebraic variety. arXiv:1309.0049
8. Josse, A., Pène, F.: Degree and class of caustics by reflection for a generic source. C. R. Math. Acad. Sci. Paris **351**(7–8), 295–297 (2013)
9. Josse, A., Pène, F.: On the normal class of curves and surfaces. arXiv:1402.7266
10. Kleiman, S.L.: Tangency and duality. In: Proceedings of Vancouver 1984 Conference in Algebraic Geometry. CMS Conference Proceedings, vol. 6, pp. 163–225 (1986)
11. Kleiman, S.L., Laksov, D.: Schubert calculus. Amer. Math. Monthly **79**, 1061–1082 (1972)
12. Laumon, G.: Degré de la variété duale d'une hypersurface. Bull. SMF **104**, 51–63 (1976)
13. Mork, H.: Real algebraic curves and surfaces. Ph. D. Thesis, University of Oslo (2011)
14. Mork, H., Piene, R.: Polars of real singular plane curves. IMA Vol. Math. Appl. **146**, 99–115 (2008)
15. Piene, R.: Polar classes of singular varieties. Ann. Sci. École Norm. Sup. **11**(4), 247–276 (1978)
16. Piene, R.: Some formulas for a surface in $\mathbb{P}^3$. In: Olson, L.D. (ed.) Algebraic Geometry. LNM, vol. 687, pp. 197–235. Springer, Heidelberg (1978)
17. Piene, R.: Cycles polaires et classes de Chern pour les variétés projectives singulières. Introduction à la théorie des singularités, II, 7–34, Travaux en Cours, 37, Hermann, Paris (1988)

18. Safey El Din, M., Schost, E.: Polar varieties and computation of one point in each connected component of a smooth algebraic set. In: Proceedings of ISSAC 2003, pp. 224–231. ACM, New York (2003)

19. Soares, M.: On the geometry of Poincaré's problem for one-dimensional projective foliations. An. Acad. Bras. Cienc. **73**, 475–482 (2001)

20. Teissier, B.: Cycles évanscents, sections planes, et conditions de Whitney. Astérisque **7–8**, 285–362 (1973)

21. Teissier, B.: Sur diverses conditions numériques d'équisingularité des familles de courbes (et un principe de spécialisation de la dépendance intégrale). Centre de Math. de l'École Polytechnique, M208.0675 (1975)

22. Trifogli, C.: Focal loci of algebraic hyper surfaces: a general theory. Geometrica Ded. **70**, 1–26 (1998)

23. Urabe, T.: Duality of numerical characters of polar loci. Publ. Res. Inst. Math. Sci. Tokyo **17**, 331–345 (1981)

A Note on a Problem Proposed by Kim and Lisonek

Cristian-Silviu Radu[✉]

Research Institute for Symbolic Computation,
Johannes Kepler University, 4040 Linz, Austria
sradu@risc.jku.at

Abstract. Let ℓ a prime number and $\Phi_\ell(X,Y)$ the modular polynomial of level ℓ. Since this polynomial has integer coefficients one may compute it modulo primes. Petr Lisonek and Yung-Jung Kim compute the polynomial $\Phi_\ell(X,Y)$ modulo 2 explicitly for several ℓ and they conjecture that the coefficients under the diagonal vanish. In this note we prove their conjecture and that the same property holds modulo the primes 3 and 5.

1 Introduction

In a talk [2] given at RISC, Petr Lisonek posed the following problem. Let ℓ be a prime and

$$\Phi_\ell(X,Y) = \sum_{i=0}^{\ell+1}\sum_{k=0}^{\ell+1} b_{k,i}^{(\ell)} X^k Y^i$$

be the ℓ-th modular polynomial; i.e., the minimal monic irreducible polynomial for which $\Phi_\ell(\mathcal{J}(\ell\tau), \mathcal{J}(\tau)) = 0$ for all $\tau \in \mathbb{H} := \{\tau \in \mathbb{C} : \mathrm{Im}(\tau) > 0\}$ where $\mathcal{J}$ is Klein's invariant. In particular it is known that $\Phi_\ell(X,Y) \in \mathbb{Z}[X,Y]$. In this setting Lisonek's conjecture states that

$$b_{k,i}^{(\ell)} \equiv 0 \pmod 2, \text{ for all } 0 \le k+i \le \ell.$$

This conjecture can be found in explicit form in the master thesis (supervised by Lisonek) of Yun-Jung Kim. In this note we prove the conjecture by using modular forms.

Given a function $f : \mathbb{H} \to \mathbb{C}$ we define $f_{a,b} : \mathbb{H} \to \mathbb{C}$ by

$$f_{a,b}(\tau) := f(a\tau + b),$$

for all $\tau \in \mathbb{H}$.

Let $p \in \{2,3,5\}$. The main result of this paper is that

$$b_{k,i}^{(\ell)} \equiv 0 \pmod p \quad \text{if} \quad 0 \le k+i \le \ell,$$

for all $\ell \ne p$. For $p = 2$, this problem can by found as [1, Conjecture 1.5.5].

C.-S. Radu—The research was supported by the strategic program "Innovatives OÖ 2010 plus" by the Upper Austrian Government in the frame of project W1214-N15-DK6 of the Austrian Science Fund (FWF).

J. Gutierrez et al. (Eds.): Computer Algebra and Polynomials, LNCS 8942, pp. 151–156, 2015.
DOI: 10.1007/978-3-319-15081-9_9

2   Definitions and Lemma's

For the proof we need the following simple lemma:

Lemma 1. *Let K be a field and let $r_1, \cdots, r_n \in K$. Set $a_k := (-1)^k \sum_{1 \le i_1 < \cdots < i_k \le n} r_{i_1} \cdots r_{i_k}$, $k = 1, \cdots, n$. Then*

$$(X - r_1) \cdots (X - r_n) = X^n + a_1 X^{n-1} + a_2 X^{n-2} + \cdots + a_{n-1} X + a_n$$

and

$$(X - r_1^{-1}) \cdots (X - r_n^{-1}) = X^n + \frac{a_{n-1}}{a_n} X^{n-1} + \frac{a_{n-2}}{a_n} X^{n-2} + \cdots + \frac{a_1}{a_n} X + a_n^{-1}.$$

Proof.

$$\frac{a_k}{a_n} = \frac{(-1)^{n-k}}{r_1 \cdots r_n} \sum_{1 \le i_1 < \cdots < i_k \le n} r_{i_1} \cdots r_{i_k} = (-1)^{n-k} \sum_{1 \le i_1 < \cdots < i_{n-k} \le n} r_{i_1}^{-1} \cdots r_{i_{n-k}}^{-1}.$$

In order to be able to apply Lemma 1 we need to write $\Phi_\ell(X, \mathcal{J})$ in the form:

$$\Phi_\ell(X, \mathcal{J}) = X^{\ell+1} + f_1(\mathcal{J}) X^\ell + \cdots + f_\ell(\mathcal{J}) X + f_{\ell+1}(\mathcal{J}),.$$

By [3, Chap. 6,Sect. 4]:

$$\Phi_\ell(X, \mathcal{J}) = (X - \mathcal{J}_{\ell,0}) \prod_{i=1}^{\ell} \left(X - \mathcal{J}_{\frac{1}{\ell}, \frac{i}{\ell}} \right).$$

By Lemma 1:

$$\frac{\Phi_\ell(X, \mathcal{J})}{X^{\ell+1} f_{\ell+1}(\mathcal{J})} = X^{-\ell-1} + \frac{f_\ell(\mathcal{J})}{f_{\ell+1}(\mathcal{J})} X^{-\ell} + \cdots + \frac{f_1(\mathcal{J})}{f_{\ell+1}(\mathcal{J})} X^{-1} + f_{\ell+1}(\mathcal{J})^{-1}$$

$$= (X^{-1} - \mathcal{J}_{\ell,0}^{-1}) \prod_{i=1}^{\ell} \left(X^{-1} - \mathcal{J}_{\frac{1}{\ell}, \frac{i}{\ell}}^{-1} \right). \tag{1}$$

Let $p \in \{2, 3, 5\}$. Let $q : \mathbb{H} \to \mathbb{C}$ with $\tau \mapsto q_\tau$ be defined by

$$q_\tau := e^{2\pi i \tau}, \quad \tau \in \mathbb{H}.$$

For $Q(X_1, X_2) \in \mathbb{C}[X_1, X_2]$ with

$$Q(X_1, X_2) = \sum_{(\alpha_1, \alpha_2) \in \mathbb{N}^2} a(\alpha_1, \alpha_2) X_1^{\alpha_1} X_2^{\alpha_2}$$

we define

$$< X_1^{\beta_1} X_2^{\beta_2} > Q(X_1, X_2) := a(\beta_1, \beta_2).$$

where $(\beta_1, \beta_2) \in \mathbb{N}^2$.

By [3, Chap. 3,Sect. 1]:

$$J(\tau) = \frac{E_4(\tau)^3}{\Delta(\tau)}, \quad \tau \in \mathbb{H}. \tag{2}$$

where $E_4 : \mathbb{H} \to \mathbb{C}$ is defined by

$$E_4(\tau) := 1 + 240 \sum_{n=1}^{\infty} \left(\sum_{d|n} d^3 \right) q_\tau^n, \quad \tau \in \mathbb{H} \tag{3}$$

and $\Delta : \mathbb{H} \to \mathbb{C}$ be defined by

$$\Delta(\tau) := q_\tau \prod_{n=1}^{\infty} (1 - q_\tau^n)^{24}, \quad \tau \in \mathbb{H}. \tag{4}$$

Given two functions $f, g : \mathbb{H} \to \mathbb{C}$ which can be expanded as Taylor series in powers of q_τ and integer coefficients that is

$$f(\tau) = \sum_{n=0}^{\infty} a(n) q_\tau^n \quad \text{and} \quad g(\tau) = \sum_{n=0}^{\infty} b(n) q_\tau^n, \quad \tau \in \mathbb{H}$$

we write

$$f \equiv g \pmod{p}$$

iff $a(n) \equiv b(n) \pmod{p}$ for all $n \in \mathbb{N}$.

Then by (2)–(4) it follows that

$$J^{-1} \equiv \Delta \pmod{p}. \tag{5}$$

For $\gamma = \begin{pmatrix} a & b \\ c & d \end{pmatrix} \in \mathrm{SL}_2(\mathbb{Z})$ (the group of 2×2 matrices with integer entries and determinant one), $k \in \mathbb{Z}$ and $f : \mathbb{H} \to \mathbb{C}$ we define $(f|_k\gamma) : \mathbb{H} \to \mathbb{C}$ by

$$(f|_k\gamma)(\tau) = (c\tau + d)^{-k} f\left(\frac{a\tau + b}{c\tau + d} \right), \quad \tau \in \mathbb{H}.$$

Recall that we say that an analytic function $f : \mathbb{H} \to \mathbb{C}$ is a modular form of weight k for $\mathrm{SL}_2(\mathbb{Z})$ iff $f|_k\gamma = f$ for all $\gamma \in \mathrm{SL}_2(\mathbb{Z})$ and

$$f(\tau) = \sum_{n=0}^{\infty} a_f(n) e^{2\pi i n \tau} = \sum_{n=0}^{\infty} a_f(n) q_\tau^n, \quad \tau \in \mathbb{H}. \tag{6}$$

Let $f : \mathbb{H} \to \mathbb{C}$ be analytic on $\mathbb{H}$ and additionally:

- $f|_0\gamma = f$ for all $\gamma \in \mathrm{SL}_2(\mathbb{Z})$;
- $f(\tau) = \sum_{n=-m}^{\infty} a_f(n) e^{2\pi i \tau n}$, $\tau \in \mathbb{H}$ and $a_f(n) \in \mathbb{Z}$.

We call such an f an analytic modular function with integer coefficients. Furthermore, one can prove that the set of all analytic modular functions with integer coefficients equals to $\mathbb{Z}[\mathcal{J}]$. This can be seen as follows. Assume that $f_0 := f = c_0 q^{-a} + \ldots$; we know that $\mathcal{J}^a = q^{-a} + \ldots$, which implies that $f_1 := f - c_0 \mathcal{J}^a = c_1 q^{-a+1} + \ldots$. Repeating this procedure we finally arrive at $f_{a+1} = c_{a+1} q + \ldots$, hence f_{a+1} is analytic on $\mathbb{H}$ and can be expanded as a Taylor series in powers of q, therefore it is a modular form of weight 0, then by [4, Chap. 7, Sect. 3, Corollary 1], every modular form of weight 0 is constant, which implies that $f_{a+1} = 0$. Note that since $\mathcal{J}$ has only integer coefficients in its q-expansion and the leading coefficient 1 it follows that each c_i, $i = 0, \ldots, a$ are integers and that $f = c_0 \mathcal{J}^a + c_1 \mathcal{J}^{a-1} + \cdots + c_a \mathcal{J}^0$.

Lemma 2. *Let ℓ be a prime. Let $p(X_1, X_2, \ldots, X_{\ell+1}) \in \mathbb{C}[X_1, X_2, \ldots, X_{\ell+1}]$ be a homogeneous symmetric polynomial of degree n. Then*

$$p\left(\ell^{12} \Delta_{\ell,0}, \Delta_{\frac{1}{\ell},0}, \Delta_{\frac{1}{\ell},\frac{1}{\ell}}, \ldots, \Delta_{\frac{1}{\ell},\frac{\ell-1}{\ell}}\right)$$

is a modular form of weight $12n$.

Proof. Let $\gamma \in \mathrm{SL}_2(\mathbb{Z})$, then

$$p\left(\ell^{12} \Delta_{\ell,0}, \Delta_{\frac{1}{\ell},0}, \Delta_{\frac{1}{\ell},\frac{1}{\ell}}, \ldots, \Delta_{\frac{1}{\ell},\frac{\ell-1}{\ell}}\right)\big|_{12n}\gamma$$

$$= p\left(\ell^{12} \Delta_{\ell,0}|_{12}\gamma, \Delta_{\frac{1}{\ell},0}|_{12}\gamma, \Delta_{\frac{1}{\ell},\frac{1}{\ell}}|_{12}\gamma, \ldots, \Delta_{\frac{1}{\ell},\frac{\ell-1}{\ell}}|_{12}\gamma\right)$$

$$= p\left(\ell^{12} \Delta_{\ell,0}, \Delta_{\frac{1}{\ell},0}, \Delta_{\frac{1}{\ell},\frac{1}{\ell}}, \ldots, \Delta_{\frac{1}{\ell},\frac{\ell-1}{\ell}}\right)$$

because if $f \in L := \left\{\ell^{12} \Delta_{\ell,0}, \Delta_{\frac{1}{\ell},0}, \ldots, \Delta_{\frac{1}{\ell},\frac{\ell-1}{\ell}}\right\}$, then $f|_{12}\gamma \in L$.

By (5) we have that

$$\Delta_{\ell,0} \equiv \mathcal{J}_{\ell,0}^{-1} \pmod{p}$$

and for every $k \in \mathbb{Z}$

$$\Delta_{\frac{1}{\ell},\frac{k}{\ell}} \equiv \mathcal{J}_{\frac{1}{\ell},\frac{k}{\ell}}^{-1} \pmod{p} \tag{7}$$

which is interpreted in the following way:

First note that if $\tau \mapsto \frac{\tau+k}{\ell}$, then $q_\tau \mapsto e^{\frac{2\pi i k}{\ell}} q_\tau^{1/\ell}$. This implies that

$$\Delta_{\frac{1}{\ell},\frac{k}{\ell}}(\tau) = \sum_{n=1}^{\infty} a(n) q_\tau^{n/\ell}, \quad \tau \in \mathbb{H}$$

and

$$\mathcal{J}_{\frac{1}{\ell},\frac{k}{\ell}}(\tau)^{-1} = \sum_{n=1}^{\infty} b(n) q_\tau^{n/\ell}, \quad \tau \in \mathbb{H}$$

where $a(n), b(n) \in \mathbb{Z}[e^{\frac{2\pi i}{\ell}}]$. Then (7) means that $a(n) - b(n)$ is in the ideal in $\mathbb{Z}[e^{\frac{2\pi i}{\ell}}]$ generated by p for all $n \geq 1$.

However note that any symmetric polynomial in the variables $\Delta_{\frac{1}{\ell}, \frac{k}{\ell}}$ for $k = 0, \ldots, \ell - 1$ can be expanded in powers of q with integer coefficients.

By (1), (7) and because of $\ell^{12} \equiv 1 \pmod{p}$ (since $\ell \neq p$):

$$f_{\ell+1}(\mathcal{J})^{-1} \equiv (-1)^{\ell+1} \ell^{12} \Delta_{\ell,0} \prod_{i=1}^{\ell} \Delta_{\frac{i}{\ell}, \frac{i}{\ell}} \pmod{p}.$$

Note that the right hand side is a modular form for $\mathrm{SL}_2(\mathbb{Z})$ of weight $12(\ell+1)$ because of Lemma 2. Furthermore, the order of the right hand side is $\ell+1$, which means that it has the form $c_{\ell+1} q^{\ell+1} + O(q^{\ell+2})$, for some $c_{\ell+1} \neq 0$.

Next note that $(-1)^{\ell+1} \frac{\Delta_{\ell,0} \prod_{i=1}^{\ell} \Delta_{\frac{1}{\ell}, \frac{i}{\ell}}}{\Delta^{\ell+1}}$ is a modular form of weight 0 hence it is constant and by computing its q-series expansion we see that this constant must be 1. Hence

$$f_{\ell+1}(\mathcal{J}) \equiv \Delta^{-\ell-1} \equiv \mathcal{J}^{\ell+1} \pmod{p}.$$

By (1), (5) and because of $\ell^{12} \equiv 1 \pmod{p}$ (since $\ell \neq p$):

$$(-1)^{\ell+1-k} \frac{f_k(\mathcal{J})}{f_{\ell+1}(\mathcal{J})} \equiv \ell^{12} \sum_{0 \leq i_1 < \cdots < i_{\ell-k} \leq p-1} \Delta_{\ell,0} \Delta_{\frac{1}{\ell}, \frac{i_1}{\ell}} \cdots \Delta_{\frac{1}{\ell}, \frac{i_{\ell-k}}{\ell}}$$
$$+ \sum_{0 \leq i_1 < \cdots < i_{\ell-k+1} \leq \ell-1} \Delta_{\frac{1}{\ell}, \frac{i_1}{\ell}} \cdots \Delta_{\frac{1}{\ell}, \frac{i_{\ell-k+1}}{\ell}} \pmod{p}.$$

We note that the right hand side $R(\tau)$ is a modular form of weight $12(\ell-k+1)$ for the group $\mathrm{SL}_2(\mathbb{Z})$ because of Lemma 2. Furthermore, because each term of $R(\tau)$ (when viewed as a series in powers of $e^{2\pi i \tau/\ell} = q_\tau^{1/\ell}$) has no constant term it follows that $R(\tau)$ has no constant term. Denote by $M_k(\mathbb{Z})$ the set of modular forms f of weight k for $\mathrm{SL}_2(\mathbb{Z})$ with integer coefficients (that is the coefficients $a_f(n)$ in (6) are integers).

Let $A^{(k)}(\mathbb{Z})$ be defined as the set with the property: $f \in A^{(k)}(\mathbb{Z})$ iff f is an analytic modular function with integer coefficients and the order of f is greater than or equal to $-k$ (that is $a_f(n) = 0$ for $n < -k$).

In particular

$$A^{(k)}(\mathbb{Z}) = \{a_0 + a_1 \mathcal{J} + \cdots + a_k \mathcal{J}^k : a_0, a_1, \cdots, a_k \in \mathbb{Z}\}.$$

If $f \in M_{12k}(\mathbb{Z})$, then $\frac{f}{\Delta^k} \in A^{(k)}(\mathbb{Z})$. Similarly if $g \in A^{(k)}(\mathbb{Z})$, then $\Delta^k g \in M_{12k}(\mathbb{Z})$. Since $\Delta \equiv \mathcal{J}^{-1} \pmod{p}$ it follows that every $f \in M_{12k}(\mathbb{Z})$ satisfies for suitable $a_0, \ldots, a_k \in \mathbb{Z}$:

$$f \equiv a_0 \mathcal{J}^{-k} + a_1 \mathcal{J}^{-k+1} + \cdots + a_k \pmod{p}.$$

Thus since $\frac{f_{\ell-k+1}(\mathcal{J})}{f_{\ell+1}(\mathcal{J})} \equiv R(\tau)$ has no constant term modulo p and also $\mathcal{J}^{-1}$ has no constant term and it follows that for suitable $a_{1,k}, \ldots, a_{k-1,k} \in \mathbb{Z}$:

$$\frac{f_{\ell-k+1}(\mathcal{J})}{f_{\ell+1}(\mathcal{J})} \equiv a_{1,k} \mathcal{J}^{-k} + a_{2,k} \mathcal{J}^{-k+1} + \cdots a_{k-1,k} \mathcal{J}^{-1} \pmod{p}.$$

which implies

$$f_{\ell-k+1}(\mathcal{J}) \equiv a_{1,k}\mathcal{J}^{\ell+1-k} + a_{2,k}\mathcal{J}^{\ell-k+2} + \cdots + a_{k-1,k}\mathcal{J}^{\ell} \pmod{p}.$$

This implies that $b_{k,i}^{(\ell)} =< \mathcal{J}^i > f_{\ell-k+1}(\mathcal{J}) \equiv 0 \pmod{p}$ for $i = 0, 1, \cdots, \ell - k$ and hence $b_{k,i}^{(\ell)} =< X^k \mathcal{J}^i > \Phi_\ell(X, \mathcal{J}) \equiv 0 \pmod{p}$ for $i = 0, 1, \cdots, \ell - k$ or equivalently that $b_{k,i}^{(\ell)} \equiv 0 \pmod{p}$ if $0 \leq k + i \leq \ell$, which is exactly what we wanted to prove.

Acknowledgments. I would like to thank the anonymous referee for the suggestions and corrections which led to improvements of this paper.

References

1. Kim, Y.-J.: Algorithms for Kloosterman Sums. Master's thesis, Simon Fraser University (2011)
2. Lisonek, P.: Classical Modular Polynomials over GF(2). RISC Seminar Talk, 09 (2012)
3. Schoeneberg, B.: Elliptic Modular Functions. Springer, New York (1974)
4. Serre, J.P.: A Course in Arithmetic. Springer, New York (1996)

Fast Algorithms for Refined Parameterized Telescoping in Difference Fields

Carsten Schneider$^{(\boxtimes)}$

Research Institute for Symbolic Computation (RISC),
Johannes Kepler University Linz, Linz, Austria
Carsten.Schneider@risc.jku.at

Abstract. Parameterized telescoping (including telescoping and creative telescoping) and refined versions of it play a central role in the research area of symbolic summation. In 1981 Karr introduced $\Pi\Sigma$-fields, a general class of difference fields, that enables one to consider this problem for indefinite nested sums and products covering as special cases, e.g., the $(q-)$hypergeometric case and their mixed versions. This survey article presents the available algorithms in the framework of $\Pi\Sigma$-extensions and elaborates new results concerning efficiency.

1 Introduction

This article deals with the following refined parameterized telescoping problem: given $f_1(k), \ldots, f_n(k)$, that are represented in a field or ring $\mathbb{F}$ and that evaluate for $k \in \mathbb{N}$ to elements from a field $\mathbb{K}$; find constants $c_1, \ldots, c_n \in \mathbb{K}$ (not all zero) and $g(k), \psi(k) \in \mathbb{F}$ such that the refined parameterized telescoping equation

$$g(k+1) - g(k) + \psi(k) = c_1 \, f_1(k) + \cdots + c_n \, f_n(k) \tag{1}$$

holds for all $k \geq \delta$ (for some $\delta \in \mathbb{N}$) and such that ψ is as simple as possible. Here $\psi = 0$ is considered as the simplest and most desirable case. If one succeeds in this task, one can sum (1) over k from δ to m and obtains the relation

$$g(m+1) - g(\delta) + \sum_{k=\delta}^{m} \psi(k) = c_1 \sum_{k=\delta}^{m} f_1(k) + \cdots + c_n \sum_{k=\delta}^{m} f_n(k). \tag{2}$$

The special case $n = 1$ (here we can set $c_1 = 1$ and $f(k) = f_1(k)$) gives refined telescoping: given $f \in \mathbb{F}$, find $g, \psi \in \mathbb{F}$ such that

$$g(k+1) - g(k) + \psi(k) = f(k) \tag{3}$$

and such that ψ is as simple as possible. If one restricts to $\psi = 0$, we consider standard telescoping. This problem has been considered heavily for rational, $(q-)$hypergeometric and mixed terms; see, e.g., [5,13,25,32,39,41].

C. Schneider—Supported by the Austrian Science Fund (FWF) grants P20347-N18 and SFB F50 (F5009-N15) and by the EU Network LHCPhenoNet PITN-GA-2010-264564.

J. Gutierrez et al. (Eds.): Computer Algebra and Polynomials, LNCS 8942, pp. 157–191, 2015.
DOI: 10.1007/978-3-319-15081-9_10

In addition, for a rational function field $\mathbb{F} = \mathbb{K}(k)$ refined telescoping has been considered in [6]: here the simplicity of ψ is determined by the degree of the denominator polynomial. Theoretical insight and additional algorithms have been derived in [37]; see also [46]. Extensions for hypergeometric terms are given in [10].

Another application is refined creative telescoping: taking $f_i(k) = F(r + i - 1, k)$ for a bivariate expression $F(r, k)$, one obtains the recurrence relation

$$g(m + 1) - g(\delta) + \sum_{k=\delta}^{m} \psi(k) = c_1 \sum_{k=\delta}^{m} F(r, k) + \cdots + c_n \sum_{k=\delta}^{m} F(r + n - 1, k). \quad (4)$$

Specializing m, e.g., to r and collecting $g(r + 1) - g(\delta) + \sum_{k=\delta}^{r} \psi(k)$ and compensating terms in $h(r)$ yields the recurrence

$$h(r) = c_1 \, S(r) + \cdots + c_n \, S(r + n - 1) \quad (5)$$

for the sum $S(r) = \sum_{k=0}^{r} F(r, k)$. Zeilberger [44,66] observed first that creative telescoping (with $\psi = 0$) can be handled algorithmically using Gosper's algorithm; for a sophisticated Mathematica package we refer to [41]. Recently, new complexity aspects were derived yielding new tactics to compute recurrence relations for hypergeometric terms more efficiently [19,20]. Similarly, creative telescoping for the q-case and mixed case have been considered [13,32,39]. For the holonomic case we refer the reader to [21,33,54]. Moreover, parameterized telescoping (with $\psi = 0$) and its application have been considered for the hypergeometric case [38,42].

A powerful generalization of (q–)hypergeometric and mixed expressions is the class of indefinite nested sums and products covering in addition, e.g., harmonic sums [17,65] and their generalized versions [3,4,35]. Such expressions can be represented in $\Pi\Sigma$-fields, a general class of difference fields introduced by Karr [27,28]. Many aspects of parameterized telescoping (extending the results mentioned above) have been elaborated in this setting. Here one is faced with three problems:

1. Reformulate the indefinite nested product-sum expressions $f_i(k)$ of (1) in a suitable $\Pi\Sigma^*$-field, i.e., in a function field $\mathbb{F} = \mathbb{K}(t_1) \ldots (t_e)$ where the generators t_i represent the occurring sums and products; for details see Definition 2 below.
2. Solve the underlying problem in this field or in a suitable extension of it.
3. Reformulate the result in terms of sums and products to get a result for (1).

Steps 1 and 3 have been worked out, e.g., in [60,61]; for a recent survey on this part dealing with telescoping and creative telescoping as introduced above, but also considering recurrence solving, we refer to [64]. In this article we are concerned with Step 2 and present up-to-date and new algorithms that solve parameterized telescoping problems efficiently. After a short summery of $\Pi\Sigma^*$-field theory in Sect. 2, the following algorithmic and theoretical aspects are considered.

An algorithmic framework to solve first-order parameterized equations (Sect. 3).
The first algorithm of parameterized telescoping in the setting of $\Pi\Sigma$-fields
has been introduced in [27]. In short, given a $\Pi\Sigma$-field $\mathbb{F}$ in which the $f_i(k)$
are represented, find all $g \in \mathbb{F}$ ($\psi = 0$) and constants $c_j \in \mathbb{K}$ such that (1)
holds. As it turns out, one actually has to solve a more general problem within
Karr's algorithm, namely parameterized first-order linear difference equations
(FPLDE). In Sect. 3 we will present a streamlined and simplified version of Karr's
algorithm [48,56]. Here an important ingredient is that results of Karr [27] and
Bronstein's extension [18] of Abramov's denominator bounding algorithm [7]
can be combined [50]. In this presentation we do not restrict to $\Pi\Sigma^*$-fields
$\mathbb{F} = \mathbb{G}(t_1)\dots(t_e)$ where the field generators t_i represent indefinite nested sums
and products and $\mathbb{G} = \mathbb{K}$ is the constant field. But we work in a rather general
framework: $\mathbb{G}$ is a difference field (modeling extra objects) that provides certain
algorithmic building blocks. In this way, all the algorithms in this article are
applicable to indefinite nested product-sum expressions where also unspecified
sequences [29,30] and radicals [31] like $\sqrt[d]{k}$ can arise.

An improved algorithm to solve parameterized telescoping (Sect. 4). With this
preparation, we derive a simplified and efficient algorithm in Sect. 4 that solves
parameterized telescoping; see also [59, Sect. 5]. If one deals with sum extensions,
the parameterized telescoping algorithm is simplified further.

Further improvement by searching first-entry solutions (Sect. 5). So far, the existing algorithms aim at finding all available solutions $c_j \in \mathbb{K}$ and $g \in \mathbb{F}$ ($\psi = 0$)
of (1). However, at least one of the c_j should be non-zero. We will present an
optimized algorithm that determines exactly one such solution with $c_1 \neq 0$ (if it
exists); such a solution will be also called first-entry solution. An obvious application is telescoping (i.e., $n = 1$ and $c_1 \neq 0$). Another important application is
creative telescoping. If there exists a recurrence (5), one can also assume that
one with $c_1 \neq 0$ exists (if it exists for $c_1 = 0$, one can shift backwards in r and
gets a recurrence where the coefficient of $S(r)$ is non-zero). Hence w.l.o.g. the
improved algorithm is applicable.

An efficient algorithm for refined parameterized telescoping (Sect. 6). Analyz-
ing the derived algorithm for first-entry solutions, a slight modification solves
the following refined parameterized telescoping problem in a $\Pi\Sigma^*$-field $\mathbb{F} =$
$\mathbb{G}(t_1)\dots(t_e)$: find $c_1,\dots,c_n \in \mathbb{K}$ with $c_1 \neq 0$ and $g \in \mathbb{F}$, $\psi \in \mathbb{G}(t_1)\dots(t_i)$
such that (1) holds and such that i is minimal; we call such a solution also
reduced solution. Note that the derived algorithm strongly simplifies the algorithm presented in [51] and leads to a much more efficient version. In addition,
the algorithm can be combined with algorithmic ideas of [57] that generalize the
refined telescoping versions of [6]: A reduced solution can be improved further
by searching for a $\psi \in \mathbb{G}(t_1)\dots(t_i)$ such that the degrees of the numerator and
denominator polynomials in t_i are minimal. The benefit of these tools will be
illustrated for the special case of refined telescoping and creative telescoping.

Exploiting structural properties (Sect. 7). The presented algorithms can be used
to transform any $\Pi\Sigma^*$-field to a reduced version [62]. As a consequence,

we obtain a constructive version of Karr's structural theorem, which can be considered as the discrete analogue of Liouville's Theorem [34] for indefinite integration. This in turn allows additional speed ups to solve parameterized telescoping in such fields. Finally, in Sect. 8 we relate the introduced algorithms to the difference field theory of depth-optimal $\Pi\Sigma^*$-extensions [53,59].

We conclude the introduction by remarking that all the presented algorithms play an important role in concrete problem solving like in the fields of combinatorics [12,47], numerics [23], number theory [36,40] or particle physics [14,15]. In particular, if one solves linear recurrence relations in terms of d'Alembertian solutions [9,43,44,48], a subclass of Liouvillian solutions [26,45], one obtains highly nested indefinite nested sums. It is then a necessary task to simplify these sums by fast parameterized telescoping algorithms. All the presented algorithms are part of the summation package Sigma [58,64].

2   A Short Summary of $\Pi\Sigma^*$-Field Theory and $\Pi\Sigma^*$-Extensions

We start with some basic definitions and notations. All fields and rings are computable and contain as subfield (resp. subring) the rational numbers $\mathbb{Q}$; $\mathbb{N}$ denotes the non-negative integers. For a set $\mathbb{A}$ (in particular for a ring and field) we define $\mathbb{A}^* = \mathbb{A}\backslash\{0\}$. For a polynomial $f = \sum_{i=0}^{d} f_i\, t^i \in \mathbb{A}[t]$ with $f_i \in \mathbb{A}$, we define $\mathrm{coeff}(f,i) = f_i$; if $f_d \neq 0$, $\deg(f) = d$. By convention, $\deg(0) = -1$. For $m \in \mathbb{Z}$ we define $\mathbb{A}[t]_m := \{f \in \mathbb{A}[t] \mid \deg(f) \leq m\}$. Moreover, we define the rational part of $\mathbb{A}(t)$ as $\mathbb{A}(t)_{(frac)} = \{\frac{p}{q} \mid p,q \in \mathbb{A}[t], \deg(p) < \deg(q)\}$.

For a vector $\mathbf{f} = (f_1,\ldots,f_n) \in \mathbb{A}^n$ and $h \in \mathbb{A}$, we define $\mathbf{f}\wedge h = (f_1,\ldots,f_n,h)$; and for a function $\sigma : \mathbb{A} \to \mathbb{A}$, we define $\sigma(\mathbf{f}) = (\sigma(f_1),\ldots,\sigma(f_n))$. The zero-vector in $\mathbb{A}^n$ is also denoted by $\mathbf{0}$. For a linear independent set (basis) $\{b_1,\ldots,b_\nu\}$ of a vector space we assume that the elements are ordered (by the given indices).

A *difference ring (resp. difference field)* $(\mathbb{A},\sigma)$ is a ring $\mathbb{A}$ (resp. field) equipped with an automorphism $\sigma : \mathbb{A} \to \mathbb{A}$. The *constants* are given by $\mathrm{const}_\sigma\mathbb{A} := \{c \in \mathbb{A} \mid \sigma(c) = c\}$. Note that $\mathrm{const}_\sigma\mathbb{A}$ is a subring (resp. subfield) of $\mathbb{A}$ and $\mathbb{Q}$ is contained in it as a subring (resp. subfield). Throughout this article we assume that $\mathrm{const}_\sigma\mathbb{A}$ *always* forms a field also called *constant field*.

Subsequently, we will deal with difference fields (resp. rings) that are given by iterative application of certain difference field (resp. ring) extensions. In general, a difference field $(\mathbb{F},\sigma)$ is a *difference field (resp. ring) extension* of a difference field $(\mathbb{G},\sigma')$ if $\mathbb{G}$ is a subfield (resp. subring) of $\mathbb{F}$ and $\sigma(f) = \sigma'(f)$ for all $f \in \mathbb{G}$. If it is clear from the context, we do not distinguish between σ and σ' anymore. Throughout this article we assume that $\mathbb{K}$ is the constant field (of the arising difference fields) and $(\mathbb{G},\sigma)$ is a difference field (not necessarily the constant field) where certain algorithmic properties are available. Moreover, $(\mathbb{A}(t),\sigma)$ is a difference field extension of $(\mathbb{A},\sigma)$ where $\mathbb{A}(t)$ is a rational function field. In particular, there is the following chain of difference field extensions: $\mathbb{K} \leq \mathbb{G} \leq \mathbb{A} \leq \mathbb{A}(t) \leq \mathbb{F}$.

$\tau : \mathbb{F} \to \mathbb{F}'$ is called a σ-*isomorphism* between two difference fields $(\mathbb{F}, \sigma)$ and $(\mathbb{F}', \sigma')$ if τ is a field isomorphism and $\tau(\sigma(f)) = \sigma'(\tau(f))$ for all $f \in \mathbb{F}$. In particular, let $(\mathbb{F}, \sigma)$ and $(\mathbb{F}', \sigma')$ be difference field extensions of $(\mathbb{G}, \sigma)$. Then a σ-isomorphism $\tau : \mathbb{F} \to \mathbb{F}'$ is a a $\mathbb{G}$-*isomorphism* if $\tau(a) = a$ for all $a \in \mathbb{G}$.

Example 1. 1. $(\mathbb{Q}, \sigma)$ is a difference field with $\sigma = \mathrm{id}_{\mathbb{Q}}$.
2. Take the rational function field $\mathbb{Q}(k)$ and define $\sigma : \mathbb{Q}(k) \to \mathbb{Q}(k)$ by $\sigma(f) = f(k+1)$ where $f(k+1)$ is the shifted version of $f(k)$. Then σ is a field automorphism with $\sigma|_{\mathbb{Q}} = \mathrm{id}_{\mathbb{Q}}$, i.e., $(\mathbb{Q}(k), \sigma)$ is a difference field extension of $(\mathbb{Q}, \sigma)$.
3. Let $(\mathbb{F}, \sigma)$ be a difference field and let t be transcendental over $\mathbb{F}$, i.e., $\mathbb{F}(t)$ is a rational function field. Take $\alpha, \beta \in \mathbb{F}$ with $\alpha \neq 0$. Then there is exactly one way how the field automorphism $\sigma : \mathbb{F} \to \mathbb{F}$ is extended to $\sigma : \mathbb{F}(t) \to \mathbb{F}(t)$ such that $\sigma(t) = \alpha t + \beta$. Namely, for $f = \sum_i f_i t^i \in \mathbb{F}[t]$ it follows that $\sigma(f) = \sum_i \sigma(f_i)(\alpha t + \beta)^i$ and for $f, g \in \mathbb{F}[t]$ with $g \neq 0$, it follows that $\sigma(\frac{f}{g}) = \frac{\sigma(f)}{\sigma(g)}$.
4. For instance, given the rational function field $\mathbb{Q}(k)(p)(h)$, consider the difference field extensions $(\mathbb{Q}(k)(p), \sigma)$ of $(\mathbb{Q}(k), \sigma)$ determined by $\sigma(p) = (k+1) p$ and $(\mathbb{Q}(k)(p)(h), \sigma)$ of $(\mathbb{Q}(k)(p), \sigma)$ determined by $\sigma(h) = h + \frac{1}{k+1}$. Note that p and h represent the factorial $k!$ and the harmonic numbers $H_k = \sum_{i=1}^{k} \frac{1}{i}$ with their shift behaviors $(k+1)! = (k+1)k!$ and $H_{k+1} = H_k + \frac{1}{k+1}$, respectively.

In the following we deal with exactly this type of extensions with the constraint that during the extension the constant field remains unchanged.

Definition 1. *Consider the difference field extension $(\mathbb{F}(t), \sigma)$ of $(\mathbb{F}, \sigma)$ with t transcendental over $\mathbb{F}$ and $\sigma(t) = \alpha t + \beta$ where $\alpha \in \mathbb{F}^*$ and $\beta \in \mathbb{F}$.*

1. *This extension is called Π-extension if $\beta = 0$ and $\mathrm{const}_{\sigma}\mathbb{F}(t) = \mathrm{const}_{\sigma}\mathbb{F}$.*
2. *This extension is called Σ^*-extension[1] if $\alpha = 1$ and $\mathrm{const}_{\sigma}\mathbb{F}(t) = \mathrm{const}_{\sigma}\mathbb{F}$.*
3. *This extension is called $\Pi\Sigma^*$-extension if it is a Π- or Σ^*-extension.*

In the following we are interested in a tower of such extensions $(\mathbb{F}_0, \sigma) < (\mathbb{F}_1, \sigma) < \cdots < (\mathbb{F}_e, \sigma)$, i.e., we start with a given difference field $(\mathbb{F}_0, \sigma) := (\mathbb{G}, \sigma)$ and construct iteratively the $\Pi\Sigma^*$-extension $(\mathbb{F}_i, \sigma)$ of $(\mathbb{F}_{i-1}, \sigma)$ where $\mathbb{F}_i = \mathbb{F}_{i-1}(t_i)$, i.e., t_i is transcendental over $\mathbb{F}_{i-1}$, σ is extended from $\mathbb{F}_{i-1}$ to $\mathbb{F}_i$ subject the shift relation $\sigma(t_i) = \alpha_i t_i + \beta_i$ $(\alpha_i \in \mathbb{F}_{i-1}^*, \beta_i = 0$ or $\alpha_i = 1, \beta_i \in \mathbb{F}_{i-1})$, and $\mathrm{const}_{\sigma}\mathbb{F}_i = \mathrm{const}_{\sigma}\mathbb{F}_{i-1}$. This gives the difference field $(\mathbb{F}, \sigma)$ where $\mathbb{F} = \mathbb{G}(t_1) \ldots (t_e)$ is a rational function field and $\mathrm{const}_{\sigma}\mathbb{F} = \mathrm{const}_{\sigma}\mathbb{G}$. Throughout this article, it is assumed that the generators $t_1, \ldots, t_e$ of such an extension are given explicitly.

---

[1] Karr's Σ-extensions [27] are given by generators with $\sigma(t) = \alpha t + \beta$ with extra conditions on α. For simplicity, we work with Σ^*-extensions that are relevant in symbolic summation.

Definition 2. *A difference field extension* $(\mathbb{F}, \sigma)$ *of* $(\mathbb{G}, \sigma)$ *with* $\mathbb{F} = \mathbb{G}(t_1) \dots (t_e)$ *and* $\mathbb{K} := \mathrm{const}_\sigma \mathbb{G}$ *is called* (nested) $\Pi\Sigma^*$-extension (resp. Π-/Σ^*-extension), *if it is a tower of (single)* $\Pi\Sigma^*$-extensions *(resp.* Π-/Σ^*-extensions*). If* $\mathbb{G} = \mathbb{K}$, *such a difference field is called* $\Pi\Sigma^*$-field over $\mathbb{K}$.

Summarizing, the generators $t_1, \dots, t_e$ represent indefinite nested sums and products whose shift-behaviors are modeled by σ. E.g., the difference field $(\mathbb{Q}(k)(p)(h), \sigma)$ from Example 1.4 is a $\Pi\Sigma^*$-field over $\mathbb{Q}$ representing $k!$ and H_k. We emphasize that the construction of Σ^*-extensions is directly connected to telescoping.

Theorem 1 *[27, 28]. Consider the difference field extension* $(\mathbb{F}(t), \sigma)$ *of* $(\mathbb{F}, \sigma)$ *with* t *being transcendental over* $\mathbb{F}$ *and* $\sigma(t) = t + \beta$. *Then this is a* Σ^*-extension *iff there is no* $g \in \mathbb{F}$ *with* $\sigma(g) = g + \beta$.

If $(\mathbb{F}, \sigma)$ is a $\Pi\Sigma^*$-field, the existence of such an element $g \in \mathbb{F}$ with $\sigma(g) = g + \beta$ can be decided constructively with Karr's telescoping algorithm [27]. However, the algorithms are not tuned for large fields. In this article we aim at developing refined telescoping algorithms in order to construct large $\Pi\Sigma^*$-fields efficiently. To demonstrate the underlying construction process, consider the following example.

Example 2. Consider $S(k) = \sum_{i=1}^k F(i)$ with $F(i) = (i^2 + 1)\, i!\, H_i^2$. We rephrase $F(k)$ in the $\Pi\Sigma^*$-field from Example 1.4: replacing $k!$ and H_k by p and h, respectively, we get $\tilde{f} = (k^2 + 1)\, p\, h^2$. In particular, $F(k+1) = \frac{1}{k+1}(k^2 + 2k + 2)\, k!(H_k(k+1)+1)^2$ is given by

$$f = \sigma(\tilde{f}) = \tfrac{1}{k+1}(k^2 + 2k + 2)p(h(k+1)+1)^2. \tag{6}$$

Using our summation algorithm (for the concrete execution steps see Example 8) we prove that there does not exist a $g \in \mathbb{Q}(k)(p)(h)$ with $\sigma(g) = g + f$. Consequently, we can construct the Σ^*-extension $(\mathbb{Q}(k)(p)(h)(t), \sigma)$ of $(\mathbb{Q}(k)(p)(h), \sigma)$ with $\sigma(t) = t + f$ by Theorem 1. In particular, like p and h represent $k!$ and H_k, respectively, t represents the sum $S(k)$ with the appropriate shift-relation.

More generally, consider an indefinite nested sum, say $S(k) = \sum_{i=1}^k F(i)$ where the summand $F(n)$ with $n \in \mathbb{N}$ evaluates to elements of the field $\mathbb{K}$. Now suppose that $F(k)$ is already represented in a $\Pi\Sigma^*$-field $(\mathbb{F}, \sigma)$ over $\mathbb{K}$ with $\tilde{f} \in \mathbb{F}$, i.e., $f := \sigma(\tilde{f}) \in \mathbb{F}$ represents $F(k+1)$. More precisely, as worked out in [64] one can attach a mapping such that each element from $\mathbb{F}$ represents an indefinite nested product-sum expression. Then by the telescoping algorithms given below one can decide algorithmically if there is a $g \in \mathbb{F}$ such that $\sigma(g) = g + f$ holds. If yes, one can construct an indefinite nested product-sum expression $G(k)$ with $G(k+1) - G(k) = F(k+1)$. Since also $S(k+1) - S(k) = F(k+1)$, it follows that $G(k) = S(k) + c$ for some $c \in \mathbb{K}$. Looking at the initial value $k = 1$ gives $c := G(1) - S(1) \in \mathbb{K}$. In other words, $g + c$ represents the sum $S(k)$ in the given field $\mathbb{F}$. Otherwise, if there does not exist a $g \in \mathbb{F}$ with $\sigma(g) = g + f$, we can construct the Σ^*-extension $(\mathbb{F}(t), \sigma)$ of $(\mathbb{F}, \sigma)$ with $\sigma(t) = t + f$ by Theorem 1. In this field the generator t represents $S(k)$. Since $(\mathbb{F}(t), \sigma)$ is again a $\Pi\Sigma^*$-field over $\mathbb{K}$, we can repeat this process iteratively exploiting the available telescoping algorithm. In a nutshell, we can represent expressions of indefinite nested sums in a $\Pi\Sigma^*$-field; for further details see [64]. The Π-case can be treated similarly, however further aspects have to be considered; see [11,24,55].

Depending on the ground field $(\mathbb{G}, \sigma)$ in Definition 2, the class of indefinite nested sums and products can be enhanced. Besides the case $\mathrm{const}_\sigma\mathbb{G} = \mathbb{G}$ (the usual case), the following classes have been considered so far.

Example 3.   1. The *free difference field* $(\mathbb{G}, \sigma)$ over $\mathbb{K}$: here we are given a rational function field $\mathbb{G} = \mathbb{K}(\ldots, x_{-1}, x_0, x_1, \ldots)$ with $\sigma(c) = c$ for all $c \in \mathbb{K}$, and $\sigma(x_i) = x_{i+1}$. In this field one can model indefinite nested sums and products over unspecified sequences; see [29,30].
2. The *radical difference field* $(\mathbb{G}, \sigma)$ over $\mathbb{K}$ of order $d \in \mathbb{N}^*$: starting with the $\Pi\Sigma^*$-field $(\mathbb{K}(x), \sigma)$ over $\mathbb{K}$ with $\sigma(x) = x + 1$ one takes the infinite field extension $\mathbb{K}(x)(\ldots, y_{-1}, y_0, y_1, \ldots)$ subject to the relations $y_k^d = x$ and $\sigma(y_k) = y_{k+1}$ for all $k \in \mathbb{Z}$. With this field one can model indefinite nested sums and products involving objects like $\sqrt[d]{k}$; see [31].

3   Solving Parameterized First-Order Equations

As motivated in the introduction, we aim at solving *parameterized telescoping* (PT) equations in a difference field $(\mathbb{A}, \sigma)$. The classical version ($\psi = 0$) can be formulated as follows. Given $\mathbf{f} = (f_1, \ldots, f_n) \in \mathbb{A}^n$, find all $c_j \in \mathrm{const}_\sigma\mathbb{A} =: \mathbb{K}$ and $q \in \mathbb{A}$ (note that q takes over the role of g used in (1)) such that

$$\sigma(q) - q = c_1 f_1 + \cdots + c_n f_n \tag{7}$$

holds. Here all proposed telescoping algorithms rely on the fact that one can solve the more general case of *first-order parameterized linear difference equations* (FPLDE) in a difference field $(\mathbb{A}, \sigma)$: Given $\mathbf{0} \neq \mathbf{a} = (a_0, a_1) \in \mathbb{A}^2$ and $\mathbf{f} = (f_1, \ldots, f_n) \in \mathbb{A}^n$, find all $c_i \in \mathrm{const}_\sigma\mathbb{A} =: \mathbb{K}$ and $q \in \mathbb{A}$ such that

$$a_1\sigma(q) + a_0 q = c_1 f_1 + \cdots + c_n f_n \tag{8}$$

holds. More generally, given a subspace W of $\mathbb{A}$ over $\mathbb{K}$, we are interested in the following solution sets (for FPLDE and PT):

$$V(\mathbf{a}, \mathbf{f}, W) := \{(c_1, \ldots, c_n, q) \in \mathbb{K}^n \times W \mid (8) holds\},$$
$$V(\mathbf{f}, W) := V((-1, 1), \mathbf{f}, W) = \{(c_1, \ldots, c_n, q) \in \mathbb{K}^n \times W \mid (7) holds\}.$$

Note that $V := V(\mathbf{a}, \mathbf{f}, W)$ is a subspace of $\mathbb{K}^n \times \mathbb{A}$ over $\mathbb{K}$ and its dimension is less than or equal to $n+1$: there is at most one homogeneous solution $(0 \ldots, 0, h)$ and there are at most n linearly independent particular solutions; for a proof see [49] which is based on [22, Theorem XII]. Note: if V consists only of the zero vector, its basis is the empty set by convention.

Summarizing, if $W = \mathbb{A}$, we aim at solving the following problems.

Problem FPLDE in $(\mathbb{A}, \sigma)$. Given a difference field (resp. ring) $(\mathbb{A}, \sigma)$, and given $\mathbf{0} \neq \mathbf{a} = (a_0, a_1) \in \mathbb{A}^2$ and $\mathbf{f} = (f_1, \ldots, f_n) \in \mathbb{A}^n$; find a basis of $V(\mathbf{a}, \mathbf{f}, \mathbb{A})$.

Problem PT in $(\mathbb{A}, \sigma)$. Given a difference field (resp. ring) $(\mathbb{A}, \sigma)$ and a vector $\mathbf{f} = (f_1, \ldots, f_n) \in \mathbb{A}^n$; find a basis of $V(\mathbf{f}, \mathbb{A}) = V((-1, 1), \mathbf{f}, \mathbb{A})$.

3.1   A General Strategy

Based on [27] the following strategy has been proposed in [48,51,56] to solve Problem FPLDE for a $\Pi\Sigma^*$-extension $(\mathbb{A}(t), \sigma)$ of $(\mathbb{A}, \sigma)$ with $\sigma(t) = \alpha t + \beta$ and $\mathbb{K} = \mathrm{const}_\sigma \mathbb{A}$: given $\mathbf{0} \neq \mathbf{a} = (a_0, a_1) \in \mathbb{A}(t)^2$ and $\mathbf{f} \in \mathbb{A}(t)^n$, find a basis of $V := V(\mathbf{a}, \mathbf{f}, \mathbb{A}(t))$.

A simple special case. If $a_0 \, a_1 = 0$, a basis of V can be obtained by solving a linear system of equations over $\mathbb{A}(t)$. Thus (under the assumption that one can solve linear systems in $\mathbb{A}(t)$), it suffices to consider the case $a_0 \, a_1 \neq 0$.

Step 1: denominator bounding. Next, we suppose that we can solve the following denominator bound problem.

Problem DenB. Given a $\Pi\Sigma^*$-extension $(\mathbb{A}(t), \sigma)$ of $(\mathbb{A}, \sigma)$, $\mathbf{a} \in (\mathbb{A}(t)^*)^2$, and $\mathbf{f} \in \mathbb{A}(t)^n$. Find $d \in \mathbb{A}[t]^*$ such that for all $(c_1, \ldots, c_n, q) \in V$ we have that $q \, d \in \mathbb{A}[t]$.

In short, d contains all arising denominators of the solution set. Since the $\mathbb{K}$-vector space V has finite dimension, it follows that such a d exists. Subsequently a d with this property is called *denominator bound* or *universal denominator*.

Suppose that one succeeds in computing a denominator bound d. Then the remaining (and often most challenging) task is to calculate the possible numerators of the rational solutions, i.e., we are interested in finding all solutions of the $\mathbb{K}$-vector space

$$
\begin{aligned}
V' &= V((\tfrac{a_0}{d}, \tfrac{a_1}{\sigma(d)}), \mathbf{f}, \mathbb{A}[t]) \\
&= \{(c_1, \ldots, c_n, p) \in \mathbb{K}^n \times \mathbb{A}[t] \mid a_1 \sigma(\tfrac{p}{d}) + a_0 \tfrac{p}{d} = c_1 f_1 + \cdots + c_n f_n\}.
\end{aligned}
\tag{9}
$$

Since the dimension of V is bounded by $n + 1$, it follows immediately that also the dimension of V' is bounded by $n + 1$. Moreover, if $\{(e_{i1}, \ldots, e_{in}, p_i)\}_{1 \leq i \leq \mu} \subseteq \mathbb{K}^n \times \mathbb{A}[t]$ is a basis of V' with dimension μ, then it is easy to see that $\{(e_{i1}, \ldots, e_{in}, \tfrac{p_i}{d})\}_{1 \leq i \leq \mu}$ is a basis of V. In a nutshell, given d, it remains to derive a basis of V' and the construction of a basis of V can be obtained. Subsequently, we clear denominators and obtain $\mathbf{a}' = (a_0', a_1') \in (\mathbb{A}[t]^*)^2$, $\mathbf{f}' = (f_1', \ldots, f_n') \in \mathbb{A}[t]^n$ such that $V' = V(\mathbf{a}', \mathbf{f}', \mathbb{A}[t])$.

To this end, we aim at finding a basis of $P := V(\mathbf{a}', \mathbf{f}', \mathbb{A}[t])$. To accomplish this task, the general tactic proceeds as follows.

Step 2: degree bounding. We suppose that we can solve the degree bound problem.

Problem DegB. Given a $\Pi\Sigma^*$-extension $(\mathbb{A}(t), \sigma)$ of $(\mathbb{A}, \sigma)$, $\mathbf{a} \in (\mathbb{A}[t]^*)^2$, and $\mathbf{f} \in \mathbb{A}[t]^n$. Find $m \in \mathbb{N} \cup \{-1\}$ such that $V(\mathbf{a}, \mathbf{f}, \mathbb{A}[t]) = V(\mathbf{a}, \mathbf{f}, \mathbb{A}[t]_m)$.

Since the dimension of P is bounded, such an m exists; in the following m is also called *degree bound* of P. For the following considerations it will be crucial that m satisfies the following additional property:

$$m \geq \max(\deg(f_1'), \dots, \deg(f_n')) - \max(\deg(a_0'), \deg(a_1')). \tag{10}$$

If $m = -1$, i.e., $\mathbb{A}[t]_{-1} = \{0\}$, a basis of $V(\mathbf{a}', \mathbf{f}', \{0\})$ can be calculated by linear algebra. Otherwise, if $m \geq 0$, we are in the position to compute a basis of $V(\mathbf{a}', \mathbf{f}', \mathbb{A}[t]_m)$ provided that we can solve Problem FPLDE in $(\mathbb{A}, \sigma)$. Namely, if $m = 0$ ($\mathbb{A}[t]_m = \mathbb{A}[t]_0 = \mathbb{A}$), we are in the base case and can calculate a basis. Otherwise, if $m \geq 1$, we utilize the following reduction introduced in [27].

Step 3: degree reduction. We search for all solutions $c_i \in \mathbb{K}$ and $q = q_0 + q_1 t + \dots + q_m t^m$ for (8). Here the crucial idea is to compute a set (more precisely a basis of a vector space) that contains all the possible choices of the leading coefficient q_m, to plug in this sub-result and to compute the remaining coefficients q_i with $i < m$ by recursion. More precisely, let $(c_1, \dots, c_n, g\, t^m + h) \in P$ with $h \in \mathbb{A}[t]_{m-1}$ and $g(= q_m) \in \mathbb{A}$. Thus

$$a_1' \sigma(g\, t^m + h) + a_0'(g\, t^m + h) = c_1 f_1' + \dots + c_n f_n'. \tag{11}$$

Now define $l := \max(\deg(a_0'), \deg(a_1'))$ and observe that the degree of the arising terms is bounded by $m + l$; this is guaranteed by (10). Thus by coefficient comparison w.r.t. t^{m+l} and using $\sigma(t) = \alpha t + \beta$ we get the following constraint on $g(= q_m)$:

$$\text{coeff}(a_1', l)\, \alpha^m\, \sigma(g) + \text{coeff}(a_0', l)\, g = c_0\, \text{coeff}(f_1', l+m) + \dots + c_n\, \text{coeff}(f_n', l+m). \tag{12}$$

Step 3.1: a solution for the leading coefficient. Now we solve this FPLDE problem in the ground field $(\mathbb{A}, \sigma)$, i.e., we compute a basis of $\tilde{V} = V(\tilde{\mathbf{a}}, \tilde{\mathbf{f}}, \mathbb{A})$ with

$$\tilde{\mathbf{f}} := (\text{coeff}(f_1', m+l), \dots, \text{coeff}(f_n', m+l)) \in \mathbb{A}^n,$$
$$\mathbf{0} \neq \tilde{\mathbf{a}} := (\text{coeff}(a_0', l), \alpha^m\, \text{coeff}(a_1', l)) \in \mathbb{A}^2. \tag{13}$$

Special case: finding the homogeneous solution. If it turns out that $\tilde{V} = \{0\}$, it follows that there is no way to find a $\mathbf{0} \neq \mathbf{c} \in \mathbb{K}^n$ such that there is a $g \in \mathbb{A}[t]_m$ with (8). However, there might still exist a solution of the homogeneous version, i.e., $a_1' \sigma(h) + a_0' h = 0$. But since $\tilde{V} = \{0\}$, the highest possible term (being of degree m) is 0 and consequently $\deg(h) < m$. Thus by recursion we compute a basis of $V(\mathbf{a}', (0), \mathbb{A}[t]_{m-1})$. If its basis is $\{\}$, i.e., there is no nonzero homogeneous solution, $P = \{0\} \subseteq \mathbb{K}^n \times \mathbb{A}[t]$. Thus we return the empty basis $\{\}$ for P. Otherwise, we can extract an $h \in \mathbb{A}[t]_{m-1}^*$ and $\{(0, \dots, 0, h)\} \subseteq \mathbb{K}^n \times \mathbb{A}[t]_{m-1}$ is a basis of P.

If $\tilde{V} \neq \{0\}$, let $\{(c_{i1}, \ldots, c_{in}, g_i)\}_{1 \leq i \leq \lambda} \subseteq \mathbb{K}^n \times \mathbb{A}$ be a basis with $\lambda \geq 1$. Then there are $d_1, \ldots, d_\lambda$ such that $g = d_1 g_1 + \cdots + d_\lambda g_\lambda$ and $c_j = d_1 c_{1j} + \cdots + d_\lambda c_{\lambda j}$ for $1 \leq j \leq n$. In vector notation this reads as

$$g = \mathbf{d}\, \mathbf{g} \text{ and } \mathbf{c} = \mathbf{d}\, \mathbf{C} \tag{14}$$

for $\mathbf{d} = (d_1, \ldots, d_\lambda) \in \mathbb{K}^\lambda$ and $\mathbf{C} = (c_{ij}) \in \mathbb{K}^{\lambda \times n}$. Moving the occurring g in (11) to the right hand side and replacing g and $\mathbf{c}$ by the right hand sides given in (14) yield

$$a_1' \sigma(h) + a_0' h = \mathbf{c}\, \mathbf{f}' - (a_1'\, \sigma(g\, t^m) + a_0'\, g\, t^m) = \mathbf{d}\, \mathbf{C}\, \mathbf{f}' - (a_1' \sigma(\mathbf{d}\, g\, t^m) + a_0'\, \mathbf{d}\, g\, t^m) = \mathbf{d}\, \phi$$

for

$$\phi := \mathbf{C}\, \mathbf{f}' - (a_1'\, \sigma(\mathbf{g}\, t^m) + a_0'\, \mathbf{g}\, t^m) \in \mathbb{A}[t]_{l+m-1}^\lambda. \tag{15}$$

Step 3.2: the solution of the remaining coefficients by recursion. In other words, we obtain a first-order parameterized linear difference equation, but this time the desired solution is reduced in its degree, i.e., $h \in \mathbb{A}[t]_{m-1}$. In short, we need a basis of the solution space $V(\mathbf{a}', \phi, \mathbb{A}[t]_{m-1})$. Now we apply the degree reduction recursively, and obtain a basis $\{(d_{i1}, \ldots, d_{i\lambda}, h_i)\}_{1 \leq i \leq \mu} \subseteq \mathbb{K}^\lambda \times \mathbb{A}[t]_{m-1}$ of the corresponding solution space $V(\mathbf{a}', \phi, \mathbb{A}[t]_{m-1})$. In vector notation the underlying difference equations read as

$$a_1'\, \sigma(\mathbf{h}) + a_0'\, \mathbf{h} = \mathbf{D}\, \phi \tag{16}$$

for $\mathbf{D} = (d_{ij}) \in \mathbb{K}^{\mu \times \lambda}$ and $\mathbf{h} := (h_1, \ldots, h_\mu) \in \mathbb{A}[t]_{m-1}^\mu$.
Step 3.3: merging the sub-solutions. Compute

$$\mathbf{E} = (e_{ij}) := \mathbf{D}\, \mathbf{C} \in \mathbb{K}^{\mu \times n} \text{ and } \mathbf{p} = (p_1, \ldots, p_\mu) := \mathbf{D}\, \mathbf{g}\, t^m + \mathbf{h} \in \mathbb{A}[t]_m^\mu. \tag{17}$$

Then it follows that

$$a_1'\sigma(\mathbf{p}) + a_0'\, \mathbf{p} \stackrel{(17)}{=} a_1'\, \sigma(\mathbf{h}) + a_0'\, \mathbf{h} + \mathbf{D}(a_1'\, \sigma(\mathbf{g}\, t^m) + a_0'\, \mathbf{g}\, t^m)$$
$$\stackrel{(16)}{=} \mathbf{D}(\phi + a_1'\, \sigma(\mathbf{g}\, t^m) + a_0'\, \mathbf{g}\, t^m) \stackrel{(15)}{=} \mathbf{D}\, \mathbf{C}\, \mathbf{f}' \stackrel{(17)}{=} \mathbf{E}\, \mathbf{f}',$$

i.e., $B := \{(e_{i1}, \ldots, e_{in}, p_i)\}_{1 \leq i \leq \mu}$ is a subset of $P = V(\mathbf{a}', \mathbf{f}', \mathbb{A}[t]_m)$. By further arguments (see [49, Theorem 6.2]) it follows that B is a basis of P.

Example 4. Consider the $\Pi\Sigma^*$-field $(\mathbb{Q}(k), \sigma)$ over $\mathbb{Q}$ with $\sigma(k) = k + 1$. Using the above strategy we can calculate a basis of $V = V(\mathbf{a}, \mathbf{f}, \mathbb{Q}(k))$ with $\mathbf{a} = (-1, 1 + k) \in \mathbb{Q}[k]^2$ and $\mathbf{f} = (2k, 0) \in \mathbb{Q}[k]^2$ as follows. A denominator bound of V is $d = 1$; here one can use the algorithms mentioned in Remark 1. Hence $V = V(\mathbf{a}', \mathbf{f}', \mathbb{Q}[k])$ with $\mathbf{a}' = \mathbf{a}$ and $\mathbf{f}' = \mathbf{f}$. Moreover, a degree bound is $m = 1$ (see again Remark 1); note that the required property (10) holds. Thus $V = V(\mathbf{a}', \mathbf{f}', \mathbb{Q}[k]_1)$. We are now in the position to start the degree reduction process with $m = 1$ (step 3). By coefficient comparison (see (13) with $l = 1$) we get that $\tilde{\mathbf{a}} = (0, 1) \in \mathbb{Q}^2$ and $\tilde{\mathbf{f}} = (2, 0) \in \mathbb{Q}^2$. Next, we calculate a basis of

$\tilde{V} = V((0,1),(2,0),\mathbb{Q})$: By solving the corresponding linear system we get the basis $\{(1/2,0,1),(0,1,0)\}$. Hence we extract the matrix $\mathbf{C} = \begin{pmatrix} 1/2 & 0 \\ 0 & 1 \end{pmatrix}$ and $\mathbf{g} = (1,0)$. Next, we compute $\phi = \mathbf{C}\mathbf{f}' - ((k+1)\sigma(\mathbf{g}\,k) - \mathbf{g}\,k) = (0,0)$; compare (15). What remains to calculate is a basis of $V((-1,k+1),(0,0),\mathbb{Q})$. Here we activate the degree reduction process for $m = 0$. Taking, e.g., the basis $\{(1,0,0),(0,1,0)\}$, we obtain the matrix $\mathbf{D} = \begin{pmatrix} 1 & 0 \\ 0 & 1 \end{pmatrix}$ and the vector $\mathbf{h} = (0,0)$. Finally, we get the matrix $\mathbf{E} = \begin{pmatrix} 1/2 & 0 \\ 0 & 1 \end{pmatrix}$ and the vector $\mathbf{p} = (k,0)$; see (17). To this end, we derive the basis $\{(1/2,0,k),(0,1,0)\}$ of V.

We remark that for the rational case $(\mathbb{K}(t),\sigma)$ with $\sigma(t) = t+1$ (and also the q-rational case) a direct approach is more efficient (and thus implemented in Sigma): Plugging in the ansatz $q = q_0 + q_1 t + \cdots + q_m t^m$ with $q_i \in \mathbb{K}$ into (11) yields a linear system for the unknowns $q_i, c_i \in \mathbb{K}$ and solving it provides a basis of $V(\mathbf{a}',\mathbf{f}',\mathbb{K}(k))$.

3.2   Turning the Strategy to Algorithms

The reduction method above can be summarized as follows.

Proposition 1. *Let* $(\mathbb{A}(t),\sigma)$ *be a* $\Pi\Sigma^*$*-extension of* $(\mathbb{A},\sigma)$*. If one can solve linear systems in* $\mathbb{A}(t)$*, can solve Problems DenB and DegB in* $(\mathbb{A}(t),\sigma)$*, and can solve Problem FPLDE in* $(\mathbb{A},\sigma)$*, then one can solve Problem FPLDE in* $(\mathbb{A}(t),\sigma)$*.*

The following properties are needed to apply this tactic in a nested $\Pi\Sigma^*$-extension.

Definition 3. *A* $\Pi\Sigma^*$*-extension* $(\mathbb{G}(t_1)\ldots(t_e),\sigma)$ *of* $(\mathbb{G},\sigma)$ *is called FPLDE-solvable, if there are algorithms that solve linear systems with multivariate polynomials over* $\mathbb{G}$*, that solve Problems DenB and DegB in the* $\Pi\Sigma^*$*-extensions* $(\mathbb{G}(t_1)\ldots(t_i),\sigma)$ *of* $(\mathbb{G}(t_1)\ldots(t_{i-1}),\sigma)$ *with* $1 \leq i \leq e$*, and that solve Problem FPLDE in* $(\mathbb{G},\sigma)$*.*

Then by recursive application of the method above we obtain the following theorem.

Theorem 2. *Let* $(\mathbb{F},\sigma)$ *be an FPLDE-solvable* $\Pi\Sigma^*$*-extension of* $(\mathbb{G},\sigma)$*. Algorithm SolveFPLDE (using DegreeReductionFPLDE) solves Problem FPLDE in* $(\mathbb{F},\sigma)$*.*

Algorithm SolveFPLDE $(\mathbf{a},\mathbf{f},\mathbb{F})$
Input: a $\Pi\Sigma^*$-extension $(\mathbb{F},\sigma)$ of $(\mathbb{G},\sigma)$ with $\mathbb{F} = \mathbb{G}(t_1)\ldots(t_e)$ which is FPLDE-solvable; $\mathbf{0} \neq \mathbf{a} = (a_0,a_1) \in \mathbb{F}^2$, $\mathbf{f} = (f_1,\ldots,f_n) \in \mathbb{F}^n$.
Output: a basis of $V(\mathbf{a},\mathbf{f},\mathbb{F})$ over $\mathbb{K} := \mathrm{const}_\sigma\mathbb{G}$.

1. IF $a_0\,a_1 = 0$, compute a basis B of $V(\mathbf{a},\mathbf{f},\mathbb{F})$ by solving a linear system and RETURN B.

2. IF $e = 0$, compute a basis B of $V(\mathbf{a}, \mathbf{f}, \mathbb{F})$ and RETURN B.
   Denote $\mathbb{A} := \mathbb{G}(t_1 \ldots, t_{e-1})$, $t := t_e$.
3. Get a denominator bound $d \in \mathbb{A}[t]^*$ of $V(\mathbf{a}, \mathbf{f}, \mathbb{A}(t))$.
4. Clear denominators, i.e., get $\mathbf{a}' \in (\mathbb{A}[t]^*)^2, \mathbf{f}' \in \mathbb{A}[t]^2$ s.t. $V((\frac{a_0}{d}, \frac{a_1}{\sigma(d)}), \mathbf{f}, \mathbb{A}[t]) = V(\mathbf{a}', \mathbf{f}', \mathbb{A}[t])$.
5. Get a degree bound $m \geq -1$ of $V(\mathbf{a}', \mathbf{f}', \mathbb{A}[t])$ with (10).
6. Get $B' := \mathsf{DegreeReductionFPLDE}(m, \mathbf{a}', \mathbf{f}', \mathbb{A}(t))$,
   say, $B' := \{(e_{i1}, \ldots, e_{in}, p_i)\}_{1 \leq i \leq \mu}$.
7. RETURN $\{(e_{i1}, \ldots, e_{in}, \frac{p_i}{d})\}_{1 \leq i \leq \mu}$.

Algorithm $\mathsf{DegreeReductionFPLDE}(m, \mathbf{a}', \mathbf{f}', \mathbb{A}(t))$
Input: a $\Pi\Sigma^*$-extension $(\mathbb{A}(t), \sigma)$ of $(\mathbb{G}, \sigma)$ with $\sigma(t) = \alpha\, t + \beta$ which is FPLDE-solvable; $\mathbf{a}' = (a_0', a_1') \in (\mathbb{A}[t]^*)^2$, $\mathbf{f}' = (f_1', \ldots, f_n') \in \mathbb{A}[t]^n$; $m \in \mathbb{N} \cup \{-1\}$ such that (10) holds.
Output: a basis of $V(\mathbf{a}', \mathbf{f}', \mathbb{A}[t]_m)$ over $\mathbb{K} := \mathrm{const}_\sigma \mathbb{G}$.

1. IF $m = -1$, compute a basis B of $V(\mathbf{a}, \mathbf{f}', \{0\})$ by linear algebra and RETURN B.
2. IF $m = 0$, get $B := \mathsf{SolveFPLDE}(\mathbf{a}', \mathbf{f}', \mathbb{A})$ and RETURN B.
3. Define $l := \max(\deg(a_0'), \deg(a_1'))$, and take $\tilde{\mathbf{f}} \in \mathbb{A}^n$, $\mathbf{0} \neq \tilde{\mathbf{a}} \in \mathbb{A}^2$ as in (13).
4. Get $\tilde{B} := \mathsf{SolveFPLDE}(\tilde{\mathbf{a}}, \tilde{\mathbf{f}}, \mathbb{A})$, say $\tilde{B} = \{(c_{i1}, \ldots, c_{in}, g_i)\}_{1 \leq i \leq \lambda} \subseteq \mathbb{K}^n \times \mathbb{A}$
5. IF $\tilde{B} = \{\}$ THEN execute $\mathsf{DegreeReductionFPLDE}(m-1, \mathbf{a}', (0), \mathbb{A}(t))$
   and check if there is a $h \neq 0$ with $a_1'\sigma(h) + a_0'(h) = 0$.
   IF yes, RETURN $\{(0, \ldots, 0, h\}$ ELSE RETURN $\{\}$.
6. Take $\mathbf{C} := (c_{ij}) \in \mathbb{K}^{\lambda \times n}$ and $\mathbf{g} = (g_1, \ldots, g_\lambda) \in \mathbb{A}^\lambda$, and set $\phi \in \mathbb{A}[t]_{l+m-1}^\lambda$ as given in (15).
7. Get $G := \mathsf{DegreeReductionFPLDE}(m-1, \mathbf{a}', \phi, \mathbb{A}(t))$,
   say $G = \{(d_{1i}, \ldots, d_{i\lambda}, h_i)\}_{1 \leq i \leq \mu}$.
8. If $G = \{\}$, RETURN $\{\}$.
9. Take $\mathbf{D} = (d_{ij}) \in \mathbb{K}^{\mu \times \lambda}$ and $\mathbf{h} := (h_1, \ldots, h_\mu) \in \mathbb{A}[t]_{m-1}^\mu$, and define $(e_{ij}) \in \mathbb{K}^{\mu \times n}$ and $(p_1, \ldots, p_\mu) \in \mathbb{A}[t]_m^\mu$ as given in (17).
10. RETURN $\{(e_{i1}, \ldots, e_{in}, p_i)\}_{1 \leq i \leq \mu}$.

In this article we aim at refinements and improvements of the presented reduction tactic for the special case of parameterized telescoping. Here we assume that the given $\Pi\Sigma^*$-extension $(\mathbb{F}, \sigma)$ of $(\mathbb{G}, \sigma)$ is FPLDE-solvable. To make this assumption more concrete, we present certain classes of difference fields in which the problems mentioned in Definition 3 can be solved by available algorithms, i.e., the telescoping algorithms of the next sections are applicable.

As worked out by M. Karr [27] this is the case if one restricts to the case that $\mathbb{K} := \mathrm{const}_\sigma \mathbb{G} = \mathbb{G}$ and one requires that $\mathbb{K}$ is σ-computable.

Definition 4. *A field $\mathbb{K}$ is σ-computable if the following properties hold.*

1. *One can perform the usual operations, in particular linear system solving with multivariate rational functions over $\mathbb{K}$ and deciding if $k \in \mathbb{Z}$ for any $k \in \mathbb{K}$,*

2. one can factorize multivariate polynomials over $\mathbb{K}$, and
3. for any $f_i \in \mathbb{K}^*$ one can compute a $\mathbb{Z}$-basis of $\{(n_1, \ldots, n_r) \in \mathbb{Z}^r \mid f_1^{n_1} \ldots f_r^{n_r} = 1\}$.

More precisely, if $\mathbb{K}$ is σ-computable, then any $\Pi\Sigma^*$-field over $\mathbb{K}$ is FPLDE-solvable: Problem FPLDE in $\mathbb{K}$ reduces to a simple linear algebra problem (property 1). Problems DenB and DegB can be solved by exploiting all three properties. In this regard, the following remarks are in place.

Remark 1. Originally, Karr [27] solved Problem FPLDE if $(\mathbb{K}(t_1) \ldots (t_e), \sigma)$ is a $\Pi\Sigma^*$-field over a σ-computable $\mathbb{K}$. Namely, he solved Problem DegB for a $\Pi\Sigma^*$-field over $\mathbb{K}$; for detailed proofs and extensions see [48,52]. In order to deal with denominators rather complicated reduction techniques (extending the degree reduction strategy above) have been utilized. In [18] Bronstein generalized Abramov's algorithm [5] which solves partially Problem DenB. Finally, in [50] algorithms of Karr [27] have been utilized to obtain a full solution. In a nutshell, this simplified version presented above is implemented in the summation package Sigma.

Example 5. Given the $\Pi\Sigma^*$-field $(\mathbb{Q}(k)(p), \sigma)$ over $\mathbb{Q}$ from Example 1.4, we calculate a basis of $V = V(\mathbf{f}, \mathbb{Q}(k)(p)) = V(\mathbf{a}, \mathbf{f}, \mathbb{Q}(k)(p))$ with $\mathbf{a} = (-1, 1)$ and $\mathbf{f} = (2kp, -\frac{2}{k+1}) \in \mathbb{Q}(k)(p)^2$. Since $\mathbb{Q}$ is σ-computable, this can be accomplished by executing SolveFPLDE$((-1, 1), \mathbf{f}, \mathbb{Q}(k)(p))$. We get the denominator bound $d = 1$ and set $\mathbf{a}' = (a_0', a_1') = (-1, 1)$ and $\mathbf{f}' = \mathbf{f}$. Moreover, we determine the degree bound $m = 1$. This means that $V = V(\mathbf{a}', \mathbf{f}', \mathbb{Q}(k)[p]_1)$. Hence we start the degree reduction with DegreeReductionFPLDE$(1, \mathbf{a}', \mathbf{f}', \mathbb{Q}(k)(p))$, and we define $l := \max(\deg(a_0'), \deg(a_1')) = 0$.

We calculate $\tilde{\mathbf{a}} = (-1, k+1)$ and $\tilde{\mathbf{f}} = (2k, 0)$ by (12). Next, we calculate the basis $\{(\frac{1}{2}, 0, k), (0, 1, 0)\}$ of $V(\tilde{\mathbf{a}}, \tilde{\mathbf{f}}, \mathbb{Q}(k))$ with SolveFPLDE$(\tilde{\mathbf{a}}, \tilde{\mathbf{f}}, \mathbb{Q}(k))$; for the details of this calculation we refer to Example 4. This gives $\mathbf{C} = \begin{pmatrix} 1/2 & 0 \\ 0 & 1 \end{pmatrix}$ and $\mathbf{g} = (k, 0)$. Then we calculate $\phi = (0, -\frac{2}{k+1})$ using (15).

We repeat the degree reduction to determine a basis of $V((-1, 1), \phi, \mathbb{Q}(k)[p]_0)$ for $m = 0$. There we calculate the basis $\{(1, 0, 0), (0, 0, 1)\}$ of $V((-1, 1), \phi, \mathbb{Q}(k))$ by executing SolveFPLDE$((-1, 1), \phi, \mathbb{Q}(k))$.

Finally, take $\mathbf{D} = \begin{pmatrix} 1 & 0 \\ 0 & 0 \end{pmatrix}$ and $\mathbf{h} = (0, 1)$ and derive $\mathbf{E} = \begin{pmatrix} 1/2 & 0 \\ 0 & 0 \end{pmatrix}$ and $\mathbf{p} = (p, 1)$ as given in (17). This produces the basis $\{(\frac{1}{2}, 0, p), (0, 0, 1)\}$ of V.

Example 6. We repeat the calculation steps of Example 5, but this time with the vector $\mathbf{f} = (\frac{(-k-2)p}{2(k+1)}, -\frac{1}{k+1})$. Again a denominator bound is $d = 1$, and we set $\mathbf{a}' = (a_0', a_1') = (-1, 1)$ and $\mathbf{f}' = \mathbf{f}$. Moreover, we get the degree bound $m = 1$. We thus activate the degree reduction process with $l = \max(\deg(a_0'), \deg(a_1')) = 0$.

For $m = 1$, we get $\tilde{\mathbf{a}} = (-1, k+1)$ and $\tilde{\mathbf{f}} = (\frac{-k-2}{2(k+1)}, 0)$ by (12). Next, we calculate the basis $\{(0, 1, 0)\}$ of $V(\tilde{\mathbf{a}}, \tilde{\mathbf{f}}, \mathbb{Q}(k))$ with SolveFPLDE$(\tilde{\mathbf{a}}, \tilde{\mathbf{f}}, \mathbb{Q}(k))$, and we extract the matrix

$$\mathbf{C} = (0,1) \tag{18}$$

and the vector $\mathbf{g} = (0)$. This yields $\phi = (\frac{-1}{k+1})$ using (15).

Repeating the degree reduction for $m = 0$ we calculate for $V((-1,1), \phi, \mathbb{Q}(k))$ the basis $\{(0,1)\}$ and extract the matrix $\mathbf{D} = (0)$ and the vector $\mathbf{h} = (1)$. This finally gives $\mathbf{E} = (0,0)$ and $\mathbf{p} = (1)$ using (17), i.e., we end up at the basis $\{(0,0,1)\}$ of V.

As worked out in [30] these algorithmic ideas can be generalized if the difference field $(\mathbb{G}, \sigma)$ satisfies the following (rather technical) properties. In this context the following functions are used: for $f \in \mathbb{A}^*$ and $k \in \mathbb{Z}$ we define

$$f_{(k,\sigma)} := \begin{cases} f\sigma(f)\dots\sigma^{k-1}(f) & \text{if } k > 0 \\ 1 & \text{if } k = 0 \\ \frac{1}{\sigma^{-1}(f)\dots\sigma^{-k}(f)} & \text{if } k < 0, \end{cases} \quad f_{\{k,\sigma\}} := \begin{cases} f_{(0,\sigma)} + f_{(1,\sigma)} + \dots + f_{(k-1,\sigma)} & \text{if } k > 0 \\ 0 & \text{if } k = 0 \\ -(f_{(-1,\sigma)} + \dots + f_{(k,\sigma)}) & \text{if } k < 0. \end{cases}$$

Definition 5. *A difference field $(\mathbb{G}, \sigma)$ is σ-computable if the following holds.*

1. *There is an algorithm that factors multivariate polynomials over $\mathbb{G}$ and that solves linear systems with multivariate rational functions over $\mathbb{G}$.*
2. *$(\mathbb{G}, \sigma^r)$ is torsion free for all $r \in \mathbb{Z}$, i.e., for all $r \in \mathbb{Z}$, for all $k \in \mathbb{Z}^*$ and all $g \in \mathbb{G}^*$ the equality $\left(\frac{\sigma^r(g)}{g}\right)^k = 1$ implies $\frac{\sigma^r(g)}{g} = 1$.*
3. *Π-Regularity. Given $f, g \in \mathbb{G}$ with f not a root of unity, there is at most one $n \in \mathbb{Z}$ such that $f_{(n,\sigma)} = g$. There is an algorithm that finds, if possible, this n.*
4. *Σ-Regularity. Given $k \in \mathbb{Z}\backslash\{0\}$ and $f, g \in \mathbb{G}$ with $f = 1$ or f not a root of unity, there is at most one $n \in \mathbb{Z}$ such that $f_{\{n,\sigma^k\}} = g$. There is an algorithm that finds, if possible, this n.*
5. *Orbit-Problem. There is an algorithm that solves the orbit problem: Given $(\mathbb{G}, \sigma)$ and $f_1, \dots, f_m \in \mathbb{G}^*$, find a basis of the following $\mathbb{Z}$-module:*

$$M(f_1, \dots, f_m; \mathbb{G}) := \{ (e_1, \dots, e_m) \in \mathbb{Z}^m \mid \exists g \in \mathbb{F}^* : f_1^{e_1} \cdots f_m^{e_m} = \frac{\sigma(g)}{g} \}. \tag{19}$$

6. *FPLDE Problem. There is an algorithm that solves Problem FPLDE in $(\mathbb{G}, \sigma)$.*

More precisely, there is the following result.

Theorem 3 ([30]). *Let $(\mathbb{G}, \sigma)$ be a σ-computable difference field. Then any $\Pi\Sigma^*$-extension $(\mathbb{F}, \sigma)$ of $(\mathbb{G}, \sigma)$ is FPLDE-solvable.*

In particular, for the following ground fields $(\mathbb{G}, \sigma)$ Problem FPLDE can be solved.

Example 7. 1. We can take $(\mathbb{K}, \sigma)$ with $\text{const}_\sigma\mathbb{K} = \mathbb{K}$ which is σ-computable[2]; e.g., $\mathbb{K}$ can be a rational function field over an algebraic number field; see [55].

---

[2] $(\mathbb{K}, \sigma)$ is σ-computable (see Definition 5) iff $\mathbb{K}$ is σ-computable (see Definition 4); we refer to [27,30].

2. We can take the free difference field $(\mathbb{G}, \sigma)$ over $\mathbb{K}$ as given in Example 3.1. Then $(\mathbb{G}, \sigma)$ is σ-computable if $(\mathbb{K}, \sigma)$ is σ-computable; see [30].
3. We can take the radical difference field $(\mathbb{G}, \sigma)$ of order d over $\mathbb{K}$ as given in Example 3.2. Then $(\mathbb{G}, \sigma)$ is σ-computable if $(\mathbb{K}, \sigma)$ is σ-computable; see [31].

We remark that all the presented ideas of this section can be generalized to solve mth-order linear difference equations using results from [8, 18, 43, 48, 56].

4   Special Case: Parameterized Telescoping

In the following we consider Problem PT in a $\Pi\Sigma^*$-extension $(\mathbb{F}, \sigma)$ of $(\mathbb{G}, \sigma)$ with $\mathbb{F} = \mathbb{G}(t_1)\dots(t_e)$ which is FPLDE-solvable. Namely, given $\mathbf{f} = (f_1, \dots, f_n) \in \mathbb{F}^n$ we aim at computing a basis of $V(\mathbf{f}, \mathbb{F}) = V((-1, 1), \mathbf{f}, \mathbb{F})$. Of course, one option is to execute Algorithm SolveFPLDE$((-1, 1), \mathbf{f}, \mathbb{F})$. Subsequently, we present a refined algorithm that is more efficient and will serve as a basis for further improvements.

If $\mathbb{F} = \mathbb{G}$, we are in the base case (there one should use, if available, an optimized PT-solver and not a general FPLDE-solver). Otherwise, denote the top generator by $t := t_e$ and consider the $\Pi\Sigma^*$-extension $(\mathbb{A}(t), \sigma)$ of $(\mathbb{A}, \sigma)$ with $\mathbb{A} = \mathbb{G}(t_1)\dots(t_{e-1})$ and $\sigma(t) = \alpha t + \beta$. Looking at the general strategy in Sect. 3.1, we first have to solve Problem DenB, i.e., we compute a denominator bound $d \in \mathbb{A}[t]^*$. If $d \neq 1$, we reduce the problem to calculate a basis of (9) and end up at Problem FPLDE (which is harder to solve than Problem PT). This situation is avoided partially as follows.

Consider the rational part $\mathbb{A}(t)_{(frac)}$ and polynomial part $\mathbb{A}[t]$ as $\mathbb{K}$-subspaces of $\mathbb{A}(t)$. Then for the direct sum $\mathbb{A}(t) = \mathbb{A}[t] \oplus \mathbb{A}(t)_{(frac)}$ we can utilize the following lemma; for a more general version see [57, Lemma 3.1].

Lemma 1. *Let* $(\mathbb{A}(t), \sigma)$ *be a* $\Pi\Sigma^*$*-extension of* $(\mathbb{A}, \sigma)$*, and let* $p, g_1 \in \mathbb{A}[t]$ *and* $r, g_2 \in \mathbb{A}(t)_{(frac)}$*. Then* $\sigma(g_1 + g_2) - (g_1 + g_2) = p + r$ *iff* $\sigma(g_1) - g_1 = p$ *and* $\sigma(g_2) - g_2 = r$*.*

Thus we can separate the telescoping problem, i.e., finding a basis of $V = V(\mathbf{f}, \mathbb{A}(t))$ for both the rational and polynomial part. Namely, by performing polynomial division with remainder on each component of $\mathbf{f}$ we get $\mathbf{r} \in \mathbb{A}(t)_{(frac)}^n$ and $\mathbf{p} \in \mathbb{A}[t]^n$ such that

$$\mathbf{f} = \mathbf{r} + \mathbf{p}. \tag{20}$$

In general, the bases of $V(\mathbf{r}, \mathbb{A}(t)_{(frac)})$ and $V(\mathbf{p}, \mathbb{A}[t])$ can be computed independently (e.g., in parallel), and the results can be combined by system solving to derive a basis of $V = V(\mathbf{f}, \mathbb{A}(t))$. For later considerations, we propose the following tactic.

Step 1: Solve the rational part. Get a basis $\{(c_{i1}, \dots, c_{in}, g_i)\}_{1 \le i \le \nu} \subseteq \mathbb{K}^n \times \mathbb{A}(t)_{(frac)}$ of $V_1 = V(\mathbf{r}, \mathbb{A}(t)_{(frac)})$. If $\nu = 0$, i.e., $V_1 = \{\mathbf{0}\}$, it follows that $V = \{\mathbf{0}\}^n \times \mathbb{K}$, i.e., $\{(0, \dots, 0, 1)\}$ is a basis of our original solution space V. Otherwise, define $\mathbf{C} = (c_{ij}) \in \mathbb{K}^{\nu \times n}$ and $\mathbf{g} = (g_1, \dots, g_\nu) \in \mathbb{A}(t)_{(frac)}^\nu$ and proceed as follows.

Step 2: Solve a refined version of the polynomial part. Set

$$\mathbf{f}' := \mathbf{C}\,\mathbf{p} \in \mathbb{A}[t]^{\nu}. \tag{21}$$

Note that $\nu \leq n$, i.e., the polynomial part might get simpler. Then compute a basis $\{(d_{i1}, \ldots, d_{i\nu}, h_i)\}_{1 \leq i \leq \mu} \subseteq \mathbb{K}^{\nu} \times \mathbb{A}[t]_m$ of $V_2 = V(\mathbf{f}', \mathbb{A}[t])$; note that $\mu \geq 1$, since $(0, \ldots, 0, 1)$ is a solution.

Step 3: Combine the rational and polynomial part. Define $\mathbf{D} = (d_{ij}) \in \mathbb{K}^{\mu \times \nu}$ and $\mathbf{h} = (h_1, \ldots, h_{\mu}) \in \mathbb{A}[t]^{\mu}$. By Step 1, $\sigma(\mathbf{g}) - \mathbf{g} = \mathbf{C}\,\mathbf{r}$, thus $\sigma(\mathbf{D}\,\mathbf{g}) - \mathbf{D}\,\mathbf{g} = \mathbf{D}\,\mathbf{C}\,\mathbf{r}$, and by Step 2 we have that $\sigma(\mathbf{h}) - \mathbf{h} = \mathbf{D}\,\mathbf{f}' = \mathbf{D}\,\mathbf{C}\,\mathbf{p}$. Adding the last two equations shows that $\sigma(\mathbf{D}\,\mathbf{g} + \mathbf{h}) - (\mathbf{D}\,\mathbf{g} + \mathbf{h}) = \mathbf{D}\,\mathbf{C}(\mathbf{r} + \mathbf{p}) = \mathbf{D}\,\mathbf{C}\,\mathbf{f}$. Consequently define

$$(e_{ij}) := \mathbf{D}\,\mathbf{C} \in \mathbb{K}^{\mu \times n} \quad \text{and} \quad (q_1, \ldots, q_{\mu}) := \mathbf{D}\mathbf{g} + \mathbf{h} \in \mathbb{A}(t)^{\mu}. \tag{22}$$

Then $B = \{(e_{i1}, \ldots, e_{in}, q_i)\}_{1 \leq i \leq \mu} \subseteq V(\mathbf{f}, \mathbb{A}(t))$. By further linear algebra arguments and Lemma 1 it follows that B forms a basis of $V = V(\mathbf{f}, \mathbb{A}(t))$.

Details of Step 1: Solving the rational part: Find a basis of $V_1 := V(\mathbf{r}, \mathbb{A}(t)_{(frac)})$. Derive a denominator bound $d \in \mathbb{A}[t]^*$ for V_1 (see DenB). Then we have to compute a basis of $V' := V((-\frac{1}{d}, \frac{1}{\sigma(d)}), \mathbf{r}, \mathbb{A}[t]_{\deg(d)-1})$; note that the numerator degree is bounded by $\deg(d) - 1$. Hence clear denominators and get $\mathbf{a}' \in \mathbb{A}[t]^2$, $\mathbf{f}' \in \mathbb{A}[t]^n$ with $V' = V(\mathbf{a}', \mathbf{f}', \mathbb{A}[t]_{\deg(d)-1})$. Thus DegreeReductionF-PLDE$(\deg(d) - 1, \mathbf{a}', \mathbf{f}', \mathbb{A}(t))$ gives a basis $\{(c_{i1}, \ldots, c_{in}, p_i)\}_{1 \leq i \leq \nu} \subseteq \mathbb{K}^n \times \mathbb{A}[t]$ of V'. As a consequence $\{(c_{i1}, \ldots, c_{in}, \frac{p_i}{d})\}_{1 \leq i \leq \nu}$ is a basis of V_1.

Details of Step 2: Solving the polynomial part: Find a basis of $V_2 = V(\mathbf{f}', \mathbb{A}[t])$ with $\mathbf{f}' = (f'_1, \ldots, f'_{\nu}) \in \mathbb{A}[t]^{\nu}$. Here Problem DegB can be read off by the following result given in [27]; for further details and proofs see Corollaries 3 and 6 in [52].

Theorem 4. *Let $(\mathbb{A}(t), \sigma)$ be a $\Pi\Sigma^*$-extension and $\mathbf{f}' = (f'_1, \ldots, f'_{\nu}) \in \mathbb{A}[t]^{\nu}$. Then a degree bound of $V(\mathbf{f}', \mathbb{A}[t])$ is*

$$m := \begin{cases} \max_{1 \leq i \leq \nu} \deg(f'_i) + 1 \ if \sigma(t) - t \in \mathbb{A} \\ \max_{1 \leq i \leq \nu} \deg(f'_i) \qquad if \sigma(t)/t \in \mathbb{A}. \end{cases} \tag{23}$$

Thus taking the corresponding $m \geq 0$, we can make the ansatz $q = g\,t^m + h$ for some $g \in \mathbb{A}$ and $h \in \mathbb{A}[t]_{m-1}$, and we can activate the degree reduction strategy; see Step 3 in Sect. 3.1. Note that $\mathbf{f}' \in \mathbb{A}[t]^{\nu}_m$. Thus the highest possible degree in $\sigma(g\,t^m + h) - (g\,t^m + h) = c_1 f'_1 + \cdots + c_{\nu} f'_{\nu}$ is m and doing coefficient comparison on this degree and using $\sigma(t) = \alpha t + \beta$ give the constraint $\alpha^m \sigma(g) - g = c_1 \operatorname{coeff}(f'_1, m) + \cdots + c_{\nu} \operatorname{coeff}(f'_{\nu}, m)$; compare (12) for the general situation. Thus we have to compute a basis of the solution space $\tilde{V} = V((-1, \alpha^m), \tilde{\mathbf{f}}, \mathbb{A})$ with

$$\tilde{\mathbf{f}} := (\operatorname{coeff}(f'_1, m), \ldots, \operatorname{coeff}(f'_{\nu}, m)) \in \mathbb{A}^{\nu}. \tag{24}$$

If $\alpha = 1$, this is again a PT problem. Otherwise, a FPLDE-solver (e.g., our SolveFPLDE) has to be activated. Given a basis $\{(c_{i1}, \ldots, c_{i\nu}, g_i)\}_{1 \leq i \leq \lambda} \subseteq \mathbb{K}^{\nu} \times \mathbb{A}$, we continue to extract the remaining part $h \in \mathbb{A}[t]_{m-1}$ by recursion. I.e., take

$$\phi := \mathbf{C}\mathbf{f} - [\sigma(\mathbf{g}\,t^m) - \mathbf{g}\,t^m] \in \mathbb{A}[t]^{\lambda}_{m-1} \tag{25}$$

and compute a basis $\{(d_{1i}, \ldots, d_{i\lambda}, h_i)\}_{1 \leq i \leq \mu} \subseteq \mathbb{K}^\lambda \times \mathbb{A}[t]_{m-1}$ of $V(\phi, \mathbb{A}[t]_{m-1})$. Finally, take $\mathbf{D} := (d_{ij}) \in \mathbb{K}^{\mu \times \lambda}$ and define

$$(e_{ij}) := \mathbf{D}\,\mathbf{C} \in \mathbb{K}^{\mu \times \nu} \text{ and } (p_1, \ldots, p_\mu) := \mathbf{D}\,\mathbf{g}\,t^m + (h_1, \ldots, h_\mu) \in \mathbb{A}[t]_m^\mu. \quad (26)$$

Then $\{(e_{i1}, \ldots, e_{i\nu}, p_i)\}_{1 \leq i \leq \mu}$ is a basis of $V(\mathbf{f}', \mathbb{A}[t]_m)$. This refined reduction can be summarized as follows.

Algorithm SolvePTRat$(\mathbf{f}, \mathbb{F})$
Input: a $\Pi\Sigma^*$-ext. $(\mathbb{F}, \sigma)$ of $(\mathbb{G}, \sigma)$ with $\mathbb{F} = \mathbb{G}(t_1) \ldots (t_e)$ which is FPLDE-solvable; $\mathbf{f} \in \mathbb{F}^n$.
Output: a basis of $V(\mathbf{f}, \mathbb{F})$ over $\mathbb{K} := \text{const}_\sigma \mathbb{G}$.

1. IF $e = 0$, compute a basis B of $V(\mathbf{f}, \mathbb{G})$ and RETURN B.
   Denote $\mathbb{A} := \mathbb{G}(t_1 \ldots, t_{e-1})$, $t := t_e$.
2. Compute $\mathbf{r} \in \mathbb{A}(t)_{(frac)}^n$ and $\mathbf{p} \in \mathbb{A}[t]^n$ such that $\mathbf{f} = \mathbf{r} + \mathbf{p}$.
3. Get a basis of $V(\mathbf{r}, \mathbb{A}(t)_{(frac)})$, say $B_1 = \{(c_{i1}, \ldots, c_{in}, g_i)\}_{1 \leq i \leq \nu}$; see "Details of Step 1".
4. IF $B_1 = \{\}$, RETURN $\{(0, \ldots, 0, 1)\}$.
5. Define $\mathbf{C} := (c_{ij}) \in \mathbb{K}^{\nu \times n}$, $\mathbf{g} := (g_1, \ldots, g_\nu) \in \mathbb{A}(t)_{(frac)}^\nu$; set $\mathbf{f}' := (f_1', \ldots, f_\nu') = \mathbf{C}\,\mathbf{p} \in \mathbb{A}[t]^\nu$.
6. Define $m \in \mathbb{N} \cup \{-1\}$ as given in (23).
7. Get $B_2 := \text{DegreeReductionRat}(m, \mathbf{f}', \mathbb{A}(t))$, say $B_2 = \{(d_{i1}, \ldots, d_{i\nu}, h_i)\}_{1 \leq i \leq \mu}$.
8. Take $\mathbf{D} := (d_{ij}) \in \mathbb{K}^{\mu \times \nu}$ and $\mathbf{h} = (h_1, \ldots, h_\mu) \in \mathbb{A}[t]_m^\mu$; define $(e_{ij}) \in \mathbb{K}^{\mu \times n}$, $(q_1, \ldots, q_\mu) \in \mathbb{A}(t)^\mu$ as given in (22).
9. RETURN $\{(e_{i1}, \ldots, e_{in}, q_i)\}_{1 \leq i \leq \mu}$.

Algorithm DegreeReductionRat$(m, \mathbf{f}', \mathbb{A}(t))$
Input: $m \in \mathbb{N} \cup \{-1\}$, a $\Pi\Sigma^*$-extension $(\mathbb{A}(t), \sigma)$ of $(\mathbb{G}, \sigma)$ with $\sigma(t) = \alpha t + \beta$ which is FPLDE-solvable; $\mathbf{f}' = (f_1', \ldots, f_\nu') \in \mathbb{A}[t]_m^\nu$.
Output: a basis of $V(\mathbf{f}', \mathbb{A}[t]_m)$ over $\mathbb{K} := \text{const}_\sigma \mathbb{G}$.

1. IF $m = -1$, compute a basis B of $V(\mathbf{f}', \{0\})$ by linear algebra and RETURN B.
2. IF $m = 0$, RETURN SolvePTRat$(\mathbf{f}', \mathbb{A})$.
3. Define $\tilde{\mathbf{f}} \in \mathbb{A}^\nu$ as in (24).
4. Get $\tilde{B} := \begin{cases} \text{SolvePTRat}(\tilde{\mathbf{f}}, \mathbb{A}) & \text{if } \alpha = 1 \\ \text{SolveFPLDE}((-1, \alpha^m), \tilde{\mathbf{f}}, \mathbb{A}) & \text{if } \alpha \neq 1 \end{cases}$, say $\tilde{B} = \{(c_{i1}, \ldots, c_{i\nu}, g_i)\}_{1 \leq i \leq \lambda}$.
5. IF $\tilde{B} = \{\}$, RETURN $\{(0, \ldots, 0, 1)\}$.   (*possible if $\alpha \neq 1$*)
6. Take $\mathbf{C} := (c_{ij}) \in \mathbb{K}^{\lambda \times \nu}$, $\mathbf{g} := (g_1, \ldots, g_\lambda) \in \mathbb{A}^\lambda$, and define $\phi \in \mathbb{A}[t]_{m-1}^\lambda$ as given in (25).
7. Get $G := \text{DegreeReductionRat}(m-1, \phi, \mathbb{A}(t))$, say $G = \{(d_{1i}, \ldots, d_{i\lambda}, h_i)\}_{1 \leq i \leq \mu}$.
8. Take $\mathbf{D} := (d_{ij}) \in \mathbb{K}^{\mu \times \lambda}$ and $\mathbf{h} := (h_1, \ldots, h_\mu) \in \mathbb{A}[t]_{m-1}^\mu$; then define $(e_{ij}) \in \mathbb{K}^{\mu \times \nu}$ and $(p_1, \ldots, p_\mu) \in \mathbb{A}[t]_m^\mu$ as given in (26).
9. RETURN $\{(e_{i1}, \ldots, e_{i\nu}, p_i)\}_{1 \leq i \leq \mu}$.

In concrete applications one is usually given $(\mathbb{G}, \sigma)$ (e.g., as a $\Pi\Sigma^*$-field build by Π-extensions) and deals with Σ^*-extensions on top where the generators (describing sums) occur only in the numerators. This motivates the following definition.

Definition 6. $(\mathbb{G}(t_1)\ldots(t_e), \sigma)$ *is called* polynomial Σ^*-extension *of* $(\mathbb{G}, \sigma)$ *if for all* $1 \leq i \leq e$, $(\mathbb{G}(t_1)\ldots(t_i), \sigma)$ *is a* Σ^*-ext. *of* $(\mathbb{G}, \sigma)$ *with* $\sigma(t_i) - t_i \in \mathbb{G}[t_1, \ldots, t_{i-1}]$.

In such difference fields we have the following property; for a more general version and its corresponding proof see [61, Theorem 2.7].

Theorem 5. *Let* $(\mathbb{G}(t_1)\ldots(t_e), \sigma)$ *be a polynomial* Σ^*-extension *of* $(\mathbb{G}, \sigma)$. *Then for all* $g \in \mathbb{G}(t_1)\ldots(t_e)$: $\sigma(g) - g \in \mathbb{G}[t_1, \ldots, t_e]$ *iff* $g \in \mathbb{G}[t_1, \ldots, t_e]$.

Here the (easy) direction from right to left implies that $(\mathbb{G}[t_1, \ldots, t_e], \sigma)$ forms a difference ring. The other direction implies that a solution of a telescoping problem does not introduce sums in the denominator provided that the summand has no sums in the denominator. As a consequence, a denominator bound of $V(\mathbf{f}, \mathbb{G}(t_1)\ldots(t_{i-1})(t_i))$ with $\mathbf{f} \in \mathbb{G}(t_1)\ldots(t_{i-1})[t_i]^n$ is always 1. In other words, Algorithms SolvePTRat and DegreeReductionRat can be simplified to Algorithms SolvePTPoly and DegreeReductionPoly, respectively. In particular (since Problem DenB and DegB can be obtained without any cost), we end up at the following

Corollary 1. *Let* $(\mathbb{G}(t_1)\ldots(t_e), \sigma)$ *be a polynomial* Σ^*-extension *of* $(\mathbb{G}, \sigma)$. *Then one can solve Problem PT in* $(\mathbb{G}[t_1, \ldots, t_e], \sigma)$ *if one can solve Problem PT in* $(\mathbb{G}, \sigma)$.

Algorithm SolvePTPoly$(\mathbf{f}, \mathbb{G}(t_1)\ldots(t_e))$
Input: a polynomial Σ^*-extension $(\mathbb{G}(t_1)\ldots(t_e), \sigma)$ of $(\mathbb{G}, \sigma)$ where Problem PT is solvable in $(\mathbb{G}, \sigma)$; $\mathbf{f} = (f_1, \ldots, f_n) \in \mathbb{G}[t_1, \ldots, t_e]^n$.
Output: a basis of $V(\mathbf{f}, \mathbb{G}(t_1)\ldots(t_e)) = V(\mathbf{f}, \mathbb{G}[t_1, \ldots, t_e])$ over $\mathbb{K} := \mathrm{const}_\sigma \mathbb{G}$.

1. IF $e = 0$, compute a basis B of $V(\mathbf{f}, \mathbb{G})$ and RETURN B.
2. Define $m := \max_{1 \leq i \leq n} \deg_{t_e}(f_i) + 1$.   (*Note: $\mathbf{f} \in \mathbb{G}[t_1, \ldots, t_{e-1}][t_e]^n_m$ and $m \geq 0$*)
3. Get $B := \mathsf{DegreeReductionPoly}(m, \mathbf{f}, \mathbb{G}(t_1, \ldots, t_e))$ and RETURN B.

Algorithm DegreeReductionPoly$(m, \mathbf{f}', \mathbb{G}(t_1, \ldots, t_e))$
Input: $m \in \mathbb{N}$; a polynomial Σ^*-extension $(\mathbb{G}(t_1)\ldots(t_e), \sigma)$ of $(\mathbb{G}, \sigma)$ where Problem PT is solvable in $(\mathbb{G}, \sigma)$; $\mathbf{f}' = (f_1', \ldots, f_\nu') \in \mathbb{G}[t_1, \ldots, t_{e-1}][t_e]^\nu_m$.
Output: a basis of $V(\mathbf{f}', \mathbb{G}[t_1, \ldots, t_{e-1}][t_e]_m)$ over $\mathbb{K} := \mathrm{const}_\sigma \mathbb{G}$
Denote $\mathbb{A} := \mathbb{G}[t_1 \ldots, t_{e-1}]$, $t := t_e$.

1. IF $m = 0$, get $B := \mathsf{SolvePTPoly}(\mathbf{f}', \mathbb{G}(t_1)\ldots(t_{e-1}))$ and RETURN B.
2. Define $\tilde{\mathbf{f}} \in \mathbb{A}^\nu$ as in (24).
3. Get $\tilde{B} := \mathsf{SolvePTPoly}(\tilde{\mathbf{f}}, \mathbb{G}(t_1)\ldots(t_{e-1}))$,
   say $\tilde{B} = \{(c_{i1}, \ldots, c_{i\nu}, g_i)\}_{1 \leq i \leq \lambda} \subseteq \mathbb{K}^\nu \times \mathbb{A}$.

4. Let $\mathbf{C} := (c_{ij}) \in \mathbb{K}^{\lambda \times \nu}$, $\mathbf{g} = (g_1, \ldots, g_\lambda) \in \mathbb{A}^\lambda$, and set $\phi \in \mathbb{A}[t]^\lambda_{m-1}$ as in (25).
5. Get $G := \mathsf{DegreeReductionPoly}(m-1, \phi, \mathbb{G}(t_1)\ldots(t_e))$,
   say $G = \{(d_{1i}, \ldots, d_{i\lambda}, h_i)\}_{1 \le i \le \mu}$.
6. Let $\mathbf{D} = (d_{ij}) \in \mathbb{K}^{\mu \times \lambda}$, $\mathbf{h} := (h_1, \ldots, h_\mu) \in \mathbb{A}[t]^\mu_{m-1}$, and define $(e_{ij}) := \mathbf{D}\mathbf{C} \in \mathbb{K}^{\mu \times \nu}$ and $(p_1, \ldots, p_\mu) := \mathbf{D}\mathbf{g}t^m + \mathbf{h} \in \mathbb{A}[t]^\mu_m$.
7. RETURN $\{(e_{i1}, \ldots, e_{i\nu}, p_i)\}_{1 \le i \le \mu}$.

Example 8. Consider f given in (6) within the $\Pi\Sigma^*$-field $(\mathbb{Q}(k)(p)(h), \sigma)$ from Example 2. Note that $(\mathbb{G}(h), \sigma)$ with $\mathbb{G} = \mathbb{Q}(k)(p)$ is trivially a polynomial Σ^*-extension of $(\mathbb{G}, \sigma)$. Since $f \in \mathbb{G}[h]$, we can calculate a basis of $V((f), \mathbb{G}(h))$ by executing $\mathsf{SolvePTPoly}((f), \mathbb{G}(h))$. We obtain the degree bound $m = 3$ and start the degree reduction with $\mathsf{DegreeReductionPoly}(3, \phi_3, \mathbb{G}(h)))$ where $\phi_3(= \mathbf{f}') = (f)$.

- We determine $\tilde{\mathbf{f}}_3(= \tilde{\mathbf{f}}) = (0)$ using (24), and take the basis $\tilde{B}_3 = \{(1,0), (0,1)\}$ of $V(\tilde{\mathbf{f}}_3, \mathbb{G})$. Hence we get $\mathbf{C}_3 = (1, 0)$ and $\mathbf{g}_3 = (0, 1)$. This gives $\mathbf{f}_2 = \phi = (f, -\sigma(h^3) + h^3) = (f, -\frac{3h^2}{k+1} - \frac{3h}{(k+1)^2} - \frac{1}{(k+1)^3})$ using (25).
- We repeat the degree reduction for $m = 2$. Taking the coefficients of h^2 from the entries in $\mathbf{f}_2$ produces the vector $\tilde{\mathbf{f}}_2 = ((1+k)(2+2k+k^2)p, -(3/(1+k)))$ using (24); here ϕ_2 and $\tilde{\mathbf{f}}_2$ take over the role of $\mathbf{f}'$ and $\tilde{\mathbf{f}}$, respectively. We calculate the basis $\tilde{B}_2 = \{(-1, 0, -k(k+1)p), (0, 0, 1)\}$ of $V(\tilde{\mathbf{f}}_2, \mathbb{G})$ by executing $\mathsf{SolvePTRat}(\phi_2, \mathbb{G})$. This gives $\mathbf{g}_2 = (-k(k+1)p, 1)$ and $\mathbf{C}_2 = \begin{pmatrix} -1 & 0 \\ 0 & 0 \end{pmatrix}$. Next, we calculate $\phi_1 = (2hkp + \frac{kp}{k+1}, -\frac{2h}{k+1} - \frac{1}{(k+1)^2})$ using (25).
- We active the degree reduction with $m = 1$. This time we get the leading coefficient vector $\tilde{\mathbf{f}}_1 = (2kp, -\frac{2}{k+1})$ using (24). We obtain the basis $\tilde{B}_1 = \{(\frac{1}{2}, 0, p), (0, 0, 1)\}$ of $V(\tilde{\mathbf{f}}_1, \mathbb{G})$ by executing $\mathsf{SolvePTRat}(\phi_1, \mathbb{G})$; the calculation steps agree with those of $\mathsf{SolveFPLDE}((-1, 1), \phi_1, \mathbb{G})$ as given in Example 5. This gives $\mathbf{C}_1 = \begin{pmatrix} \frac{1}{2} & 0 \\ 0 & 0 \end{pmatrix}$ and $\mathbf{g}_1 = (p, 1)$. Finally, using (25) we calculate

$$\phi_0 = (\frac{(-k-2)p}{2(k+1)}, -\frac{1}{k+1}). \tag{27}$$

- A basis of $V(\phi_0, \mathbb{G})$ is $\{(0, 0, 1)\}$; for the calculation steps see Example 6.

To this end, we combine the solutions. We get $\mathbf{D}_0 = (0, 0)$ and $\mathbf{h}_0 = (1)$. This yields $\mathbf{E}_1 = \mathbf{C}_0\mathbf{D}_0 = (0, 0)$ and $p_1 = (1)$, i.e., we get the basis $\{(0, 0, 1)\}$ of $V(\phi_1, \mathbb{G}[h]_1)$. Similarly, we get the basis $\{(0, 0, 1)\}$ of $V(\phi_2, \mathbb{G}[h]_2)$ and the basis $(0, 1)$ of $V(\phi_3, \mathbb{G}[h]_3) = V((f), \mathbb{G}(h))$.

5   Parameterized First-Entry Telescoping Solutions

The developed algorithms can be optimized further for first-entry solutions.

Definition 7. *Let $(\mathbb{A}, \sigma)$ be a difference ring with constant field $\mathbb{K} := \text{const}_\sigma \mathbb{A}$. Let W be a $\mathbb{K}$-subspace of $\mathbb{A}$ and let $\mathbf{f} \in \mathbb{A}^n$. An element $(c_1, \ldots, c_n, g) \in V(\mathbf{f}, W)$ is called* first-entry solution *if $c_1 \ne 0$. A* first-entry solution set *of $V(\mathbf{f}, W)$ is the empty set if there does not exist a first-entry solution of $V(\mathbf{f}, W)$. Otherwise, the first-entry solution set consists of exactly one such first entry solution element.*

Trivially, a first-entry solution set can be determined by computing a basis of the full solution space $V(\mathbf{f}, \mathbb{G}(t_1)\ldots(t_e))$ and taking –if possible– a vector where the first entry is non-zero. Note that exactly such a solution is needed if one solves the telescoping problem or if one hunts for a creative telescoping solution (see the introduction). For the following considerations the basis representation is refined as follows.

Definition 8. *Let $\mathbb{K}$ be a field being a subring of $\mathbb{A}$. A matrix $(c_{ij}) \in \mathbb{K}^{\lambda \times n}$ is first-row reduced if for the first column $(c_{11},\ldots,c_{\lambda 1})$ we have that $c_{21} = \cdots = c_{2\lambda} = 0$, i.e., only the first entry may be non-zero. A linearly independent set $\{(c_{i1},\ldots,c_{in},g_i)\}_{1 \le i \le \lambda} \subseteq \mathbb{K}^n \times \mathbb{A}$ over $\mathbb{K}$ is first-entry reduced if $(c_{ij}) \in \mathbb{K}^{\lambda \times n}$ is first-row reduced.*

Example 9. The matrices $\mathbf{C_3} = (1,0)$, $\mathbf{C_2} = \begin{pmatrix} -1 & 0 \\ 0 & 0 \end{pmatrix}$, and $\mathbf{C_3} = (1,0)$ from Example 8 are first-row reduced, i.e., in the first column only the first entry is non-zero.

Subsequently, we simplify the already derived reduction technique for Problem PT (resp. Algorithm SolvePTRat) as much as possible such that exactly a first-entry solution set is produced. For further considerations, suppose we are given a $\Pi\Sigma^*$-extension $(\mathbb{G}(t_1)\ldots(t_e), \sigma)$ of $(\mathbb{G}, \sigma)$ being FPLDE-solvable; let $\mathbf{f} = (f_1,\ldots,f_n)$.

If $f_1 = 0$, $\{(1,0,\ldots,0)\}$ is a first-entry solution. Moreover, if $e = 0$, we calculate by assumption a basis of $V = V(\mathbf{f}, \mathbb{G})$ and extract a first-entry solution set (ideally, one should use here improved algorithms to determine a first-entry solution set).

Now let $\mathbb{A} := \mathbb{G}(t_1,\ldots,t_{e-1})$ and $t := t_e$ with $\sigma(t) = \alpha t + \beta$. We proceed as in Sect. 4. Write $\mathbf{f}$ in the form (20) with $\mathbf{r} \in \mathbb{A}(t)_{(frac)}^n$ and $\mathbf{p} \in \mathbb{A}[t]^n$. Consider the solution space $V_1 = V(\mathbf{r}, \mathbb{A}(t)_{(frac)})$ of the rational part. If $V_1 = \{0\}$, then there is no way to get a first-entry solution for V, and $\{\}$ is the first-entry solution set. Otherwise, let $B_1 = \{(c_{i1},\ldots,c_{in},g_i)\}_{1 \le i \le \nu} \subseteq \mathbb{K}^n \times \mathbb{A}(t)_{(frac)}$ be a first-entry reduced basis of V_1, take the first-row reduced matrix $\mathbf{C} := (c_{ij}) \in \mathbb{K}^{\nu \times n}$ and define $\mathbf{f}' \in \mathbb{A}[t]^\nu$ as in (21). Finally, take a first-entry reduced basis $B_2 = \{(d_{i1},\ldots,d_{i\nu},h_i)\}_{1 \le i \le \mu} \subseteq \mathbb{K}^\nu \times \mathbb{A}[t]_m$ of $V_2 = V(\mathbf{f}', \mathbb{A}[t])$. Hence we get the first-row reduced matrix $\mathbf{D} = (d_{ij}) \in \mathbb{K}^{\mu \times \nu}$ and define the matrix

$$\mathbf{E} = (e_{ij}) := \mathbf{D}\,\mathbf{C} \in \mathbb{K}^{\mu \times n} \tag{28}$$

and $q_i \in \mathbb{A}(t)$ as in (22). In particular, we obtain a basis $B = \{(e_{i1},\ldots,e_{in},q_i)\}_{1 \le i \le \mu}$ of V. Moreover, since $\mathbf{C}$ and $\mathbf{D}$ are first-row reduced, $\mathbf{E}$ is first-row reduced. In particular, the first entry of the first column of $\mathbf{E}$ is non-zero if and only if the first entry of the first column for both, $\mathbf{C}$ and $\mathbf{D}$, are non-zero. In other words, B contains a first entry solution if and only if B_1 and B_2 have a first entry solution. With this knowledge, the reduction can be simplified as follows. If the first column of $\mathbf{C}$ is the zero-vector, a first-entry solution of V does not exist. Hence $\{\}$ is the first-entry solution set. Otherwise, B has a first-entry solution if and only if B_2 has a first-entry solution. In particular, if we are given a first-entry solution of V_2, then we obtain immediately a first-entry solution of V. The corresponding modifications of Algorithm SolvePTRat are summarized in Algorithm FirstEntryPT.

In particular, we do not have to compute a full bases of $V_2 = V(\mathbf{f}', \mathbb{A}[t])$, but we only need a first-entry solution set of V_2. Here the following refinements are in place. If $f_1' = 0$, then $\{(1, 0, \ldots, 0)\}$ is a first-entry solution set. Otherwise, let m be the degree bound of V_2 as given in (23). Note that $m = -1$ is only possible if $\sigma(t) = \alpha t$ and $\mathbf{f}' = \mathbf{0}$. But this case is already covered with $f_1' = 0$. Thus we may assume that $m \geq 0$. Then the coefficient of the highest possible term t^m of the polynomial solutions is contained in the solution space $\tilde{V} = V((-1, \alpha^m), \tilde{\mathbf{f}}, \mathbb{A})$ with (24). If $\tilde{V} = \{\mathbf{0}\}$, it follows that there is no first-entry solution of V_2 and $\{\}$ is the first-entry solution set. Otherwise, let $\{(c_{i1}, \ldots, c_{i\nu}, g_i)\}_{1 \leq i \leq \lambda} \subseteq \mathbb{K}^\nu \times \mathbb{A}$ be a first-entry reduced basis of $\tilde{V}$, and take $\mathbf{C} := (c_{ij}) \in \mathbb{K}^{\lambda \times \nu}$ and $\mathbf{g} := (g_1, \ldots, g_\lambda) \in \mathbb{A}^\lambda$. Finally, define (25) and take a basis $\{(d_{1i}, \ldots, d_{i\lambda}, h_i)\}_{1 \leq i \leq \mu} \subseteq \mathbb{K}^\lambda \times \mathbb{A}[t]_{m-1}$ of $W = V(\phi, \mathbb{A}[t]_{m-1})$. Take $\mathbf{D} := (d_{ij}) \in \mathbb{K}^{\mu \times \lambda}$ and define $\mathbf{E} = (e_{ij}) := \mathbf{D}\mathbf{C} \in \mathbb{K}^{\mu \times \nu}$ and $p_i \in \mathbb{A}[t]_m$ as in (26). Then $B_2 = \{(e_{i1}, \ldots, e_{i\nu}, p_i)\}_{1 \leq i \leq \mu}$ is a basis of V_2. As above we can conclude that V_2 has a first-entry solution if and only if W has a first-entry solution. In particular, given an explicit first-entry solution of W and a basis of $\tilde{V}$ (which contains a first-entry solution), one can construct a first-entry solution of V_2. In summary, it suffices to calculate a first-entry solution set of W (instead of calculating a full basis of W).

Example 10. In Example 8 we calculated a full basis of $V((f), \mathbb{Q}(k)(p)(h))$ where f is given in (6). Now we calculate only a first-entry solution set. In the beginning, the calculations are the same as in Example 8. Since the matrices $\mathbf{C_i}$ $i = 3, 2, 1$ given in Example 8 are first-entry reduced and the corresponding bases $\tilde{B}_i$ contain a first-entry solution, nothing changes in these steps. Finally, we enter the problem to compute a first-entry solution for $V_0 = V(\phi_0, \mathbb{G})$ with (27). Hence we continue the reduction as given in Example 6 with $\mathbf{f} := \phi_0$. When we enter here the degree reduction we calculate the first-row reduced matrix (18). Exactly here our proposed method delivers a shortcut. Since there is no first-entry solution, V_0 and thus also V has no first entry solution. Thus we return the first entry solution set $\{\}$.

In the previous example just a small part of the reduction could be avoided. However, if one has more field generators (e.g., $e = 100$), big parts of the full reduction process can be skipped. The refined reduction can be summarized as follows.

Algorithm FirstEntryPT$(\mathbf{f}, \mathbb{F})$
Input: a $\Pi\Sigma^*$-extension $(\mathbb{F}, \sigma)$ of $(\mathbb{G}, \sigma)$ with $\mathbb{F} = \mathbb{G}(t_1) \ldots (t_e)$ and $\mathbb{K} := \mathrm{const}_\sigma \mathbb{G}$ which is FPLDE-solvable; $\mathbf{f} \in \mathbb{F}^n$.
Output: a first-entry solution set of $V(\mathbf{f}, \mathbb{F})$.

1. IF $f_1 = 0$, RETURN $\{(1, 0, \ldots, 0)\}$.   (*shortcut*)
2. IF $e = 0$, compute a first-entry solution set B of $V(\mathbf{f}, \mathbb{G})$ and RETURN B. Denote $\mathbb{A} := \mathbb{G}(t_1 \ldots, t_{e-1})$, $t := t_e$.

3. Compute $\mathbf{r} \in \mathbb{A}(t)^n_{(frac)}$ and $\mathbf{p} \in \mathbb{A}[t]^n$ such that $\mathbf{f} = \mathbf{r} + \mathbf{p}$.
4. Get a first-entry reduced basis of $V(\mathbf{r}, \mathbb{A}(t)_{(frac)})$,
   say $B_1 = \{(c_{i1}, \ldots, c_{in}, g_i)\}_{1 \le i \le \nu}$.
5. IF $B_1 = \{\}$ OR $c_{11} = \cdots = c_{\nu 1} = 0$, RETURN $\{\}$.   (*shortcut*)
6. Take $\mathbf{C} := (c_{ij}) \in \mathbb{K}^{\nu \times n}$, $\mathbf{g} := (g_1, \ldots, g_\nu) \in \mathbb{A}(t)^\nu_{(frac)}$, and let $\mathbf{f}' :=$
   $(f'_1, \ldots, f'_\nu) = \mathbf{C}\,\mathbf{p} \in \mathbb{A}[t]^\nu$.
7. Define $m \in \mathbb{N} \cup \{-1\}$ as given in (23).
8. Get $B_2 := \mathsf{DegreeReductionFirstEntry}(m, \mathbf{f}', \mathbb{A}(t))$. IF $B_2 = \{\}$, RETURN $\{\}$.
9. Otherwise, let $B_2 = \{(d_{11}, \ldots, d_{1\nu}, h_1)\} \subseteq \mathbb{K}^\nu \times \mathbb{A}[t]_m$ and take $\mathbf{D} :=$
   $(d_{11}, \ldots, d_{1\nu}) \in \mathbb{K}^{1 \times \nu}$, $\mathbf{h} := (h_1) \in \mathbb{A}[t]^1_m$. Define $(e_{11}, \ldots, e_{1n}) := \mathbf{D}\mathbf{C} \in$
   $\mathbb{K}^{1 \times n}$ and $(q_1) := \mathbf{D}\,\mathbf{g} + \mathbf{h} \in \mathbb{A}(t)^1$.
10. RETURN $\{(e_{11}, \ldots, e_{1n}, q_1)\}$.

Algorithm DegreeReductionFirstEntry$(m, \mathbf{f}', \mathbb{A}(t))$
Input: $m \in \mathbb{N} \cup \{-1\}$, a $\Pi\Sigma^*$-extension $(\mathbb{A}(t), \sigma)$ of $(\mathbb{G}, \sigma)$ with $\sigma(t) = \alpha\,t + \beta$
which is FPLDE-solvable; $\mathbf{f}' = (f'_1, \ldots, f'_\nu) \in \mathbb{A}[t]^\nu_m$.
Output: a first-entry solution of $V(\mathbf{f}', \mathbb{A}[t]_m)$ over $\mathbb{K} := \mathrm{const}_\sigma \mathbb{G}$.

1. IF $f'_1 = 0$, RETURN $\{(1, 0, \ldots, 0)\}$.   (*shortcut - note: this covers also the
                                                      case $m = -1$*)
2. IF $m = 0$, RETURN $\mathsf{FirstEntryPT}(\mathbf{f}', \mathbb{A})$.
3. Define $\tilde{\mathbf{f}} \in \mathbb{A}^\nu$ as in (24).
4. Get $\tilde{B} := \begin{cases} \mathsf{SolvePTRat}(\tilde{\mathbf{f}}, \mathbb{A}) & \text{if } \alpha = 1 \\ \mathsf{SolveFPLDE}((-1, \alpha^m), \tilde{\mathbf{f}}, \mathbb{A}) & \text{if } \alpha \ne 1. \end{cases}$'
   say $\tilde{B} = \{(c_{i1}, \ldots, c_{i\nu}, g_i)\}_{1 \le i \le \lambda}$.
   IF the bases is not first-entry reduced, reduce it.
5. IF $\tilde{B} = \{\}$ OR $c_{11} = \cdots = c_{\lambda 1} = 0$, RETURN $\{\}$.   (*shortcut*)
6. Take $C := (c_{ij}) \in \mathbb{K}^{\lambda \times \nu}$, $\mathbf{g} := (g_1, \ldots, g_\lambda) \in \mathbb{A}^\lambda$, and let $\phi \in \mathbb{A}[t]^\lambda_{m-1}$ as
   in (25).
7. Get $G := \mathsf{DegreeReductionFirstEntry}(m - 1, \phi, \mathbb{A}(t))$. IF $G = \{\}$, RETURN $\{\}$.
8. Otherwise, let $G = \{(d_{11}, \ldots, d_{1\lambda}, h_1)\} \subseteq \mathbb{K}^\lambda \times \mathbb{A}[t]_{m-1}$. Take $\mathbf{D} :=$
   $(d_{11}, \ldots, d_{1\lambda}) \in \mathbb{K}^{1 \times \lambda}$, $\mathbf{h} := (h_1) \in \mathbb{A}[t]^1_{m-1}$ and define $(e_{11}, \ldots, e_{1\nu}) :=$
   $\mathbf{D}\,C \in \mathbb{K}^{1 \times \nu}$ and $(p_1) := \mathbf{D}\,\mathbf{g}\,t^m + \mathbf{h} \in \mathbb{A}[t]^1_m$.
9. RETURN $\{(e_{11}, \ldots, e_{1\nu}, p_1)\}$.

Remark 2. In Algorithm DegreeReductionFirstEntry we execute $\mathsf{SolvePTRat}(\tilde{\mathbf{f}}, \mathbb{A})$ in
Line 4 if $\alpha = 1$. If the found basis does not contain a first-entry solution, the
algorithm stops. Thus one can modify $\mathsf{SolvePTRat}(\tilde{\mathbf{f}}, \mathbb{A})$ such that it stops as soon
as possible when it is clear that a first-entry solution does not exist. More pre-
cisely, the modified version is similar to FirstEntryPT and DegreeReductionFirstEntry
with the difference that a full basis is returned whenever it contains a first-entry
solution. Moreover, if $\alpha \ne 1$, the algorithms SolveFPLDE and DegreeReductionF-
PLDE can and should be modified by similar refinements.

The following observations and properties will be crucial for the next section.
Let $\mathbf{f_e} \in \mathbb{G}(t_1) \ldots (t_e)^{n_e}$ with $e \ge 1$ and consider the reduction process as

carried out in $\mathsf{FirstEntryPT}(\mathbf{f_e}, \mathbb{G}(t_1)\ldots(t_e))$. Then in Line 1 it might find a solution, and in Line 5 it might return $\{\}$. Otherwise it calls $\mathsf{DegreeReduction}$-$\mathsf{FirstEntry}(m, \mathbf{f}', \mathbb{G}(t_1)\ldots(t_e))$ for some $m \in \mathbb{N} \cup \{-1\}$. If $f_1' = 0$ (in particular if $m = -1$), we find again a solution in Line 1. If $m \geq 1$, we might enter in a special case in Line 5 and return $\{\}$. Otherwise we apply $\mathsf{DegreeReductionFirstEntry}(m - 1, \phi, \mathbb{G}(t_1)\ldots(t_e))$. In a nutshell, we run in certain shortcuts (returning $\{\}$ or $\{(1, 0, \ldots, 0)\}$) or we call $\mathsf{DegreeReductionFirstEntry}$ m-times, until we enter in the case $m = 0$ and execute $\mathsf{FirstEntryPT}(\mathbf{f_{e-1}}, \mathbb{A})$ in Line 2 for some $\mathbf{f_{e-1}} \in \mathbb{A}^{n_{e-1}} = \mathbb{G}(t_1)\ldots(t_{e-1})^{n_{e-1}}$. If one enters in these shortcuts for a particular given reduction, $\mathbf{f_e}$ is called base-vector (of the given reduction). Otherwise, the resulting vector $\mathbf{f_{e-1}}$ is a reduction-vector of $\mathbf{f_e}$ and we write $\mathbf{f_e} \to \mathbf{f_{e-1}}$.

Example 11. Define $\mathbf{f_2} = (f)$ with f given in (6) and $\mathbf{f_1} = \phi_0$ given in (27). Then $\mathbf{f_1}$ is a reduction vector of $\mathbf{f_2}$ (in short $\mathbf{f_2} \to \mathbf{f_1}$). In particular, $\mathbf{f_1}$ is a base vector.

Note that such a vector $\mathbf{f_{e-1}}$ is not uniquely determined. Depending on the choice of used basis representations (i.e., $B_1, \tilde{B}$ during the reduction), different reduction vectors arise, in particular shortcuts might or might not apply; see Lemma 3.

As worked out above, if there is a reduction vector $\mathbf{f_{e-1}}$ of $\mathbf{f_e}$, the following holds: $V(\mathbf{f_e}, \mathbb{A}(t_e))$ has a first-entry solution if and only if $V(\mathbf{f_{e-1}}, \mathbb{A})$ has a first-entry solution. Applying this observation iteratively, we end up at the the following lemma[3].

Lemma 2. *Let* $(\mathbb{G}(t_1)\ldots(t_e), \sigma)$ *be a* $\Pi\Sigma^*$*-extension of* $(\mathbb{G}, \sigma)$*; let* $\mathbf{f_i} \in \mathbb{G}(t_1)\ldots(t_i)^{n_i}$ *with* $s \leq i \leq e$ *be a chain of reduction vectors, i.e.,*

$$\mathbf{f_e} \to \mathbf{f_{e-1}} \to \cdots \to \mathbf{f_s}. \tag{29}$$

Then $V(\mathbf{f_e}, \mathbb{G}(t_1)\ldots(t_e))$ *has a first-entry solution iff* $V(\mathbf{f_s}, \mathbb{G}(t_1)\ldots(t_s))$ *has one.*

Lemma 3. *Let* $(\mathbb{G}(t_1)\ldots(t_e), \sigma)$ *be a* $\Pi\Sigma^*$*-ext. of* $(\mathbb{G}, \sigma)$*. Consider a chain of reduction vectors* (29) *with* $\mathbf{f_i} \in \mathbb{G}(t_1)\ldots(t_i)^{n_i}$ *for* $s \leq i \leq e$ *where* $\mathbf{f_s}$ *is a base vector. Moreover take another chain of reduction vectors* $\bar{\mathbf{f}}_{\mathbf{e}} \to \bar{\mathbf{f}}_{\mathbf{e-1}} \to \cdots \to \bar{\mathbf{f}}_{\bar{\mathbf{s}}}$ *with* $\bar{\mathbf{f}}_{\mathbf{i}} \in \mathbb{G}(t_1)\ldots(t_i)^{\bar{n}_i}$ *for* $\bar{s} \leq i \leq e$ *where* $\bar{\mathbf{f}}_{\bar{\mathbf{s}}}$ *is a base vector. Suppose that* $\mathbf{f_e} = \bar{\mathbf{f}}_{\mathbf{e}}$.

1. *For all* j *with* $e \geq j \geq \max(s, \bar{s})$ *we have that* $n_j = \bar{n}_j$ *and* $\bar{\mathbf{f}}_{\mathbf{j}} = \mathbf{T_j} \mathbf{f_j}$ *for first-row reduced invertible matrices* $\mathbf{T_j} \in \mathbb{K}^{n_j \times n_j}$.
2. *Moreover, if* $V(\mathbf{f_e}, \mathbb{G}(t_1)\ldots(t_e))$ *has no first-entry solution, then* $s = \bar{s}$.

---

[3] To execute the steps above one needs the property that $(\mathbb{G}, \sigma)$ is FPLDE-solvable. However, in the following we are only interested in the the exploration of the possible reduction processes with the corresponding reduction vectors without the need to calculate them explicitly. We therefore drop the FPLDE-solvability in the statements below.

Proof. (1) For $j = e$ the statement clearly holds. Now suppose that for j with $e \geq j > \max(s, \bar{s})$ we have $n_j = \bar{n}_j$ and that there is a first-row reduced invertible matrix $\mathbf{T_j} \in \mathbb{K}^{n_j \times n_j}$ such that $\bar{\mathbf{f}}_j = \mathbf{T_j}\, \mathbf{f}_j$. Following the reduction with $n := n_j$ as given in FirstEntryPT we calculate $\mathbf{r}, \bar{\mathbf{r}} \in \mathbb{A}(t)^n_{(frac)}$ and $\mathbf{p}, \bar{\mathbf{p}} \in \mathbb{A}[t]^n$ s.t. $\mathbf{f}_j = \mathbf{r} + \mathbf{p}$ and $\bar{\mathbf{f}}_j = \bar{\mathbf{r}} + \bar{\mathbf{p}}$. Note that $\bar{\mathbf{r}} = \mathbf{T_j}\,\mathbf{r}$ and $\bar{\mathbf{p}} = \mathbf{T_j}\,\mathbf{p}$. Now let $B_1 = \{(c_{i1}, \ldots, c_{in}, g_i)\}_{1 \leq i \leq \nu}$ be the derived basis of $V(\mathbf{r}, \mathbb{A}(t)_{(frac)})$ and take $\mathbf{C} \in \mathbb{K}^{\nu \times n}$ and $\mathbf{f}' \in \mathbb{A}[t]^\nu$ as given in Line 6. Similarly, let $\bar{B}_1$ be the derived basis of $V(\bar{\mathbf{r}}, \mathbb{A}(t)_{(frac)})$ and take the corresponding $\bar{\mathbf{C}} \in \mathbb{K}^{\bar{\nu} \times n}$ and $\bar{\mathbf{f}}' \in \mathbb{A}[t]^{\bar{\nu}}$. Note that $(c_1, \ldots, c_n, g) \in V(\mathbf{r}, \mathbb{A}(t))$ iff $((c_1, \ldots, c_n)\mathbf{T_j}^{-1}) \wedge g \in V(\bar{\mathbf{r}}, \mathbb{A}(t))$. From this one can conclude that the dimension of B_1 equals the dimension of $\bar{B}_1$, i.e., $\nu = \bar{\nu}$. By $\sigma(\bar{\mathbf{g}}) - \bar{\mathbf{g}} = \bar{\mathbf{C}}\,\bar{\mathbf{r}} = \bar{\mathbf{C}}\,\mathbf{T_j}\,\mathbf{r}$ and using the fact that B_1 is a basis of $V(\mathbf{r}, \mathbb{A}(t))$, it follows that $\bar{\mathbf{C}}\,\mathbf{T_j} = \mathbf{T}\,\mathbf{C}$ and $\bar{\mathbf{g}} = \mathbf{T}\,\mathbf{g}$ for some $\mathbf{T} \in \mathbb{K}^{\nu \times \nu}$. Therefore $\mathbf{T}$ is a basis transformation from B_1 to B_2 and is invertible. Moreover, since $\mathbf{C}$, $\bar{\mathbf{C}}$ and $\mathbf{T_j}$ are first-row reduced (by construction and assumption), $\mathbf{T}$ is first-row reduced. Finally, $\bar{\mathbf{f}}' = \bar{\mathbf{C}}\,\bar{\mathbf{p}} = \bar{\mathbf{C}}\,\mathbf{T_j}\,\mathbf{p} = \mathbf{T}\,\mathbf{C}\,\mathbf{p} = \mathbf{T}\,\mathbf{f}'$. In summary, $\mathbf{f}'$ and $\bar{\mathbf{f}}'$ differ by a first-row reduced invertible matrix. Now enter in the degree reduction as given in DegreeReductionFirstEntry. Note that the degree bound m is in both cases the same (see Theorem 4). Completely analogously it follows that one obtains for the corresponding $\mathbf{f}'$ and $\bar{\mathbf{f}}'$ the vectors ϕ and $\bar{\phi}$ which have the same length and which are the same up to the multiplication of a first-row reduced invertible matrix. Hence after m degree reductions steps one gets the reduction vectors $\mathbf{f_{j-1}}, \bar{\mathbf{f}}_{j-1}$ of $\mathbf{f_j}, \bar{\mathbf{f}}_j$, respectively: both have the length n_{j-1} and differ only by the multiplication of a first-entry reduced invertible matrix $\mathbf{T_{j-1}}$. This proves part (1).

(2) Now suppose that $V(\mathbf{f_e}, \mathbb{G}(t_1) \ldots (t_e))$ has no first-entry solution. Then we show that a $\mathbf{f_i}$ has a reduction vector in any reduction iff $\bar{\mathbf{f}}_i$ has a reduction vector in any reduction. This shows that $s = \bar{s}$. Namely, by Lemma 2 it follows that for all $s \leq i \leq e$, $V(\mathbf{f_i}, \mathbb{G}(t_1) \ldots (t_i))$ has no first-entry solution, and similarly for all $\bar{s} \leq i \leq e$, $V(\bar{\mathbf{f}}_i, \mathbb{G}(t_1) \ldots (t_i))$ has no first-entry solution. Thus we never enter in the shortcuts of Line 1 in Algorithm FirstEntryPT and of Line 1 in Algorithm DegreeReductionFirstEntry. Moreover by part (1), we exit in Line 5 of FirstEntryPT with the input vector $\mathbf{f_i}$ iff we exit with the input vector $\bar{\mathbf{f}}_i$. Furthermore, using the fact that also the input vectors of DegreeReductionFirstEntry are the same up to first-entry reduced invertible matrices, it follows that the exit in Line 5 within the degree reduction occurs iff it happens for both vectors. Hence for both chains to a base vector the length is the same. □

As a consequence, the length of the chain of reduction vectors is the same if there is no first-entry solution. However, different choices of basis representations in Line 4 of FirstEntryPT and Line 4 of DegreeReductionFirstEntry might deliver more compact reduction vectors (e.g., more zero-entries or less monomials in an entry) which in turn speeds up the underlying rational function arithmetic. In addition, if there is a first-entry solution, the shortcuts in Line 1 of Algorithm FirstEntryPT and in Line 1 of Algorithm DegreeReductionFirstEntry might occur differently.

In Sect. 6 the presented reduction algorithm will be slightly modified to find sum representations where the summand is expressed in the smallest possible subfield. In order to prove the correctness of this refined telescoping algorithm (see Theorem 6) we need the following technical properties.

Lemma 4. *Let $(\mathbb{G}(t_1)\ldots(t_e),\sigma)$ be a $\Pi\Sigma^*$-ext. of $(\mathbb{G},\sigma)$ and $\mathbf{f_i} \in \mathbb{G}(t_1)\ldots(t_i)^{n_i}$ with $s \leq i \leq e$ such that (29) where $\mathbf{f_s}$ is a base-vector. If there is no first-entry solution of $V(\mathbf{f_e},\mathbb{G}(t_1)\ldots(t_e))$, the first entry of $\mathbf{f_s}$ is from $\mathbb{G}(t_1)\ldots(t_s)\backslash \mathbb{G}(t_1)\ldots(t_{s-1})$.*

Proof. Suppose that $\mathbf{f_s}$ is the base vector with $s \geq 1$ and suppose that there is no first-entry solution of $V(\mathbf{f_e},\mathbb{G}(t_1)\ldots(t_e))$. By Lemma 2 it follows that there is no first-entry solution of $V(\mathbf{f_s},\mathbb{G}(t_1)\ldots(t_s))$. Moreover, suppose that the first entry a of $\mathbf{f_s}$ is from $\mathbb{G}(t_1)\ldots(t_{s-1})$. With these properties we will show that there is a particular choice of basis representations such that $\mathbf{f_s}$ has a reduction vector. Since by Lemma 3 the chain of reduction vectors to a basis vector has always the same length, there cannot be a reduction process such that $\mathbf{f_s}$ is a basis vector, a contradiction. We start the reduction following Algorithm FirstEntryPT$(\mathbf{f_s},\mathbb{G}(t_1)\ldots(t_s))$, i.e., $\mathbf{f_s} = \mathbf{f} = (f_1,\ldots,f_n)$. Note that $f_1 = a = 0$ would return a first-entry solution, which is not possible. Moreover, we are not in the ground field, since $s \geq 1$. Therefore, take $\mathbf{f} = \mathbf{r} + \mathbf{p}$ with $\mathbf{r} \in \mathbb{G}(t_1)\ldots(t_s)^n_{(frac)}$ and $\mathbf{p} \in \mathbb{G}(t_1)\ldots(t_{s-1})[t_s]^n$. Since $f_1 = a$ is free of t_s, the first entry of $\mathbf{r}$ is 0 and the first entry of $\mathbf{p}$ is a. Hence we can choose the basis B_1 with the vector $(1,0,\ldots,0)$ and thus in Line 6 we can take $\mathbf{C}$ where the first row is $(1,0,\ldots,0)$ and we can take $\mathbf{g}$ where the first entry is 0. Thus we obtain $\mathbf{f}' = \mathbf{C}\mathbf{p}$ where the first entry is $f_1' := a$. Now we proceed the reduction process as given in DegreeReductionFirstEntry$(\mathbf{f}',\mathbb{G}(t_1)\ldots(t_s))$. Since $f_1' = a \neq 0$ (see above), Line 1 is not considered. Moreover, if $m = 0$, we obtain a reduction vector $\mathbf{f}'$ of $\mathbf{f_s}$, a contradiction. Thus $m \geq 1$ and we enter Lines 3 and 4 of Algorithm DegreeReductionFirstEntry. Since $\mathrm{coeff}(f_1',m) = \mathrm{coeff}(a,m) = 0$, we can take $\tilde{B}$ with the element $(-1,0,\ldots,0)$. Since $\tilde{B}$ is non-empty, we carry out Line 3 and obtain ϕ where the first entry is a. Thus we enter DegreeReductionFirstEntry$(m - 1,\phi,\mathbb{G}(t_1)\ldots(t_s))$. Hence the degree reduction is applied iteratively to $m = 0$; a contradiction (see above). □

Lemma 5. *Let $(\mathbb{A}(t),\sigma)$ be a $\Pi\Sigma^*$-extension of $(\mathbb{G},\sigma)$ with $\mathbf{f} \in \mathbb{A}(t)^n$ and $h \in \mathbb{A}$.*

1. *If $\psi \in \mathbb{A}^\mu$ is a reduction-vector of $\mathbf{f}$, $\psi \wedge h$ is a reduction-vector of $\mathbf{f} \wedge h$.*
2. *If $V(\mathbf{f} \wedge h,\mathbb{A}(t))$ has a first-entry solution and $V(\mathbf{f},\mathbb{A}(t))$ has no first-entry solution, then there is a reduction-vector $\psi \in \mathbb{A}^\mu$ of $\mathbf{f}$.*

Proof. (1) Let $\psi \in \mathbb{A}^u$ be a reduction-vector of $\mathbf{f} \in \mathbb{A}[t]^n$ and suppose that $h \in \mathbb{A}$. Consider this reduction process which leads to ψ. First, write $\mathbf{f} = \mathbf{r} + \mathbf{p}$ where the entries of $\mathbf{r}$ are from $\mathbb{A}(t)_{(frac)}$ and the entries of $\mathbf{p}$ are from $\mathbb{A}[t]$. Let $B_1 = \{(c_{i1},\ldots,c_{in},g_i)\}_{1\leq i\leq\nu}$ be the basis of $V(\mathbf{r},\mathbb{A}(t)_{(frac)})$ which will lead us to the reduction-vector ψ. In particular, let $\mathbf{f}' \in \mathbb{A}[t]^\nu$ as defined in (21) using the basis B_1. From there we activate the degree reduction to reach the reduction vector ψ. Now consider $\mathbf{f} \wedge h$ and perform the following reduction. We get $\mathbf{f} \wedge h = (\mathbf{r} \wedge 0) + (\mathbf{p} \wedge h)$. Thus we can choose the basis

$B'_1 = \{(c_{i1}, \ldots, c_{in}, 0, g_i)\}_{1 \le i \le \nu} \cup \{(0, \ldots, 0, 1, 0)\}$, and using this particular basis we obtain the vector $\mathbf{f}' \wedge h$ to activate the degree reduction. Observe that for both cases $\mathbf{f}'$ and $\mathbf{f}' \wedge h$ the degree bound $m \ge 0$ is the same (see Theorem 4). Next, let $\tilde{\mathbf{f}}$ be the coefficient vector of $\mathbf{f}'$ w.r.t. the term t^m as defined in (24). Then $\tilde{\mathbf{f}} \wedge 0$ is the corresponding coefficient vector of $\mathbf{f}' \wedge h$. Moreover, let $\tilde{B} = \{(c_{i1}, \ldots, c_{i\nu}, g_i)\}_{1 \le i \le \lambda}$ be the basis of $V(\tilde{\mathbf{f}}, \mathbb{A})$ whose choice will bring us to the reduction-vector ψ. More precisely, we get the vector ϕ as defined in (25) and continue to compute a basis of $V(\phi, \mathbb{A}[t]_{m-1})$. Similarly, we can choose the basis $\tilde{B}' := \{(c_{i1}, \ldots, c_{i\nu}, 0, g_i)\}_{1 \le i \le \lambda} \cup \{0, \ldots, 0, -1, 0\}$ for $V(\tilde{\mathbf{f}} \wedge 0, \mathbb{A})$, obtain $\phi \wedge h$ and continue to compute a basis of $V(\phi \wedge h, \mathbb{A}[t]_{m-1})$. Applying this argument iteratively, we get the reduction-vector $\psi \wedge h$ of $\mathbf{f} \wedge h$ when applied to $V(\mathbf{f} \wedge h, \mathbb{A}(t))$.

(2) Now suppose that $V(\mathbf{f} \wedge h, \mathbb{A}(t))$ has a first-entry solution, but $V(\mathbf{f} \wedge h, \mathbb{A}(t))$ has not. Note that we do not exit in Line 1 of FirstEntryPT, otherwise also $V(\mathbf{f}, \mathbb{A}(t))$ has a first-entry solution. Since h is free of t, we can choose the basis B'_1 as in part (1). Note that we cannot exit in Line 5 by assumption of the existence of a first entry solution. Similarly, we cannot exist in Line 1 in Algorithm DegreeReductionFirstEntry. Thus we enter the degree reduction process with a degree bound $m \ge 0$ and a certain vector $\mathbf{f}' \wedge h$. Now activate the reduction for the vector $\mathbf{f}$. We therefore can take the basis B_1 as in part (1), get the same degree bound m (see Theorem 4), and start the degree reduction process with $\mathbf{f}'$. Again we cannot exist in Line 1 in Algorithm DegreeReductionFirstEntry (otherwise it would apply for the reduction vector $\mathbf{f}' \wedge h$). Thus in both cases we carry out the degree reduction. In particular, choosing the basis accordingly as in part (1), we can perform the degree reduction from m to $m-1$ simultaneously. Applying this argument iteratively, shows that the degree reductions of $\mathbf{f}' \wedge h$ and $\mathbf{f}'$ can be brought to reduction vectors $\psi \wedge h$ and ψ, respectively. $\square$

Applying Lemma 5.1 iteratively gives the following

Corollary 2. *Let* $(\mathbb{G}(t_1) \ldots (t_e), \sigma)$ *be a* $\Pi\Sigma^*$-*ext. of* $(\mathbb{G}, \sigma)$ *and* $\mathbf{f_i} \in \mathbb{G}(t_1) \ldots (t_i)^{n_i}$ *with* $n_i \ge 1$, *and* $h \in \mathbb{G}(t_1) \ldots (t_s)$. *If* (29) *is a chain of reduction vectors then also*

$$\mathbf{f_e} \wedge h \to \mathbf{f_{e-1}} \wedge h \to \cdots \to \mathbf{f_s} \wedge h. \tag{30}$$

6   Refined Parameterized Telescoping: Reduced Solutions

Finally, we turn to refined parameterized telescoping. A solution of (1) in the setting of difference rings (resp. fields) is called special solution.

Definition 9. *Let* $(\mathbb{A}, \sigma)$ *be a difference ring with constant field* $\mathbb{K} := \mathrm{const}_\sigma \mathbb{A}$ *and take* $\mathbf{f} = (f_1, \ldots, f_n) \in \mathbb{A}^n$. $(\psi, (c_1, \ldots, c_n, g))$ *is called a special* $(\mathbf{f}, \mathbb{A})$-*solution if* $\psi \in \mathbb{A}$ *and* $(c_1, \ldots, c_n, g) \in \mathbb{K}^n \times \mathbb{A}$ *with* $c_1 \ne 0$ *such that* $\sigma(g) - g + \psi = c_1 f_1 + \cdots + c_n f_n$.

Remark 3. A first-entry solution $(c_1, \ldots, c_n, g)$ of $V(\mathbf{f}, \mathbb{A})$ delivers a $(\mathbf{f}, \mathbb{A})$-special solution $(0, (c_1, \ldots, c_n, g))$.

More precisely, we look for a special solution $(\psi, (c_1, \ldots, c_n, g))$ which is reduced.

Definition 10. Let $(\mathbb{F}, \sigma)$ be a $\Pi\Sigma^*$-extension of $(\mathbb{G}, \sigma)$ with $\mathbb{F} = \mathbb{G}(t_1) \ldots (t_e)$ and $\mathbb{K} := \mathrm{const}_\sigma \mathbb{G}$; let $\mathbf{f} \in \mathbb{F}^n$. A special $(\mathbf{f}, \mathbb{F})$-solution $(\psi, (c_1, \ldots, c_n, g))$ is called reduced over $\mathbb{G}$ if one of the following holds:

1. $\psi = 0$.
2. $\psi \in \mathbb{G} \backslash \{0\}$ and there is no $(\mathbf{f}, \mathbb{F})$-special solution $(0, (\kappa_1, \ldots, \kappa_n, \gamma))$.
3. $\psi \in \mathbb{G}(t_1) \ldots (t_i) \backslash \mathbb{G}(t_1 \ldots, t_{i-1})$ for some $1 \leq i \leq e$ and there is no $(\mathbf{f}, \mathbb{F})$-special solution $(\phi, (\kappa_1, \ldots, \kappa_n, \gamma))$ with $\phi \in \mathbb{G}(t_1) \ldots (t_{i-1})$.

In other words, we are interested in a special solution $(\psi, (c_1, \ldots, c_n, g))$ where ψ is 0 or $\psi \neq 0$ lives in the smallest possible field, i.e., $\psi \in \mathbb{G}(t_1) \ldots (t_i)$ with minimal i.

Note that this problem contains refined telescoping (compare (3)): given $(\mathbb{F}, \sigma)$ and $f \in \mathbb{F}$, find $g, \psi \in \mathbb{F}$ with

$$\sigma(g) - g + \psi = f \tag{31}$$

such that $\psi = 0$ or $\psi \neq 0$ is taken from the smallest possible extension field of $\mathbb{G}$.

Example 12. Consider the $\Pi\Sigma^*$-field $(\mathbb{Q}(k)(p)(h), \sigma)$ as given in Example 1.4 and take f as given in (6). Then $(\psi, (-1/2, g))$ with $\psi = -\frac{(k+2)p}{2(k+1)}$ and $g = -\frac{1}{2}h^2 k(k+1)p + hp$ is a reduced solution, i.e., we have that (31) where ψ is from the smallest possible sub-field. Reinterpreting this solution in terms of $k!$ and H_k yields a solution of (3) with $f(k) = H_k^2(k^2 + 1)k!$. Thus summing (3) over k produces

$$\sum_{k=1}^m (k^2 + 1) \, k! \, H_k^2 = \sum_{k=1}^m \frac{(k+1)!}{k^2} + m \, (m+1)! \, H_m^2 - 2m! \, H_m. \tag{32}$$

Similarly, this enables one to refine creative telescoping [23, 36, 40].

Example 13. We aim at calculating a recurrence for $S(r) = \sum_{k=0}^{2r} f(r, k)$ with the summand $f(r, k) = (-1)^k \binom{2r}{k}^3 H_k$. Note that $P_r(k) = (-1)^k \binom{2r}{k}^3$ has the shift-behavior $P_r(k+1) = -\frac{(2r-k)^3}{(k+1)^3} P_r(k)$ and $P_{r+i}(k) = \prod_{j=1}^{2i} \frac{(2r+j)^3}{(2r-k+j)^3} P_r(k)$. To accomplish this task, take the rational function field $\mathbb{Q}(r)$, i.e., r is considered as a variable, and construct the $\Pi\Sigma^*$-field $(\mathbb{Q}(r)(k)(p)(h), \sigma)$ over $\mathbb{Q}(r)$ with $\sigma(k) = k + 1$, $\sigma(p) = -\frac{(2r-k)^3}{(k+1)^3} p$ and $\sigma(h) = h + \frac{1}{k+1}$. Thus p and h represent $P_r(k)$ and h, respectively. In particular, $f(r+i, k)$ with $i \in \mathbb{N}$ can be rephrased with $f_i = \left(\prod_{j=1}^{2i} \frac{(2r+j)^3}{(2r-k+j)^3} \right) p \, h \in \mathbb{Q}(r)(k)(p)(h)$.

First, we activate the classical creative telescoping approach with the function call FirstEntryPT$((f_0, f_1, \ldots, f_n), \mathbb{Q}(r)(k)(p)(h))$ for $n = 0, 1, 2, \ldots$ Eventually we find a first-entry solution for $n = 2$ which produces a summand recurrence of order two of the form (1) with $\psi = 0$. Summing this equation over k provides a recurrence of the form (5) with $\psi = 0$ whose coefficients c_0, c_1, c_2 are rather large.

Second, we execute $\mathsf{ReducedPT}((f_0, f_1, \ldots, f_n), \mathbb{Q}(r)(k)(p)(h))$. For $n = 1$ we obtain a reduced solution over $\mathbb{Q}(r)$, namely $(\frac{-108r^3 - 171r^2 - 86r - 13}{2(r+1)(2r+1)}p, (3(3r + 1)(3r+2), (r+1)^2, g))$ with $g = p(a + b\,h)$ where $a, b \in \mathbb{Q}(r)(k)$ are large rational functions. Rephrasing this equation in terms of the summation objects gives the summand recurrence (1). Finally, summing this equation over k produces the recurrence

$$(r+1)^2\, S(r+1) + 3(3r+1)(3r+2)\, S(r) = \frac{-108r^3 - 171r^2 - 86r - 13}{2(r+1)(2r+1)} \sum_{k=0}^{2r}(-1)^k \binom{2r}{k}^3;$$

note that by further simplification (e.g., using symbolic summation tools) one gets that $\sum_{k=0}^{2r}(-1)^k \binom{2r}{k}^3 = \frac{(-1)^r(3r)!}{(r!)^3}$. Finally, solving this recurrence yields/proves

$$\sum_{k=0}^{r}(-1)^k \binom{r}{k}^3 H_k = \left(H_r + 2H_{2r} - H_{3r}\right)\frac{(-1)^r(3r)!}{2(r!)^3}.$$

In order to solve this problem, we modify Algorithm $\mathsf{FirstEntryPT}$ as follows. Instead of returning a first-entry solution set, we return always a special solution: If we obtain a first-entry solution, it is returned without any changes; see Remark 3. If this is not possible, i.e., we obtain the base vector $(f_1, \ldots, f_n) \in \mathbb{A}(t_1) \ldots (t_e)^n$ in our reduction, we do not return $\{\}$, but we return the special solution $(f_1, (1, 0, \ldots, 0))$ which trivially holds: $\sigma(0) - 0 + f_1 = 1\,f_1 + 0\,f_2 + \cdots + 0\,f_n$.

Example 14. We apply this tactic for Example 12. More precisely, we refine the reduction described in Example 10. Namely, when we reach the base vector (27), we do not return the first-entry solution set $\{\}$, but we return the special solution $(\psi, (1, 0, 0))$ of $V(\phi_0, \mathbb{Q}(k)(p))$ with $\psi = \frac{(-k-2)p}{2(k+1)}$. Now we combine this solution with the derived sub-solutions of the degree reduction. This yields the special solutions $(\psi, (\frac{1}{2}, 0, hp))$, $(\psi, \{(-\frac{1}{2}, 0, -\frac{1}{2}h^2k(k+1)p + hp)\}$ and $(\psi, \{(-\frac{1}{2}, -\frac{1}{2}h^2k(k+1)p+hp)\}$ of the sets $V(\phi_1, \mathbb{Q}(k)(p)[h]_1)$, $V(\phi_2, \mathbb{Q}(k)(p)[h]_2)$ and $V(\phi_3, \mathbb{Q}(k)(p)[h]_3) = V((f), \mathbb{Q}(k)(p)(h))$, respectively. By Theorem 6 this solution is reduced over $\mathbb{Q}$.

With this mild modification we obtain the following algorithm and Theorem 6.

Algorithm $\mathsf{ReducedPT}(\mathbf{f}, \mathbb{F})$
Input: a $\Pi\Sigma^*$-extension $(\mathbb{F}, \sigma)$ of $(\mathbb{G}, \sigma)$ with $\mathbb{F} = \mathbb{G}(t_1) \ldots (t_e)$ and $\mathbb{K} := \mathrm{const}_\sigma \mathbb{G}$ which is FPLDE-solvable; $\mathbf{f} \in \mathbb{F}^n$.
Output: a special $(\mathbf{f}, \mathbb{F})$-solution $(\psi, (c_1, \ldots, c_n, g))$ being reduced over $\mathbb{G}$.

1. IF $f_1 = 0$, RETURN $(0, (1, 0, \ldots, 0))$.
2. IF $e = 0$, compute a first-entry solution set B of $V(\mathbf{f}, \mathbb{G})$.   (*Return a special solution*)

> IF $B = \{\}$, RETURN $(f_1, (1, 0, \ldots, 0))$
> ELSE take $B = \{h\}$ and RETURN $(0, h)$.

  Denote $\mathbb{A} := \mathbb{G}(t_1 \ldots, t_{e-1})$, $t := t_e$.

3. Compute $\mathbf{r} \in \mathbb{A}(t)^n_{(frac)}$ and $\mathbf{p} \in \mathbb{A}[t]^n$ such that $\mathbf{f} = \mathbf{r} + \mathbf{p}$.

4. Get a first-entry reduced basis of $V(\mathbf{r}, \mathbb{A}(t)_{(frac)})$, say $B_1 = \{(c_{i1}, \ldots, c_{in}, g_i)\}_{1 \le i \le \nu}$.

5. IF $B_1 = \{\}$ OR $c_{11} = \cdots = c_{\nu 1} = 0$,
        RETURN $(f_1, (1, 0, \ldots, 0))$.   (*NEW: return a special solution*)

6. Take $C := (c_{ij}) \in \mathbb{K}^{\nu \times n}$, $\mathbf{g} := (g_1, \ldots, g_\nu) \in \mathbb{A}(t)^\nu_{(frac)}$; define $\mathbf{f}' := (f'_1, \ldots, f'_\nu) = \mathbf{C}\,\mathbf{p} \in \mathbb{A}[t]^\nu$.

7. Define $m \in \mathbb{N} \cup \{-1\}$ as given in (23).

8. Get $B_2 := \mathsf{DegreeReductionReduced}(m, \mathbf{f}', \mathbb{A}(t))$

9. IF $B_2 = ()$, RETURN $(f_1, (1, 0, \ldots, 0))$.   (*NEW: return a special solution*)

10. Otherwise, let $B_2 = (\psi, (d_{11}, \ldots, d_{1\nu}, h_1))$ and take $\mathbf{D} := (d_{11}, \ldots, d_{1\nu}) \in \mathbb{K}^{1 \times \nu}$, $\mathbf{h} := (h_1) \in \mathbb{A}[t]^1_m$. Define $(e_{11}, \ldots, e_{1n}) := \mathbf{D}\,\mathbf{C} \in \mathbb{K}^{1 \times n}$ and $(q_1) := \mathbf{D}\,\mathbf{g} + \mathbf{h} \in \mathbb{A}(t)^1$.

11. RETURN $(\psi, (e_{11}, \ldots, e_{1n}, q_1))$.

Algorithm $\mathsf{DegreeReductionReduced}(m, \mathbf{f}, \mathbb{A}(t))$
Input: $m \in \mathbb{N} \cup \{-1\}$, a $\Pi\Sigma^*$-extension $(\mathbb{A}(t), \sigma)$ of $(\mathbb{G}, \sigma)$ with $\sigma(t) = \alpha\,t + \beta$ which is FPLDE-solvable; $\mathbf{f}' = (f'_1, \ldots, f'_\nu) \in \mathbb{A}[t]^\nu_m$.
Output: a special $(\mathbf{f}, \mathbb{A}[t])$-solution $(\psi, (c_1, \ldots, c_n, g))$ with $\psi \in \mathbb{A}$, $g \in \mathbb{A}[t]$, $c_i \in \mathbb{K}$ being reduced over $\mathbb{G}$. If this is not possible, the output is $()$.

1. IF $f'_1 = 0$, RETURN $(0, (1, 0, \ldots, 0))$. (*Note that here we cover also the case $m = -1$*)

2. IF $m = 0$, RETURN $\mathsf{ReducedPT}(\mathbf{f}', \mathbb{A})$.

3. Define $\tilde{\mathbf{f}} \in \mathbb{A}^\nu$ as in (24).

4. Get $\tilde{B} := \begin{cases} \mathsf{SolvePTRat}(\tilde{\mathbf{f}}, \mathbb{A}) & \text{if } \alpha = 1 \\ \mathsf{SolveFPLDE}((-1, \alpha^m), \tilde{\mathbf{f}}, \mathbb{A}) & \text{if } \alpha \neq 1 \end{cases}$,
   say $\tilde{B} = \{(c_{i1}, \ldots, c_{i\nu}, g_i)\}_{1 \le i \le \lambda}$.
   IF the bases is not reduced, reduce it.

5. IF $\tilde{B} = \{\}$ OR $c_{11} = \cdots = c_{\lambda 1} = 0$, RETURN $()$.

6. Take $\mathbf{C} := (c_{i,j}) \in \mathbb{K}^{\lambda \times \nu}$, $\mathbf{g} := (g_1, \ldots, g_\lambda) \in \mathbb{A}^\lambda$, and let $\phi \in \mathbb{A}[t]^\lambda_{m-1}$ as in (25).

7. Get $G := \mathsf{DegreeReductionReduced}(m - 1, \phi, \mathbb{A}(t))$. IF $G = ()$, RETURN $()$.

8. Let $G = (\psi, (d_{11}, \ldots, d_{1\lambda}, h_1)) \subseteq \mathbb{A} \times \mathbb{K}^\lambda \times \mathbb{A}[t]_{m-1}$; take $D := (d_{11}, \ldots, d_{1\lambda}) \in \mathbb{K}^{1 \times \lambda}$, $\mathbf{h} := (h_1) \in \mathbb{A}[t]^1_{m-1}$, and define $(e_{11}, \ldots, e_{1\nu}) := \mathbf{D}\,\mathbf{C} \in \mathbb{K}^{1 \times \nu}$ and $(p_1) := \mathbf{D}\,\mathbf{g}\,t^m + \mathbf{h} \in \mathbb{A}[t]^1_m$.

9. RETURN $(\psi, (e_{11}, \ldots, e_{1\nu}, p_1))$.

Theorem 6. Let $(\mathbb{F}, \sigma)$ be a $\Pi\Sigma^*$-extension of $(\mathbb{G}, \sigma)$ which is FPLDE-solvable, let $\mathbf{f} \in \mathbb{F}^n$. Then one can compute a special $(\mathbf{f}, \mathbb{F})$-solution being reduced over $\mathbb{G}$.

Proof. By construction Algorithm ReducedPT returns a special $(\mathbf{f}, \mathbb{F})$-solution, say $(\psi, (c_1, \ldots, c_n, g))$. In particular, it returns a first-entry solution of $V(\mathbf{f}, \mathbb{F})$ if it exists. Consequently, if $\psi = 0$, the special solution is reduced over $\mathbb{G}$. Moreover, if $\psi \in \mathbb{G}^*$, $\psi = 0$ is not possible, and thus the special solution is again reduced over $\mathbb{F}$. Finally suppose that $\psi \in \mathbb{G}(t_1) \ldots (t_s) \backslash \mathbb{G}(t_1) \ldots (t_{s-1})$. This implies that there does not exist a first-entry solution of $V(\mathbf{f}, \mathbb{F})$. Thus by Lemma 4 there is a chain of reduction-vectors (29) with $\mathbf{f_e} = \mathbf{f}$ where $\mathbf{f_s}$ is a base-vector and where the first entry in $\mathbf{f_s}$ is ψ. Now suppose that there is a special $(\mathbf{f}, \mathbb{F})$-solution $(h, (\kappa_1, \ldots, \kappa_n, \gamma))$ with $h \in \mathbb{G}(t_1) \ldots (t_{s-1})$ and $\kappa_1 \neq 0$. By Corollary 2 it follows that there is also the chain (30) of reduction-vectors. However, we have that $\sigma(\gamma) - \gamma = \kappa_1 f_1 + \cdots + \kappa_n f_n - h$. Consequently, there is a first-entry solution $(\kappa_1, \ldots, \kappa_n, -1, \gamma)$ of $V(\mathbf{f} \wedge h, \mathbb{F})$. Thus by Lemma 2 $V(\mathbf{f_s} \wedge h, \mathbb{G}(t_1) \ldots (t_s))$ has a first-entry solution. Moreover, since $V(\mathbf{f}, \mathbb{F})$ has no first-entry solution, $V(\mathbf{f_s}, \mathbb{G}(t_1) \ldots (t_s))$ has no first-entry solution by Lemma 2. Therefore we can apply Lemma 5.2 and it follows that there is a reduction-vector $\mathbf{f_{s-1}}$ of $\mathbf{f_s}$; a contradiction that $\mathbf{f_s}$ is a base-vector. $\qquad \square$

Remark 4. **(1)** Restricting to a polynomial Σ^*-extension $(\mathbb{F}, \sigma)$ of $(\mathbb{G}, \sigma)$ in which one can solve Problem FPLDE, the algorithms can be simplified further (compare SolvePTPoly and DegreeReductionPoly).
(2) Also the improvements of Remark 2 can be applied to the algorithms from above.
(3) Furthermore, Algorithm ReducedPT can be refined further (and is available in Sigma) as follows. In Line 5 of Algorithm DegreeReductionReduced, one does not return (), but returns $(f_1', \{(1, 0, \ldots, 0)\})$. Then one can show that one obtains a solution where the degree of the extension t is minimal. This yields an improved version of the algorithm introduced in [57, Algorithm 4.2]. Besides that, one can modify Line 5 in Algorithm ReducedPT to find optimal degree representations of t in the numerators and denominators using the ideas of [57, Section 5].

7   A Constructive Version of Karr's Structural Theorem

In [27,28] reduced $\Pi \Sigma^*$-fields are introduced to derive a discrete analogue of Liouville's Theorem [34]. More generally, for $\Pi \Sigma^*$-extensions we need the following

Definition 11. *A $\Pi \Sigma^*$-extension* $(\mathbb{G}(t_1) \ldots (t_e), \sigma)$ *of* $(\mathbb{G}, \sigma)$ *is called* reduced *if for any Σ^*-extension t_i ($1 \leq i \leq e$) with $f := \sigma(t_i) - t_i \in \mathbb{G}(t_1) \ldots (t_{i-1}) \backslash \mathbb{G}$ the following property holds: there do not exist a $g \in \mathbb{G}(t_1) \ldots (t_{i-1})$ and an $\psi \in \mathbb{G}$ with* (31).

With Algorithm ReducedPT one can immediately check if a given $\Pi \Sigma^*$-extension $(\mathbb{G}(t_1) \ldots (t_e), \sigma)$ of $(\mathbb{G}(t_1) \ldots (t_{e-1}), \sigma)$ with $f := \sigma(t_e) - t_e$ is reduced over $\mathbb{G}$. If $f \in \mathbb{G}$, it is reduced. Otherwise, let i such that $f \in \mathbb{G}(t_1) \ldots (t_i) \backslash \mathbb{G}(t_1) \ldots (t_{i-1})$. Now calculate a reduced solution $(\psi, (1, g))$ with ReducedPT$((f), \mathbb{G}(t_1) \ldots (t_{i-1}))$. Then the extension is not reduced iff $\psi \in \mathbb{G}(t_1) \ldots (t_{i-1})$. In this case it can be transformed to a reduced one using the following lemma; see [62, Lemma 21].

Lemma 6. *Let $(\mathbb{F}(t), \sigma)$ be a Σ^*-extension of $(\mathbb{F}, \sigma)$ with $\sigma(t) = t + f$, and let $\psi, g \in \mathbb{F}$ such that (31) holds. Then there is a Σ^*-extension $(\mathbb{F}(s), \sigma)$ of $(\mathbb{F}, \sigma)$ with $\sigma(s) = s + \psi$ together with an $\mathbb{F}$-isomorphism $\tau : \mathbb{F}(t) \to \mathbb{F}(s)$ with $\tau(t) = s + g$.*

Namely, we can construct the Σ^*-extension $(\mathbb{F}(s), \sigma)$ of $(\mathbb{F}, \sigma)$ with $\sigma(s) = s + \psi$ together with the $\mathbb{F}$-isomorphism $\tau : \mathbb{F}(t) \to \mathbb{F}(s)$ defined by $\tau(t) = s + g$.

Example 15. Take the $\Pi\Sigma^*$-field $(\mathbb{Q}(k)(p)(h), \sigma)$ from Example 1.4 and consider the Σ^*-extension $(\mathbb{Q}(k)(p)(h)(t), \sigma)$ of $(\mathbb{Q}(k)(p)(h), \sigma)$ with $\sigma(t) = t + f$ where f is given in (2). We get an improved field extension by using the reduced solution $(\psi, (-1/2, g))$ from Example 12. Here we take the equivalent reduced solution $(-2\psi, (1, -2g))$. By Lemma 6 we can construct the Σ^*-extension $(\mathbb{Q}(k)(p)(h)(s), \sigma)$ of $(\mathbb{Q}(k)(p)(h), \sigma)$ with $\sigma(s) = s + \frac{(k+2)p}{(k+1)}$ and get the $\mathbb{Q}(k)(p)(h)$-isomorphism $\tau : \mathbb{Q}(k)(p)(h)(t) \to \mathbb{Q}(k)(p)(h)(s)$ with $\tau(t) = s + k(k+1)p\,h^2 - 2\,h\,p$. Note that this map is also reflected in (32).

Applying this transformation iteratively (see [62, Algorithm 1]) enables one to transform any $\Pi\Sigma^*$-extension to a reduced version.

Theorem 7. *For any $\Pi\Sigma^*$-extension $(\mathbb{H}, \sigma)$ of $(\mathbb{G}, \sigma)$ there is a reduced $\Pi\Sigma^*$-extension $(\mathbb{F}, \sigma)$ of $(\mathbb{G}, \sigma)$ and an $\mathbb{F}$-isomorphism $\tau : \mathbb{H} \to \mathbb{F}$.*

1. *Such $(\mathbb{F}, \sigma)$ and τ can be given explicitly, if $(\mathbb{F}, \sigma)$ is FPDLE-computable.*
2. *If $(\mathbb{H}, \sigma)$ is a polynomial Σ^*-extension of $(\mathbb{G}, \sigma)$ and one can solve Problem PT in $(\mathbb{G}, \sigma)$, such a polynomial Σ^*-extension $(\mathbb{F}, \sigma)$ of $(\mathbb{G}, \sigma)$ and τ can be calculated.*

Given such a reduced $\Pi\Sigma^*$-extension, we are in the position to apply the following theorem. For a proof in the context of reduced $\Pi\Sigma$-fields see [28, Result, p. 315], and in the context of reduced $\Pi\Sigma^*$-extensions as above see [48, Theorem 4.2.1].

Theorem 8 (Karr's structural theorem). *Let $(\mathbb{F}, \sigma)$ be a reduced $\Pi\Sigma^*$-extension of $(\mathbb{G}, \sigma)$ with $\mathbb{F} = \mathbb{G}(t_1) \ldots (t_e)$ and $\sigma(t_i) = \alpha_i t_i + \beta_i$, and define $S := \{1 \le i \le e | \sigma(t_i) - t_i \in \mathbb{G}\}$. Let $f \in \mathbb{G}$. Then for any $g \in \mathbb{F}$ with $\sigma(g) - g = f$ we have that*

$$g = w + \sum_{i \in S} c_i\, t_i, \quad \text{with } c_i \in \mathrm{const}_\sigma \mathbb{G} \text{ and } w \in \mathbb{G}. \tag{33}$$

This theorem can simplify the calculation of a solution $g \in \mathbb{G}(t_1) \ldots (t_e)$ of $\sigma(g) - g = f \in \mathbb{F}$ dramatically: Let $S = \{i_1, \ldots, i_r\}$ and $f \in \mathbb{A}$. Then calculate a first-entry solution set of $V((f, \beta_{i_1}, \ldots, \beta_{i_r}), \mathbb{G})$. If it is the empty set, there does not exist such a g by Theorem 8. Otherwise, the computed first-entry solution $(c, c_1, \ldots, c_r, w)$ with $c \ne 0$ yields the desired solution $g = \frac{1}{c}(w - c_1 t_{i_1} - \cdots - c_r t_{i_r})$. Similarly, parameterized telescoping can be simplified using Karr's structural theorem.

8 Conclusion

We started with a general framework to solve parameterized first-order linear difference equations in a $\Pi\Sigma^*$-extension over a ground field that possesses certain computational properties. From there various refinements are derived yielding fast parameterized telescoping algorithms. In particular, a new algorithmic variant has been developed to compute reduced solutions efficiently. This enables one to simplify indefinite sums by telescoping using an extra sum whose summand is expressed in the smallest possible sub-field. In addition, recurrences can be produced using such sum extensions yielding shorter recurrences than naive creative telescoping. Finally, this algorithm enables one to construct reduced $\Pi\Sigma^*$-extensions and to exploit structural properties such as Theorem 8. Exactly such results give rise to even more efficient strategies to calculate parameterized telescoping solutions.

In [53,59] depth-optimal $\Pi\Sigma^*$-extension have been introduced which refine the notion of reduced $\Pi\Sigma^*$-extensions and which give even stronger structural results than Theorem 8; see [62]. In particular one can search for such an improved $\Pi\Sigma^*$-field that leads to sum representations with minimal nesting depth [60]. As for reduced $\Pi\Sigma^*$-extensions this construction is algorithmically by an improved telescoping algorithm. This has been accomplished by refining Algorithm SolveP-TRat in [59]. However, with the technology presented in this article it is possible to perform this refinement starting with Algorithm FirstEntryPT (instead of SolveP-TRat). This new variant (similarly as we obtained a new algorithm for reduced $\Pi\Sigma^*$-extensions) is meanwhile implemented in Sigma and gives currently the best algorithm to solve parameterized linear difference equations in large $\Pi\Sigma^*$-fields.

The algorithms presented in this article and the refinements for depth-optimal $\Pi\Sigma^*$-extension are heavily exploited in ongoing calculations coming from QCD (Quantum ChromoDynamics). In these computations highly complicated Feynman integrals [1,2,14,15] are transformed to multi-sums [16] and are simplified in terms of indefinite nested product-sum expressions using the packages introduced in [63].

Acknowledgments. I would like to thank the referee for the very careful reading and the valuable suggestions.

References

1. Ablinger, J., Blümlein, J., Hasselhuhn, A., Klein, S., Schneider, C., Wissbrock, F.: Massive 3-loop ladder diagrams for quarkonic local operator matrix elements. Nucl. Phys. B. **864**, 52–84 (2012). ArXiv:1206.2252v1 [hep-ph]
2. Ablinger, J., Blümlein, J., Klein, S., Schneider, C., Wissbrock, F.: The $O(\alpha_s^3)$ massive operator matrix elements of $O(n_f)$ for the structure function $F_2(x, Q^2)$ and transversity. Nucl. Phys. B. **844**, 26–54 (2011). ArXiv:1008.3347 [hep-ph]
3. Ablinger, J., Blümlein, J., Schneider, C.: Harmonic sums and polylogarithms generated by cyclotomic polynomials. J. Math. Phys. **52**(10), 1–52 (2011). arXiv:1007.0375 [hep-ph]

4. Ablinger, J., Blümlein, J., Schneider, C.: Analytic and algorithmic aspects of generalized harmonic sums and polylogarithms. J. Math. Phys. **54**(8), 1–74 (2013). ArXiv:1302.0378 [math-ph]
5. Abramov, S.A.: On the summation of rational functions. Zh. vychisl. mat. Fiz. **11**, 1071–1074 (1971)
6. Abramov, S.A.: The rational component of the solution of a first-order linear recurrence relation with a rational right-hand side. U.S.S.R. Comput. Math. Math. Phys. **15**, 216–221 (1975). Transl. from Zh. vychisl. mat. mat. fiz. 15, 1035–1039 (1975)
7. Abramov, S.A.: Rational solutions of linear differential and difference equations with polynomial coefficients. U.S.S.R. Comput. Math. Math. Phys. **29**(6), 7–12 (1989)
8. Abramov, S.A., Bronstein, M., Petkovšek, M., Schneider, C.: (2015, In preparation)
9. Abramov, S.A., Petkovšek, M.: D'Alembertian solutions of linear differential and difference equations. In: von zur Gathen, J. (ed.) Proceedings of ISSAC 1994, pp. 169–174. ACM Press (1994)
10. Abramov, S.A., Petkovšek, M.: Rational normal forms and minimal decompositions of hypergeometric terms. J. Symbolic Comput. **33**(5), 521–543 (2002)
11. Abramov, S.A., Petkovšek, M.: Polynomial ring automorphisms, rational (w, σ)-canonical forms, and the assignment problem. J. Symbolic Comput. **45**(6), 684–708 (2010)
12. Andrews, G.E., Paule, P., Schneider, C.: Plane partitions VI: stembridge's TSPP theorem. Adv. Appl. Math. **34**(4), 709–739 (2005). Special Issue Dedicated to Dr. David P. Robbins. Edited by D. Bressoud
13. Bauer, A., Petkovšek, M.: Multibasic and mixed hypergeometric Gosper-type algorithms. J. Symbolic Comput. **28**(4–5), 711–736 (1999)
14. Bierenbaum, I., Blümlein, J., Klein, S., Schneider, C.: Two-loop massive operator matrix elements for unpolarized heavy flavor production to $O(\epsilon)$. Nucl. Phys. B. **803**(1–2), 1–41 (2008). arXiv:hep-ph/0803.0273
15. Blümlein, J., Hasselhuhn, A., Klein, S., Schneider, C.: The $O(\alpha_s^3 n_f T_F^2 C_{A,F})$ contributions to the gluonic massive operator matrix elements. Nucl. Phys. B. **866**, 196–211 (2013)
16. Blümlein, J., Klein, S., Schneider, C., Stan, F.: A symbolic summation approach to feynman integral calculus. J. Symbolic Comput. **47**, 1267–1289 (2012)
17. Blümlein, J., Kurth, S.: Harmonic sums and Mellin transforms up to two-loop order. Phys. Rev. **D60**, 014018 (1999)
18. Bronstein, M.: On solutions of linear ordinary difference equations in their coefficient field. J. Symbolic Comput. **29**(6), 841–877 (2000)
19. Chen, S., Jaroschek, M., Kauers, M., Singer, M.F.: Desingularization explains order-degree curves for ore operators. In: Kauers, M. (ed.) Proceedings of ISSAC 2013, pp. 157–164 (2013)
20. Chen, S., Kauers, M.: Order-degree curves for hypergeometric creative telescoping. In: van der Hoeven, J., van Hoeij, M. (eds.) Proceedings of ISSAC 2012, pp. 122–129 (2012)
21. Chyzak, F.: An extension of Zeilberger's fast algorithm to general holonomic functions. Discrete Math. **217**, 115–134 (2000)
22. Cohn, R.M.: Difference Algebra. Interscience Publishers, John Wiley & Sons, New York (1965)
23. Driver, K., Prodinger, H., Schneider, C., Weideman, J.A.C.: Padé approximations to the logarithm III: alternative methods and additional results. Ramanujan J. **12**(3), 299–314 (2006)

24. Eröcal, B.: Algebraic extensions for summation in finite terms. Ph.D. thesis, RISC, Johannes Kepler University, Linz (2011)
25. Gosper, R.W.: Decision procedures for indefinite hypergeometric summation. Proc. Nat. Acad. Sci. U.S.A. **75**, 40–42 (1978)
26. Hendriks, P.A., Singer, M.F.: Solving difference equations in finite terms. J. Symbolic Comput. **27**(3), 239–259 (1999)
27. Karr, M.: Summation in finite terms. J. ACM **28**, 305–350 (1981)
28. Karr, M.: Theory of summation in finite terms. J. Symbolic Comput. **1**, 303–315 (1985)
29. Kauers, M., Schneider, C.: Application of unspecified sequences in symbolic summation. In: Dumas, J. (ed.) Proceedings of ISSAC 2006, pp. 177–183. ACM Press (2006)
30. Kauers, M., Schneider, C.: Indefinite summation with unspecified summands. Discrete Math. **306**(17), 2021–2140 (2006)
31. Kauers, M., Schneider, C.: Symbolic summation with radical expressions. In: Brown, C. (ed.) Proceedings of ISSAC 2007, pp. 219–226 (2007)
32. Koornwinder, T.H.: On Zeilberger's algorithm and its q-analogue. J. Comp. Appl. Math. **48**, 91–111 (1993)
33. Koutschan, C.: Creative telescoping for holonomic functions. In: Schneider, C., Blümlein, J. (eds.) Computer Algebra in Quantum Field Theory: Integration, Summation and Special Functions. Texts and Monographs in Symbolic Computation, pp. 171–194. Springer (2013). ArXiv:1307.4554 [cs.SC]
34. Liouville, J.: Mémoire sur l'intégration d'une classe de fonctions transcendantes. J. Reine Angew. Math. **13**, 93–118 (1835)
35. Moch, S.O., Uwer, P., Weinzierl, S.: Nested sums, expansion of transcendental functions, and multiscale multiloop integrals. J. Math. Phys. **6**, 3363–3386 (2002)
36. Osburn, R., Schneider, C.: Gaussian hypergeometric series and extensions of supercongruences. Math. Comp. **78**(265), 275–292 (2009)
37. Paule, P.: Greatest factorial factorization and symbolic summation. J. Symbolic Comput. **20**(3), 235–268 (1995)
38. Paule, P.: Contiguous relations and creative telescoping, p. 33. Unpublished manuscript (2001)
39. Paule, P., Riese, A.: A Mathematica q-analogue of Zeilberger's algorithm based on an algebraically motivated aproach to q-hypergeometric telescoping. In: Ismail, M., Rahman, M. (eds.) Special Functions, q-Series and Related Topics, vol. 14, pp. 179–210. AMS (1997)
40. Paule, P., Schneider, C.: Computer proofs of a new family of harmonic number identities. Adv. Appl. Math. **31**(2), 359–378 (2003)
41. Paule, P., Schorn, M.: A Mathematica version of Zeilberger's algorithm for proving binomial coefficient identities. J. Symbolic Comput. **20**(5–6), 673–698 (1995)
42. Koutschan, C., Paule, P., Suslov, S.K.: Relativistic coulomb integrals and Zeilberger's holonomic systems approach II. In: Barkatou, M., Cluzeau, T., Regensburger, G., Rosenkranz, M. (eds.) AADIOS 2012. LNCS, vol. 8372, pp. 135–145. Springer, Heidelberg (2014)
43. Petkovšek, M.: Hypergeometric solutions of linear recurrences with polynomial coefficients. J. Symbolic Comput. **14**(2–3), 243–264 (1992)
44. Petkovšek, M., Wilf, H.S., Zeilberger, D.: $A = B$. A. K. Peters, Wellesley (1996)
45. Petkovšek, M., Zakrajšek, H.: Solving linear recurrence equations with polynomial coefficients. In: Schneider, C., Blümlein, J. (eds.) Computer Algebra in Quantum Field Theory: Integration, Summation and Special Functions. Texts and Monographs in Symbolic Computation, pp. 259–284. Springer (2013)

46. Pirastu, R., Strehl, V.: Rational summation and Gosper-Petkovšek representation. J. Symbolic Comput. **20**(5–6), 617–635 (1995)
47. Prodinger, H., Schneider, C., Wagner, S.: Unfair permutations. Europ. J. Comb. **32**, 1282–1298 (2011)
48. Schneider, C.: Symbolic summation in difference fields. Technical report 01–17, RISC-Linz, J. Kepler University (2001). Ph.D. Thesis
49. Schneider, C.: Solving parameterized linear difference equations in $\Pi\Sigma$-fields. SFB-Report 02–19, J. Kepler University, Linz (2002)
50. Schneider, C.: A collection of denominator bounds to solve parameterized linear difference equations in $\Pi\Sigma$-extensions. An. Univ. Timişoara Ser. Mat.-Inform. **42**(2), 163–179 (2004). Extended version of Proceedings of SYNASC 2004
51. Schneider, C.: Symbolic summation with single-nested sum extensions. In: Gutierrez, J. (ed.) Proceedings of ISSAC 2004, pp. 282–289. ACM Press (2004)
52. Schneider, C.: Degree bounds to find polynomial solutions of parameterized linear difference equations in $\Pi\Sigma$-fields. Appl. Algebra Engrg. Comm. Comput. **16**(1), 1–32 (2005)
53. Schneider, C.: Finding telescopers with minimal depth for indefinite nested sum and product expressions. In: Kauers, M. (ed.) Proceedings of ISSAC 2005, pp. 285–292. ACM (2005)
54. Schneider, C.: A new Sigma approach to multi-summation. Adv. Appl. Math. **34**(4), 740–767 (2005)
55. Schneider, C.: Product representations in $\Pi\Sigma$-fields. Ann. Comb. **9**(1), 75–99 (2005)
56. Schneider, C.: Solving parameterized linear difference equations in terms of indefinite nested sums and products. J. Differ. Eqn. Appl. **11**(9), 799–821 (2005)
57. Schneider, C.: Simplifying sums in $\Pi\Sigma$-Extensions. J. Algebra Appl. **6**(3), 415–441 (2007)
58. Schneider, C.: Symbolic summation assists combinatorics. Sém. Lothar. Combin. **56**, 1–36 (2007). Article B56b
59. Schneider, C.: A refined difference field theory for symbolic summation. J. Symbolic Comput. **43**(9), 611–644 (2008). arXiv:0808.2543v1
60. Schneider, C.: A symbolic summation approach to find optimal nested sum representations. In: Carey, A., Ellwood, D., Paycha, S., Rosenberg, S. (eds.) Motives, Quantum Field Theory, and Pseudodifferential Operators, Clay Mathematics Proceedings, vol. 12, pp. 285–308. Amer. Math. Soc. (2010). ArXiv:0808.2543
61. Schneider, C.: Parameterized telescoping proves algebraic independence of sums. Ann. Comb. **14**(4), 533–552 (2010). arXiv:0808.2596
62. Schneider, C.: Structural theorems for symbolic summation. Appl. Algebra Engrg. Comm. Comput. **21**(1), 1–32 (2010)
63. Schneider, C.: Modern summation methods for loop integrals in quantum field theory: The packages Sigma, EvaluateMultiSums and SumProduction. In: Proceedings of ACAT 2013, To appear in J. Phys.: Conf. Ser., pp. 1–17 (2014). ArXiv:1310.0160 [cs.SC]
64. Schneider, C.: Simplifying multiple sums in difference fields. In: Schneider, C., Blümlein, J. (eds.) Computer Algebra in Quantum Field Theory: Integration, Summation and Special Functions. Texts and Monographs in Symbolic Computation, pp. 325–360. Springer (2013). ArXiv:1304.4134 [cs.SC]
65. Vermaseren, J.A.M.: Harmonic sums, Mellin transforms and integrals. Int. J. Mod. Phys. **A14**, 2037–2976 (1999)
66. Zeilberger, D.: The method of creative telescoping. J. Symbolic Comput. **11**, 195–204 (1991)

Some Results on the Surjectivity of Surface Parametrizations

J. Rafael Sendra[1], David Sevilla[2](✉), and Carlos Villarino[1]

[1] Department of Physics and Mathematics, University of Alcalá,
Ap. Correos 20, 28871 Alcalá de Henares, Madrid, Spain
{rafael.sendra,carlos.villarino}@uah.es
http://www2.uah.es/rsendra
[2] University Center of Mérida, Av. Santa Teresa de Jornet 38,
06800 Mérida, Badajoz, Spain
sevillad@unex.es
http://www.davidsevilla.com

Abstract. This paper deals with the decision problem of the surjectivity of a rational surface parametrization. We give sufficient conditions for a parametrization to be surjective, and we describe different families of parametrizations that satisfy these criteria. In addition, we consider the problem of computing a superset of the points not covered by the parametrization. In this context, we report on the case of parametrizations without projective base points and we analyze the particular case of rational ruled surfaces.

Keywords: Rational algebraic surface · Normality · Ruled surfaces · Base points

1 Introduction

Often, when motivating the applicability of rational surfaces, one claims that providing a different representation to the implicit one (for example a rational parametrization representation) is very useful in many particular applied problems, for instance in computer aided geometric desing (see [8,9]). Examples of this are plotting images in computer screens, curve or surface integration, intersection of varieties, etc. Nevertheless, even though this is true, one must add that, for the parametrization to be feasibly useful, it must satisfy certain properties. For instance, if we are using a parametrization for plotting, and the parametrization is given over the complex numbers, how do we know whether the variety is indeed real? What parameter values must be put into the parametrization in order to get real points? Of course, this is not a problem if the input parametrization is given with real coefficients; this problem has been studied in [1,2,12].

A similar phenomenon can arise when computing the intersection of two surfaces where one of them is given parametrically: how do we know that the parametrization is not missing essential information for the problem we are dealing

© Springer International Publishing Switzerland 2015
J. Gutierrez et al. (Eds.): Computer Algebra and Polynomials, LNCS 8942, pp. 192–203, 2015.
DOI: 10.1007/978-3-319-15081-9_11

with? This is, indeed, the central topic of this paper: surjectivity (also called normality) of rational surface parametrizations. Let us illustrate by an example the situation described above.

Example 1. We consider the surfaces S_1 and S_2 defined by

$$F_1(x, y, z) = x^2 y^2 z - 2xyz - x^2 + z$$

and

$$F_2(x, y, z) = -x^3 y^2 z + x^2 y^2 z + 2x^2 yz + x^3 - 2xyz - x^2 + z$$

respectively. If we compute a Gröbner basis of (F_1, F_2) we get $\{z^2, xz, x^2 - z\}$, and so $S_1 \cap S_2$ consists of the line $z = x = 0$. However, if we take the rational parametrization

$$\mathcal{P}(s, t) = \left(\frac{1}{t}, s + t, \frac{1}{s^2} \right)$$

of S_1 and we substitute it in F_2 we get $1/(ts^2)$ and hence no intersection point. Nevertheless, if we observe that F_1 is linear in z we can also consider the parametrization

$$\mathcal{Q}(s, t) = \left(s, t, \frac{s^2}{s^2 t^2 - 2st + 1} \right)$$

of S_1. Substituting this new parametrization in S_2 we get

$$\frac{s^3}{s^2 t^2 - 2st + 1}.$$

So the intersection of S_1 and S_2 is $\mathcal{Q}(0, t) = (0, t, 0)$, the expected line. The difference between $\mathcal{P}(s, t)$ and $\mathcal{Q}(s, t)$ is that the second one is surjective. See Remark 3 in relation to the parametrization $\mathcal{Q}$.

The problem of deciding the normality of a surface parametrization (i.e. whether it is surjective over the algebraic surface) can be attacked by means of elimination theory techniques, but a direct application of these can be too complex in terms of time. Moreover, deciding whether a given rational surface admits a normal parametrization is, at least to our knowledge, an open problem. In this article we approach the problem by, on one hand, giving sufficient conditions for the normality of an input rational surface parametrization; and, on the other, for general rational surfaces and under certain conditions, as well as for ruled surfaces, we describe explicitly a convenient superset of the complementary of the image in the surface.

In situations where we cannot assure normality, an alternative approach, proposed in [15], is to cover the algebraic surface with a finite number of affine parametrizations. In projective space this is possible by [4].

When dealing with coverings we need to assume that the parametrization has no projective base points. Only in some cases, like ruled surfaces, there has been progress in base point removal, see [6] and its reference [14]; see also [3] for the surface implicitization problem. Therefore in most situations it is assumed that the given parametrization has none base points, as we do here. As an intermediate step, one can reparametrize in such a way that all affine base points are sent to infinity, see [16] for details.

Ruled surfaces are of particular relevance in applications. We show how, given any rational parametrization of a ruled surface, one can reparametrize it so that provided the new parametrization does not have affine points, then it is normal or the set of missing points is contained in a line that can be made explicit.

Our coefficient field is algebraically closed of characteristic zero; for other fields (for example $\mathbb{R}$, of obvious interest) the curve case already suffers from complications that make the analysis very difficult, see [5,13].

The structure of the article is as follows: in Sect. 2 we show how to calculate algebraic supersets of the complement of a given parametrization. In Sect. 3 we present criteria for normality. Section 4 deals with the ruled case.

2   Computation of Critical Sets

Let us fix some notation through a few definitions. In the sequel, we use the following terminology: a parametrization takes its values in two-dimensional space in the variables s, t into three-dimensional space in the variables x, y, z.

Definition 1. *Let $\mathbb{K}$ be an algebraically closed field of characteristic zero and $S \subset \mathbb{K}^3$ an affine algebraic surface. A* parametrization *of S is a triple of rational functions that determines a rational dominant map*

$$
\begin{aligned}
\mathcal{P}: \mathbb{K}^2 &- \to & S \\
(s,t) &\mapsto & \left(\frac{p_1(s,t)}{q(s,t)}, \frac{p_2(s,t)}{q(s,t)}, \frac{p_3(s,t)}{q(s,t)} \right).
\end{aligned}
$$

We assume w.l.o.g. that $\gcd(p_1, p_2, p_3, q) = 1$. We denote as $\overline{S}$ the projective closure of S in $\mathbb{P}^3(\mathbb{K})$. The function $\mathcal{P}$ has a projective counterpart, $\overline{\mathcal{P}}$:

$$
\begin{aligned}
\overline{\mathcal{P}}: \quad \mathbb{P}^2(\mathbb{K}) &\quad - \to & \mathbb{P}^3(\mathbb{K}) \\
\mathbf{s} = (s:t:u) &\mapsto & (\overline{p_1}(\mathbf{s}) : \overline{p_2}(\mathbf{s}) : \overline{p_3}(\mathbf{s}) : \overline{q}(\mathbf{s}))
\end{aligned}
$$

where the four components are the polynomial homogenizations of the numerators and denominator of $\mathcal{P}$ such that their gcd is 1 and they have the same degree. Note that $\overline{\mathcal{P}}$ may be undefined at some points of $\mathbb{P}^2(\mathbb{K})$, since its four components may have a common zero.

Definition 2. *The common zeros of the components of $\overline{\mathcal{P}}$ are called* projective base points. *Such a point $(s:t:u)$ is also called an* affine base point *if $u \neq 0$.*

Since the gcd of the four homogeneous polynomials is 1, by Bézout's theorem it follows that there can be at most finitely many projective base points.

Definition 3. *An (affine) surface parametrization is called* normal *if it is surjective on S, that is, for every $p \in S$ there exist $s_0, t_0 \in \mathbb{K}$ such that $\mathcal{P}(s_0, t_0) = p$.*

Definition 4. *Let $\mathcal{P}$ be a parametrization that is not normal. A closed proper subset $C \subset S$ is called a* critical set *of $\mathcal{P}$ if $C \supset S \setminus \mathcal{P}(\mathbb{K}^2)$.*

Example 2. In Example 1, the line $z = x = 0$ is a critical set for $\mathcal{P}$. Any (reducible) curve on the surface that contains that line is also a critical set.

Example 3. The cone $x^2 + y^2 = z^2$ has the parametrization

$$\mathcal{P}(s,t) = \left(\frac{2st}{1+t^2}, \frac{s(1-t^2)}{1+t^2}, s \right)$$

that has the critical set $x = y + z = 0$. The complement of the image is that line minus the origin, since $\mathcal{P}(0,t) = (0,0,0)$ for any $t \neq \pm\sqrt{-1}$.

Now we give explicit descriptions of a critical set. See [15] for details. It is important to remark that the set of missing points may be finite, but we do currently know how to detect that case or compute a finite critical set when it is possible to do so, although eventually the method may generate zero-dimensional outputs (see Examples 5 and 6).

Theorem 1. *Let $\mathcal{P}$ be a non-normal parametrization of a surface S without projective base points. In the notation of Definition 1, let $n = \deg(q)$ and $l_i = \deg q - \deg p_i$ for $i = 1, 2, 3$. Necessarily $l_i \geq 0$ by the hypothesis. Let $P_{i,n-l_i}(s,t)$ be the homogeneous form of highest degree in $p_i(s,t)$ and $Q_n(s,t)$ that of $q(s,t)$. Let $\delta_{i,j}$ denote the Kronecker delta. Define*

$$C_1(s) = \left(\frac{P_{1,n-l_1}(s,1)}{Q_n(s,1)} \cdot \delta_{l_1,0}, \ \frac{P_{2,n-l_2}(s,1)}{Q_n(s,1)} \cdot \delta_{l_2,0}, \ \frac{P_{3,n-l_3}(s,1)}{Q_n(s,1)} \cdot \delta_{l_3,0} \right),$$

and

$$p = \left(\frac{P_{1,n-l_1}(1,0)}{Q_n(1,0)} \cdot \delta_{l_1,0}, \ \frac{P_{2,n-l_2}(1,0)}{Q_n(1,0)} \cdot \delta_{l_2,0}, \ \frac{P_{3,n-l_3}(1,0)}{Q_n(1,0)} \cdot \delta_{l_3,0} \right).$$

Then $S = \mathcal{P}(\mathbb{K}^2) \cup C_1(\mathbb{K}) \cup \{p\}$. In particular, the rational curve

$$C_1(\mathbb{K}) \cup \{p\} = \{\overline{\mathcal{P}}(s:t:0) \mid (s:t) \in \mathbb{P}(\mathbb{K})\} \cap S.$$

is a critical set.

Remark 1. It is worth noting that by a Gröebner basis computation we can decide if a given point in space belongs to the image of the parametrization. The same technique works if we want to test the points of a curve on the surface, whether parametrically or implicitly given.

Remark 2. In the particular case when $\mathcal{P}$ is proper (i.e. injective) with inverse $\mathcal{Q}$, the points not in the image are contained in the curves defined by the denominators of $\mathcal{P}$ and the denominators of $\mathcal{P}(\mathcal{Q})$. More precisely, one can proceed as follows:

1. Compute a representant of the inverse of $\mathcal{P}$; say

$$Q(x, y, z) = \left(\frac{A_1(x, y, z)}{B_1(x, y, z)}, \frac{A_2(x, y, z)}{B_2(x, y, z)} \right).$$

2. Compute the denominators $D_i(x, y, z)$ of $\mathcal{P}(Q(x, y, z))$.
3. The intersection of the algebraic surface and the algebraic set $V(\operatorname{lcm}(D_1, D_2, D_3, B_1, B_2))$ is a lower-dimensional algebraic set containing the set of non-reachable points. (Here $V(F_1, \ldots, F_s)$ denotes the algebraic set defined over $\mathbb{K}$ by the polynomials $\{F_1, \ldots, F_m\} \subset \mathbb{K}[x, y, z]$.)

The following example illustrates this.

Example 4. Let us consider the revolution surface $\mathcal{S}$ given parametrically by

$$\mathcal{P}(s, t) = \left(\frac{2s}{s^2 + 1} \frac{t - 1}{t + 1}, \frac{(1 - s^2)}{s^2 + 1} \frac{t - 1}{t + 1}, \frac{t + 1}{t - 1} \right)$$

obtained by rotating the hyperbola $\mathcal{C}$, of implicit equation $yz = 1$ given parametrically by $\left(0, \frac{t-1}{t+1}, \frac{t+1}{t-1} \right)$, around the z-axis. In order to analyze the missing points of the surface, we first observe that the hyperbola itself has only a missing point $P = (0, 1, 1)$ which corresponds to the limit when $t \to \infty$ (see [13]). Then it is logical to think that the circle that this point generates is a circle not reachable by $\mathcal{P}$; direct calculations confirm it. On the other hand, another candidate curve for missing points could be the limit curve $\left(0, -\frac{t-1}{t+1}, \frac{t+1}{t-1} \right)$ (obtained as $s \to \infty$) and that is, in the plane yz, the curve symmetric of the hyperbola $\mathcal{C}$ with respect the z-axis. Again, direct calculations show that no point of this curve is reachable by $\mathcal{P}$. Now, in order to see that there are no more missing points we compute a critical set by means of the inverse of $\mathcal{P}$. From a Gröbner basis for the ideal

$$\left(\operatorname{numer}(p_1 - x), \ \operatorname{numer}(p_2 - y), \ \operatorname{numer}(p_3 - z), \ w(s^2 + 1)(t^2 - 1) - 1 \right)$$

we can choose the representative of the inverse of $\mathcal{P}$

$$Q(x, y, z) = \left(\frac{z + 1}{z - 1}, \frac{xz}{yz + 1} \right).$$

The curves defined by the denominators of Q and $\mathcal{P}(Q)$ over $\mathcal{S}$ are contained in

$$z(z - 1)(yz + 1)(x^2 z^2 + y^2 z^2 + 2yz + 1) = 0, \quad x^2 z^2 + y^2 z^2 - 1 = 0.$$

The prime decomposition of this variety is $(x, yz + 1) \cup (z - 1, x^2 + y^2 - 1)$. Then, a critical set it is formed by the hyperbola $x = yz + 1 = 0$ and the circle $z - 1 = x^2 + y^2 - 1 = 0$; among the choices of the inverse, this is the best possible outcome.

Example 5. We consider the parametrization

$$P(s,t) = \left(\frac{t^2}{s^2+t^2}, \frac{s}{s^2+t^2}, \frac{st}{s^2+t^2} \right)$$

of the cylinder $x^2 + z^2 - x = 0$. A representative of the inverse of P is

$$\left(\frac{1-x}{y}, \frac{z}{y} \right).$$

Applying the procedure described in Remark 2, the critical set obtained is the point $(1,0,0)$. This point is reachable as $P(0,1)$, thus the parametrization is in fact normal.

Example 6. We consider the parametrization

$$P(s,t) = \left(\frac{t^2}{s^3+t^3}, \frac{s}{s^3+t^3}, \frac{t}{s^3+t^3} \right)$$

of the quintic surface of equation $x^2y^3 + x^2z^3 - z^4 = 0$. A representative of the inverse of P is

$$\left(\frac{xy}{z^2}, \frac{x}{z} \right).$$

From this the critical set obtained is the point $(0,0,0)$. From the parametrization, it is clear that the origin is not reachable. Thus the critical set is optimal.

In the following example the critical set given by Theorem 1 is smaller that the one generated by Remark 2.

Example 7. We consider the parametrization

$$P = \left(\frac{t^2}{s^2+1}, \frac{s^2+t}{s^2+1}, \frac{t^2+s}{s^2+1} \right).$$

Observe that it has no projective base points. The surface parametrized by $P(s,t)$ is defined by the degree 4 polynomial

$$x^4 - 4x^3y - 6x^3z + 2x^2y^2 + 8x^2yz + 7x^2z^2 - 4xy^2z - 4xyz^2 -$$
$$-4xz^3 + y^4 + 2y^2z^2 + z^4 + 3x^3 - 2x^2y - 6x^2z - xy^2 + 4xyz +$$
$$+3xz^2 - 2y^3 - 2yz^2 + 2xy + y^2 - x.$$

The inverse of P can be expressed as

$$\left\{ s = \frac{x^2 - 2xy - xz + 2yz + x - z}{x^2 - 2xz - y^2 + z^2 + 2y - 1}, \; t = \frac{x^2 - xy - 2xz + y^2 + z^2 + x - y}{x^2 - 2xz - y^2 + z^2 + 2y - 1} \right\}.$$

Thus Remark 2 provides, as a critical set, the variety

$$V(-2y + x + 1, 3x - 1 - 2z) \cup V(y - 1, x - z) \cup V(-2y + x + 1, x + 1 - 2z).$$

However, Theorem 1 reduces the critical set to $V(y - 1, x - z)$.

3   Criteria for Normality

We will describe sufficient conditions on a rational surface parametrization to be normal. Note that the composition of a normal parametrization with an affine transformation of $\mathbb{K}^3$ is again normal. So the results below are also valid applying affine transformations to the parmatrizations.

Taking into account the inverse-based method for computing critical sets described in Remark 2, we have the following criterium.

Proposition 1. *Every proper polynomial parametrization with polynomial inverse is normal.*

Example 8. We consider the surface S defined by $x^4 - 2x^2y + y^2 + x - z$. It can be parametrized by $\mathcal{P}(s,t) = (t, t^2 + s, s^2 + t)$. This parametrization is proper and its inverse can be expressed polynomially as $(y - x^2, x)$. Therefore, $\mathcal{P}(s,t)$ is normal.

Corollary 1. *A parametrization of the form*

$$\mathcal{P}(s,t) = (as + b, ct + A(s), B(s,t))$$

where $ac \neq 0$, $A \in \mathbb{K}[s]$ and $B \in \mathbb{K}[s,t]$ is normal.

Proof. Note that $\mathcal{P}(s,t)$ is polynomial and proper, and its inverse can be expressed as

$$\left(\frac{x - b}{a}, \frac{y - A(\frac{x-b}{a})}{c} \right)$$

which is polynomial.

More generally, one has the next corollary.

Corollary 2. *A parametrization of the form*

$$\mathcal{P}(s,t) = (\varphi_1(s,t), \varphi_2(s,t), B(s,t)),$$

where (φ_1, φ_2) is an automorphism of $\mathbb{K}^2$, and $B \in \mathbb{K}[s,t]$, is normal.

Proof. The inverse of $\mathcal{P}(s,t)$ can be expressed as

$$(\varphi_1(s,t), \varphi_2(s,t))^{-1}$$

which is polynomial.

Remark 3. Every irreducible surface S defined by a polynomial $F(x,y,z)$ with degree 1 with respect to one of the variables is rational. Assuming w.l.o.g. that $deg_z(F) = 1$, then F can be expressed as $F = p(x,y) - zq(x,y)$ where $\gcd(p,q) = 1$, and S can be easily parametrized as

$$\mathcal{P}(s,t) = \left(s, t, \frac{p(s,t)}{q(s,t)} \right).$$

The analysis of the normality of this parametrization is also very simple: $\mathcal{P}$ is a normal parametrization of $\mathcal{S}$ if and only if $V(p(s,t),q(s,t))$ is empty. Indeed, if $V(p(s,t),q(s,t)) = \bigcup\{(s_i,t_i)\}$, then the lines $(s_i,t_i,\lambda)_{\lambda \in \mathbb{K}}$ are in $\mathcal{S}$ and are not reachable. On the other hand, if $\mathcal{P}$ is not normal there exists a point $(x_0,y_0,z_0) \in \mathcal{S}$ not reachable by $\mathcal{P}$; this implies that $q(x_0,y_0) = p(x_0,y_0) = 0$.

In particular, the parametrization $\mathcal{Q}$ in Example 1 is normal because $V(s,s^2t^2 - 2st - 1) = \emptyset$.

The following criterium can be found in Theorem 5 in [15]. It applies to situations where there is an absence of projective base points.

Proposition 2. *Let $\mathcal{P}(s,t)$ have no projective base points. If*

$$\max\{\deg(p_1),\deg(p_2),\deg(p_3)\} > \deg(q)$$

then $\mathcal{P}(s,t)$ is normal.

Example 9. By Proposition 2, we know that the parametrization

$$\mathcal{P}(s,t) = \left(\frac{s^2}{s+2t}, \frac{t^2}{s-2t}, \frac{st+1}{s+2t} \right)$$

is normal.

Corollary 3. *A parametrization of the form*

$$\mathcal{P}(s,t) = (A_1(s), A_2(t), A_3(s,t)),$$

where A_1, A_2 are non-constant polynomials of the same degree, $A_3 \in \mathbb{K}[s,t]$ with $\deg(A_3) \le \deg(A_1)$, is normal.

Proof. Let $A_1 = a_n s^n + \cdots + a_0$, $A_2 = b_n t^n + \cdots + b_0$. Let $\mathcal{P}^H(s,t,u)$ be the projectivization of $\mathcal{P}(s,t)$. Then, $\mathcal{P}^H(s,t,0) = (a_n s^n : b_n t^n : \alpha : 0)$ for some $\alpha \in \mathbb{K}[s,t]$. So, the parametrization does not have projective base points. Now the result follows from Proposition 2.

Corollary 4. *A parametrization without affine base points of the form*

$$\mathcal{P}(s,t) = \left(\frac{p_1}{q}, \frac{p_2}{q}, \frac{p_3}{q} \right)$$

where

$$p_1 = a_n t^n + P_1^*(s,t),$$
$$p_2 = b_n s^n + P_2^*(s,t),$$

such that P_1^*, P_2^* and q have degree strictly less than n, and p_3 has degree $\le n$, is normal.

Proof. Once again, $\mathcal{P}^H(s,t,0) = (a_n s^n : b_n t^n : \alpha : 0)$ for some $\alpha \in \mathbb{K}[s,t]$ so there are no projective base points.

The next criterium is based on the notion of pseudo-normality, a concept introduced in [11]. Let I be the implicitization ideal of $\mathcal{P}(s,t)$, that is the ideal in $\mathbb{K}[W, s, t, x, y, z]$ generated by $\{q(s,t)x - p_1(s,t), q(s,t)y - p_2(s,t), q(s,t)z - p_3(s,t), qW - 1\}$. We consider the maps $\Pi_s : \mathbb{K}^5 \to \mathbb{K}^4; (s, t, x, y, z) \mapsto (t, x, y, z)$ and $\Pi_t : \mathbb{K}^4 \to \mathbb{K}^3; (t, x, y, z) \mapsto (x, y, z)$. Let V be the variety in $\mathbb{K}^5$ defined by $\{q(s,t)x - p_1(s,t), q(s,t)y - p_2(s,t), q(s,t)z - p_3(s,t)\}$. In this situation, we say that $\mathcal{P}(s,t)$ is *pseudo-normal* if

$$S \subset \pi_t(\pi_s(V)).$$

The next result can be found in Corollary 4.4. in [11].

Proposition 3. *A pseudo-normal polynomial parametrization is normal.*

To finish this section, we present a family of normal parametrizable surfaces that do not come from the previous criteria. In Fig. 1 we see one of them. As the image suggests, they correspond to surfaces having three independent planes being asymptotes of the surface.

Fig. 1. The surface $x^4 y^5 z^3 = 1$

Proposition 4. *Let $\lambda \neq 0$ and let S be the surface defined by*

$$L_1(x, y, z)^n L_2(x, y, z)^m L_3(x, y, z)^k = \lambda$$

where L_i are three linearly independent linear forms with natural exponents. Then S can be normally parametrized.

Proof. W.l.o.g. we can assume, after a suitable affine linear change, that S is given by $x^n y^m z^k = \lambda$. We consider the parametrization

$$\mathcal{P}(s,t) = \left(s^k,\ t^k,\ \sqrt[k]{\lambda}/(s^n t^m) \right).$$

Since $\lambda \neq 0$, the affine surface does not intersect the coordinate planes $x = 0$, $y = 0$, and $z = 0$. Now observe that, if $(a,b,c) \in S$, then $(a,b,c) = \mathcal{P}(\sqrt[k]{a}, \sqrt[k]{b})$.

4 Critical Set of Ruled Surfaces

Our starting point is a parametrization of a ruled surface. In [10] methods to determine if a surface is ruled are presented, including the computation of a parametrization of the surface in the form $A(t) + sB(t)$. We will assume without loss of generality that any ruled surface is given in this form.

Lemma 1. *A parametrization as above can be reparametrized into the form*

$$\mathcal{P}(s,t) = \left(\frac{r_1(s) + t p_1(s)}{q(s)},\ \frac{r_2(s) + t p_2(s)}{q(s)},\ \frac{r_3(s) + t p_3(s)}{q(s)} \right)$$

where $\gcd(p_1, p_2, p_3) = 1$, $\gcd(r_1, r_2, r_3, q) = 1$, $\deg(r_1) = \deg(r_2) = \deg(r_3) = \deg(q) = m$ *and* $\deg(p_1) = \deg(p_2) = \deg(p_3) = n$.

Proof (Sketch). It suffices to follow these steps:

- Put a common denominator.
- By a Möbius transformation, the degrees of the r, p, q are made equal.
- With the change $t \leftarrow t/\gcd(p_i)$, the p_i are made coprime.

The following theorem describes simple critical sets of ruled surface parametrizations given as in Lemma 1, and under the assumption of not having affine base points.

Theorem 2. *Let $\mathcal{P}$ be a ruled surface parametrization in the form provided by Lemma 1 without affine base points. Then the following line is a critical set of $\mathcal{P}$:*

$$\begin{cases} p_{2n}(x q_m - r_{1m}) - p_{1n}(y q_m - r_{2m}) = 0 \\ p_{3n}(x q_m - r_{1m}) - p_{1n}(z q_m - r_{3m}) = 0 \end{cases}$$

where p_{in}, r_{im} and q_m are the leading coefficients of the polynomials p_i, r_i and q respectively. Moreover, a parametrization of this line is

$$\left(\frac{r_{1m} + \lambda p_{1n}}{q_m},\ \frac{r_{2m} + \lambda p_{2n}}{q_m},\ \frac{r_{3m} + \lambda p_{3n}}{q_m} \right), \qquad \lambda \in \mathbb{K}$$

that can be obtained from $\mathcal{P}$ considering $\mathcal{Q}(0, \lambda)$ in $\mathcal{Q}(s,t) = \mathcal{P}\left(\dfrac{1}{s}, \dfrac{t}{s^{m-n}} \right)$.

Proof (Sketch). In the ring $\mathbb{K}[x, y, z, s, t, w]$ we consider the ideal

$$I = (r_1(s) + t \cdot p_1(s) - x \cdot q(s), \quad \ldots, \quad \ldots, \quad w \cdot q(s) - 1).$$

Then $\text{Image}(\mathcal{P}) = \pi(V(I))$ where $\pi \colon \mathbb{K}^6 \to \mathbb{K}^3 \colon \pi(x, y, z, s, t, w) = (x, y, z)$. Repeated application of the Extension Theorem (see in [7], also Exercise 6.3.7 in p. 283) proves the result.

Remark 4. The concept of pseudonormality introduced in [11] corresponds to the case where it is possible to extend a surface point to s, t but possibly not to w.

In the next example Theorem 2 provides a better critical set than Remark 2.

Example 10. We consider the parametrization

$$\mathcal{P}(s, t) = \left(\frac{s + t(s^2 + 1)}{s + 1}, \; \frac{s + 3 + t(s^2 - 2)}{s + 1}, \; \frac{s + 2 + t(s^2 - 3)}{s + 1} \right)$$

that parametrizes the cubic surface defined by the polynomial

$$x^3 - 8x^2 y + 7x^2 z + 16xy^2 - 32xyz + 15xz^2 + 16y^2 z - 24yz^2 +$$
$$+9z^3 - 6x^2 + 24xy - 12xz - 24yz + 18z^2 - 12x - 24y + 36z.$$

Observe that $\mathcal{P}(s, t)$ satisfies the hypotheses of Theorem 2. Hence, a critical set for $\mathcal{P}(s, t)$ is the line $x = y = z$. On the other hand a representative of the inverse of $\mathcal{P}(s, t)$ is

$$\left\{ s = -\frac{1}{12} \frac{x^2 - 4xy + 4xz - 4yz + 3z^2 - 6x + 12y - 6z - 12}{y - z}, \right.$$

$$\left. t = -\frac{1}{12}x^2 + \frac{1}{3}yx - \frac{1}{3}zx + \frac{1}{3}zy - \frac{1}{4}z^2 + \frac{1}{2}x - \frac{1}{2}z \right\}.$$

Applying the method in Remark 2 we get as a critical set the variety

$$V(y - z, \; x^3 - x^2 z - xz^2 + z^3 - 6x^2 + 12xz - 6z^2 - 12x + 12z)$$

that decomposes as the line $y = x = z$ and the conic $V(y^2 - 6y - x^2 + 6x + 12, y - z)$.

Acknowledgements. This work was developed, and partially supported, by the Spanish *Ministerio de Economía y Competitividad* under Project MTM2011-25816-C02-01. The first and third authors are members of the Research Group ASYNACS (Ref. CCEE2011/R34).

References

1. Andradas, C., Recio, T., Sendra, J.R., Tabera, L.F.: On the simplification of the coefficients of a parametrization. J. Symbolic Comput. **44**(2), 192–210 (2009)
2. Andradas, C., Recio, T., Sendra, J.R., Tabera, L.F., Villarino, C.: Proper real reparametrization of rational ruled surfaces. Comput. Aided Geom. Design **28**(2), 102–113 (2011)
3. Busé, L., Cox, D., D'Andrea, C.: Implicitization of surfaces in $\mathbb{P}^3$ in the presence of base points. J. Algebra Appl. **2**(2), 189–214 (2003)
4. Bodnár, G., Hauser, H., Schicho, J., Villamayor, O.: Plain varieties. Bull. Lond. Math. Soc. **40**(6), 965–971 (2008)
5. Bajaj, C.L., Royappa, A.V.: Finite representations of real parametric curves and surfaces. Internat. J. Comput. Geom. Appl. **5**(3), 313–326 (1995)
6. Chen, F.: Reparametrization of a rational ruled surface using the μ-basis. Comput. Aided Geom. Design **20**(1), 11–17 (2003)
7. Cox, D., Little, J., O'Shea, D.: Ideals, Varieties, and Algorithms: An Introduction to Computational Algebraic Geometry and Commutative Algebra. Undergraduate Texts in Mathematics, 3rd edn. Springer, New York (2007)
8. Farin, G., Hoschek, J., Kim, M.-S. (eds.): Handbook of Computer Aided Geometric Design. North-Holland, Amsterdam (2002)
9. Hoschek, J., Lasser, D.: Fundamentals of Computer Aided Geometric Design. A. K. Peters Ltd., Wellesley (1993). Translated from the 1992 German edition by L.L. Schumaker
10. Pérez-Díaz, S., Shen, L.-Y.: Characterization of rational ruled surfaces. J. Symbolic Comput. **50**, 450–464 (2013)
11. Pérez-Díaz, S., Sendra, J.R., Villarino, C.: A first approach towards normal parametrizations of algebraic surfaces. Internat. J. Algebra Comput. **20**(8), 977–990 (2010)
12. Recio, T., Sendra, J.R., Tabera, L.F., Villarino, C.: Generalizing circles over algebraic extensions. Math. Comp. **79**(270), 1067–1089 (2010)
13. Rafael Sendra, J.: Normal parametrizations of algebraic plane curves. J. Symbolic Comput. **33**(6), 863–885 (2002)
14. Saito, T., Sederberg, T.W.: Rational-ruled surfaces: implicitization and section curves. Graph. Models Image Process. **57**(4), 334–342 (1995)
15. Sendra, J.R., Sevilla, D., Villarino, C.: Covering of surfaces parametrized without projective base points. In: Proceedings of ISSAC 2014, pp. 375–380. ACM Press (2014)
16. Sendra, J.R., Sevilla, D., Villarino, C.: Covering rational ruled surfaces. arXiv:1406.2140v1 [math.AG]

Rational Normal Curves as Set-Theoretic Complete Intersections of Quadrics

Maria-Laura Torrente[(✉)]

Dipartimento di Matematica, Università di Genova,
Via Dodecaneso 35, 16146 Genova, Italy
torrente@dima.unige.it

Abstract. In the first part of this paper we present a short *survey* on the problem of the representation of rational normal curves as set-theoretic complete intersections. In the second part we use a method, introduced by Robbiano and Valla, to prove that the rational normal quartic is set-theoretically complete intersection of quadrics: it is an original proof of a classical result of Perron, and Gallarati-Rollero.

Keywords: Rational normal curves · Set-theoretic complete intersection · Quadratic polynomials · Gröbner bases

1 Introduction

Mathematicians have always shown great interest for rational normal curves, a special class of curves obtained as the image of the projective line. More precisely, the rational normal curve of degree n in $\mathbb{P}^n$, denoted by C^n, is defined as the image of the Veronese embedding of $\mathbb{P}^1$ in $\mathbb{P}^n$, which is given by the complete linear series of forms of degree n. In particular, if we consider the vector space spanned by such linear forms and choose as its basis the set of monomials of degree n, we get the following parametric representation of C^n:

$$x_0 = a^n,\ x_1 = a^{n-1}b, \ldots, x_{n-1} = ab^{n-1},\ x_n = b^n \tag{1}$$

Let's consider the special case $n = 2$. The parametric representation becomes

$$x_0 = a^2,\ x_1 = ab,\ x_2 = b^2$$

It is easy to verify that the Cartesian representation of C^2 is given by the unique equation $x_0 x_2 - x_1^2 = 0$, and that $x_0 x_2 - x_1^2$ generates the ideal of the curve C^2, denoted by $I(C^2)$. In this particular case everything is clear. Now, we consider the next special case $n = 3$, that is, we consider the curve C^3 classically known as the *twisted cubic*. Its parametric representation is:

$$x_0 = a^3,\ x_1 = a^2b,\ x_2 = ab^2,\ x_3 = b^3$$

The twisted cubic has degree 3 and codimension 2 in $\mathbb{P}^3$. Further, it is easy to verify that C^3 is defined by more than 2 equations. In fact, by contradiction,

© Springer International Publishing Switzerland 2015
J. Gutierrez et al. (Eds.): Computer Algebra and Polynomials, LNCS 8942, pp. 204–212, 2015.
DOI: 10.1007/978-3-319-15081-9_12

assume that it is defined by 2 equations, then, as a consequence of Bézout's theorem, C^3 would necessarily be the intersection of a plane and a surface of degree 3, contradicting the fact that C^3 is not a plane curve. So, the curve C^3 is not a *complete intersection*, i.e., the defining ideal $I(C^3)$ of C^3 cannot be generated by as many equations as its codimension. A minimal set of generators of $I(C^3)$ can be computed in a standard way by using Gröbner bases theory (see for instance [7]): $\{x_0x_2 - x_1^2, \ x_0x_3 - x_1x_2, \ x_1x_3 - x_2^2\}$. We observe that such generators are the 2×2 minors of the matrix

$$\begin{pmatrix} x_0 & x_1 & x_2 \\ x_1 & x_2 & x_3 \end{pmatrix}$$

Now, we consider the open set of the curve obtained by setting $x_0 \neq 0$; in this way, we obtain an affine chart of C^3, called Γ^3. Letting $y_i = \frac{x_i}{x_0}$, for $i = 1, \ldots, 3$, the parametric representation of Γ^3 is

$$y_1 = b, \ y_2 = b^2, \ y_3 = b^3$$

The ideal $I(\Gamma^3)$ of the affine curve Γ^3 is generated by the polynomials $f_1 = y_2 - y_1^2$ and $f_2 = y_3 - y_1y_2$, that is, $I(\Gamma^3) = (y_2 - y_1^2, y_3 - y_1y_2)$, and so Γ^3 is a complete intersection. But the curve Γ^3 can also be described by the two equations $f_1 = 0$ and $f_2^2 = 0$; on the other hand, f_1 and f_2^2 do not generate the ideal $I(\Gamma^3)$, but obviously the curve Γ^3 is contained in their zero locus. In fact we have that $I(\Gamma^3) = \sqrt{(f_1, f_2^2)}$; more precisely this is equivalent to saying that f_1 and f_2^2 define Γ^3 as a *set-theoretic complete intersection*. Now, we observe that the ideal generated by f_1 and f_2^2 is also generated by f_1 and g_2, where $g_2 = f_2^2 + y_2^2 f_1 = y_3^2 - 2y_1y_2y_3 + y_2^3$. Using another well-known result in Gröbner bases theory (see for instance [8]), it is possible to prove that the zero locus of the homogenizations of f_1 and g_2 w.r.t. x_0, which are $F_1 = x_0x_2 - x_1^2$ and $G_2 = x_0x_3^2 - 2x_1x_2x_3 + x_2^3$ respectively, is exactly the curve C^3. We conclude that C^3 is a set-theoretic complete intersection, but it is not a complete intersection.

Is it possible to give a general statement for C^n? This question has been answered in many different ways in the last seventy years. In 1941 Perron observed that all rational normal curves C^n are set-theoretical complete intersections; since then, many different contributions were devoted to improve the class of equations which define the curve C^n set-theoretically. But, why so great attention has been paid to this apparently minor issue? In our opinion, the reason is mainly due to the fact that the problem of characterizing which classes of projective varieties are set-theoretical complete intersections is not yet completely solved. Any minor contribution to this topic would be of great importance; for this reason, in the last few years many mathematicians addressed this kind of problem, giving rise to a wide literature. Among the others, we recall the papers of Barile [2–5], Badescu and Valla [1] (in particular, see Remark 4.3), Moh [9,10], Ohm [11], Robbiano [13,14], Robbiano and Valla [15,16], and Vogel [17,18].

This short survey paper, and in particular Sect. 2, is an *excursus* on the problem of the representation of the curve C^n as a set-theoretic complete intersection. In particular, throughout this paper we briefly recall the results contained in

[6,12,16,19]. The contribution of [12] is fundamental: in this paper, for the first time, it has been proved that every rational normal curve C^n is the set-theoretical complete intersection of $n-1$ hypersurfaces of degree $2,\dots,n$ respectively. Further, in the special case when n is a power of 2, and only in this case, it is proved that C^n is the set-theoretic complete intersection of $n-1$ quadrics. Starting from this result and for every integer n, in [6] the authors find another representation of C^n which lowers the degrees of the defining equations. Let s be the largest power of 2 less or equal to n; then C^n can be expressed as the set-theoretical complete intersection of $s-1$ quadrics and additional $n-s$ hypersurfaces of degree $s+1,\dots,n$ respectively. In our opinion, an alternative proof of the original results of [12], when n is a power of 2, can be obtained using the computational method presented in [16] (and reported in Sect. 2) which is a generalization of the example of the twisted cubic C^3 discussed in this introduction. To this aim, in Sect. 3, we show that this latter method works for the case of the quartic curve C^4; its extension to the case $n=2^m$ is still an open issue.

2   Classical Results

In this section we briefly recall some classical results on the problem of the representation of the rational normal curves as set-theoretic complete intersections.

Let C^n be the rational normal curve of $\mathbb{P}^n$, and $I_n = I(C^n)$ be the defining ideal of C^n in $\overline{P} = \mathbb{K}[x_0, x_1, \dots, x_n]$, the polynomial ring in the indeterminates $x_0, \dots, x_n$ over the field $\mathbb{K}$. It is well-known that I_n is generated by the 2×2 minors of the matrix

$$A = \begin{pmatrix} x_0 \ x_1 \ \dots \ x_{n-1} \\ x_1 \ x_2 \ \dots \ x_n \end{pmatrix},$$

i.e., I_n is generated by a system of $\binom{n}{2}$ quadratic forms.

A first answer to the representation problem is due to Perron (see [12]), who proved, in 1941, that the curve C^n can be represented as the set-theoretic complete intersection of $n-1$ algebraic hypersurfaces of degree $2, \dots, n$ and equations $P_1 = 0, \dots, P_{n-1} = 0$ where

$$P_i = \begin{vmatrix} x_0 \ x_1 & \dots \ x_i \\ x_1 \ x_2 & \dots \ x_{i+1} \\ \vdots \ \vdots & \quad \vdots \\ x_i \ x_{i+1} & \dots \ x_{2i} \end{vmatrix} \tag{2}$$

for $i = 1, \dots, n-1$, with $x_j = 0$ for $j > n$. The proof is direct and exploits the parametric representation of the rational normal curves: the core is to show that $(x_0 : x_1 : \dots : x_n) \in \mathbb{P}^n$ is a solution of the polynomial system $P_1 = \dots = P_n = 0$ if and only if it satisfies relation (1).

When n is a power of 2, and only in this case, by making similar considerations and using an inductive approach, Perron shows that C^n is a set-theoretic intersection of $n - 1$ quadrics of equations $Q_1^n = 0, \ldots, Q_{n-1}^n = 0$, recursively defined by:

$$Q_i^n = \begin{cases} Q_i^{n-1}(x_0, \ldots, x_{n/2}) & i = 1, \ldots, n/2 - 1 \\ Q_i^{n-1}(x_{n/2+1}, \ldots, x_n) & i = n/2, \ldots, n - 2 \\ \sum_{k=0}^{n/2}(-1)^k \binom{n/2}{k} x_k x_{n-k} & i = n - 1 \end{cases} \tag{3}$$

where the trivial case $n = 2$ is given by the quadric $Q_1^2 = x_0 x_2 - x_1^2 = 0$. In this special case the curve C^n can be represented by a system of polynomial equations whose degree, defined as the product of the degrees of the single equations, is equal to 2^{n-1}.

In the late 1970s and early 1980s, independently of the work of Perron, two different proofs of the fact that the rational normal curves are set-theoretic complete intersections were presented in [16, 19].

In 1979, exploiting some properties of suitable homogeneous polynomials, Verdi (see [19]) proves that the prime ideal $I_n = I(C^n)$ satisfies $I_n^{(n-1)!} \subseteq J_n$, where J_n is the ideal generated by the forms $V_1, \ldots, V_{n-1}$ and

$$V_i = \begin{vmatrix} x_0 & x_1 & x_2 & \ldots & x_i \\ x_1 & x_2 & & & x_{i+1} \\ x_2 & \vdots & & & 0 \\ \vdots & & & & \vdots \\ x_i & x_{i+1} & 0 & \ldots & 0 \end{vmatrix} \tag{4}$$

for $i = 1, \ldots, n - 1$, and $\deg(V_i) = i + 1$. Such a relation easily yields $\sqrt{J_n} = I_n$.

In 1983, using computational techniques derived from Gröbner bases theory and an approach already introduced in [15], Robbiano and Valla (see [16]) provide a different proof of the fact that every curve C^n is a set-theoretic complete intersection. The method they use is more general and constructive: it addresses the problem of the representation of every projective variety V of codimension r in $\mathbb{P}^n$ as a set-theoretic complete intersection, by performing the following steps:

1. Determine an affine variety W whose projective closure is V; it is well-known that this implies that $I(V) = I(W)^{\text{hom}}$, where $I(W)^{\text{hom}} \subseteq \overline{P}$ denotes the homogeneous ideal generated by the homogenizations f^{hom} of the polynomials $f \in I(W)$ w.r.t. the homogenizing indeterminate x_0 (see for instance [8], Sect. 4.3).
2. Find a representation of W as a set-theoretic complete intersection, that is, determine polynomials $g_1, \ldots, g_r$ which satisfy $\sqrt{(g_1, \ldots, g_r)} = I(W)$.
3. Lift the representation found in step 2 to the projective space to get a set-theoretic complete intersection representation of V.

Note that the instructions of step 3 are very easy to be performed if the polynomials $g_1, \ldots, g_r$ are a Gröbner basis, w.r.t. a degree compatible term ordering, of the ideal they generate $J = (g_1, \ldots, g_r)$. In this case, using classical results of Gröbner bases theory (see for instance [8], Proposition 4.3.21), the homogenizations $g_1^{\mathrm{hom}}, \ldots, g_r^{\mathrm{hom}}$ are a Gröbner basis of the homogenized ideal J^{hom}. Further, recalling that over ideals the operations of homogenization and radical commute, we get the following chain of equalities:

$$I(V) = I(W)^{\mathrm{hom}} = \left(\sqrt{(g_1, \ldots, g_r)}\right)^{\mathrm{hom}} = \sqrt{(g_1, \ldots, g_r)^{\mathrm{hom}}} = \sqrt{(g_1^{\mathrm{hom}}, \ldots, g_r^{\mathrm{hom}})}$$

which implies that V is a set-theoretic complete intersection of the algebraic hypersurfaces of equations $g_1^{\mathrm{hom}} = \ldots = g_r^{\mathrm{hom}} = 0$.

Here we show how this computational approach has been used for the case of rational normal curves (see [16], Sect. 1). We consider the affine chart defined by $x_0 \neq 0$ and the monomial affine curve Γ^n in $\mathbb{A}^n$ whose parametric equations are $x_i = b^i$, for $i = 1, \ldots, n$. It is clear that C^n is the projective closure of Γ^n, which implies that $I_n = I(C^n) = I(\Gamma_n)^{\mathrm{hom}}$. Further, it is easy to verify that Γ_n is the complete intersection of $n-1$ hypersurfaces defined by polynomials $f_i = x_{i+1} - x_1 x_i$, for $i = 1, \ldots, n-1$, from which it follows that $I(\Gamma_n) = (f_1, \ldots, f_{n-1})$. Now, let $1 \leq k \leq i \leq n-1$; we have that $x_1^k \equiv x_k \mod (f_1, \ldots, f_{i-1})$ and

$$f_i^i = (x_{i+1} - x_1 x_i)^i = \sum_{k=0}^{i} (-1)^k \binom{i}{k} x_{i+1}^{i-k} x_1^k x_i^k$$

$$\equiv x_{i+1}^i + \sum_{k=1}^{i} (-1)^k \binom{i}{k} x_{i+1}^{i-k} x_k x_i^k \mod (f_1, \ldots, f_{i-1})$$

From the above relation and the definition of radical of an ideal (see for instance [7]) it follows that $I(\Gamma_n) = (f_1, \ldots, f_{n-1}) = \sqrt{(r_1, \ldots, r_{n-1})}$, where r_i are defined as follows:

$$r_i = x_{i+1}^i + \sum_{k=1}^{i} (-1)^k \binom{i}{k} x_{i+1}^{i-k} x_k x_i^k \tag{5}$$

In $\mathbb{K}[x_1, \ldots, x_n]$, let σ be the total degree reverse lexicographical ordering induced by $x_1 >_\sigma \ldots >_\sigma x_n$ (denoted by DegRevLex in [7]); the leading term of each r_i is $\mathrm{LT}_\sigma(r_i) = x_i^{i+1}$. Since the leading terms of $r_1, \ldots, r_{n-1}$ form a regular sequence, it follows that $r_1, \ldots, r_{n-1}$ are the Gröbner basis (w.r.t. the term ordering σ) of the ideal they generate $(r_1, \ldots, r_{n-1})$ (see again [7], Corollary 2.5.10). We then conclude that the rational normal curve C^n is the set-theoretic complete intersection of the hypersurfaces of equations $r_1^{\mathrm{hom}} = 0, \ldots, r_{n-1}^{\mathrm{hom}} = 0$.

Note that the results of [12,16,19] we have been discussing here yield different representations of C^n. In particular, we observe that, though $\deg(P_i) = \deg(V_i) = \deg(r_i^{\mathrm{hom}})$ for each i, the homogeneous polynomials r_i^{hom} are in general much simpler (since they are formed by fewer terms) than the corresponding forms V_i, which in turn are simpler than the forms P_i (see also the following Example 1). Nevertheless, their common feature is the degree of the polynomial system they form, which is defined as the product of the degrees of all the polynomials and equals $2 \cdot 3 \cdot \ldots \cdot n = n!$ in each of the three cases.

In 1988, partially using the results of [12], Gallarati and Rollero (see [6]) improve the representation of C^n as a set-theoretic complete intersection from the point of view of the degree of the system. Let s be the largest power of 2 less than or equal to n. In [6] it is proved that there exists a system of polynomial equations defining C^n and whose degree is $\frac{n! \, 2^{s-1}}{s!}$, which is appreciably less than $n!$ as soon as $n \geq 5$. We give here a sketch of the proof. Let $Q_1^s, \ldots, Q_{s-1}^s$ be the set of quadrics introduced in [12] and recursively defined in (3); further, let $r_s^{\mathrm{hom}}, \ldots, r_{n-1}^{\mathrm{hom}}$ be $n-s$ homogeneous forms of degree $s+1, \ldots, n$ respectively (defined in (5)). A straightforward computation shows that $(x_0 : \ldots : x_n)$ is a solution of the polynomial system

$$\begin{cases} Q_i^s = 0 & 1 \leq i \leq s-1 \\ r_i^{\mathrm{hom}} = 0 & s \leq i \leq n-1 \end{cases} \tag{6}$$

if and only if it satisfies relation (1). But this implies that the curve C^n can be expressed as a set-theoretic complete intersection using the previous system of equations, whose degree is exactly $\frac{n! \, 2^{s-1}}{s!}$.

We end this section with an example: the aim is to compute and compare the different representations of C^n as a set-theoretic complete intersection discussed in this section.

Example 1. We consider the case of the rational normal quintic curve C^5 in $\mathbb{P}^5$. In [12] the curve C^5 is expressed by the polynomials (see formula (2)):

$$\begin{aligned} P_1 &= x_1^2 - x_0 x_2 \\ P_2 &= x_2^3 - 2x_1 x_2 x_3 + x_0 x_3^2 + x_1^2 x_4 - x_0 x_2 x_4 \\ P_3 &= x_3^4 - 3x_2 x_3^2 x_4 + x_2^2 x_4^2 + 2x_1 x_3 x_4^2 - x_0 x_4^3 + 2x_2^2 x_3 x_5 - 2x_1 x_3^2 x_5 \\ &\quad - 2x_1 x_2 x_4 x_5 + 2x_0 x_3 x_4 x_5 + x_1^2 x_5^2 - x_0 x_2 x_5^2 \\ P_4 &= x_4^5 - 4x_3 x_4^3 x_5 + 3x_3^2 x_4 x_5^2 + 3x_2 x_4^2 x_5^2 - 2x_2 x_3 x_5^3 - 2x_1 x_4 x_5^3 + x_0 x_5^4 \end{aligned} \tag{7}$$

In [19] the curve C^5 is expressed by the polynomials (see formula (4)):

$$\begin{aligned} P_1 &= x_1^2 - x_0 x_2 \\ V_2 &= x_2^3 - 2x_1 x_2 x_3 + x_0 x_3^2 \\ V_3 &= x_3^4 - 3x_2 x_3^2 x_4 + x_2^2 x_4^2 + 2x_1 x_3 x_4^2 - x_0 x_4^3 \\ P_4 &= x_4^5 - 4x_3 x_4^3 x_5 + 3x_3^2 x_4 x_5^2 + 3x_2 x_4^2 x_5^2 - 2x_2 x_3 x_5^3 - 2x_1 x_4 x_5^3 + x_0 x_5^4 \end{aligned} \tag{8}$$

In [16] the curve C^5 is expressed by the polynomials (see formula (5)):

$$\begin{aligned} P_1 &= x_1^2 - x_0 x_2 \\ V_2 &= x_2^3 - 2x_1 x_2 x_3 + x_0 x_3^2 \\ R_3 &= x_3^4 - 3x_2 x_3^2 x_4 + 3x_1 x_3 x_4^2 - x_0 x_4^3 \\ R_4 &= x_4^5 - 4x_3 x_4^3 x_5 + 6x_2 x_4^2 x_5^2 - 4x_1 x_4 x_5^3 + x_0 x_5^4 \end{aligned} \tag{9}$$

In [6] the curve C^5 is expressed by the polynomials (see formula (6)):

$$\begin{aligned}
P_1 &= x_1^2 - x_0 x_2 \\
G_2 &= x_2^2 - 2x_1 x_3 + x_0 x_4 \\
G_3 &= x_3^2 - x_2 x_4 \\
R_4 &= x_4^5 - 4x_3 x_4^3 x_5 + 6x_2 x_4^2 x_5^2 - 4x_1 x_4 x_5^3 + x_0 x_5^4
\end{aligned} \tag{10}$$

As already observed, (7), (8) and (9) are made up of homogeneous polynomials of increasing degree $2, 3, 4, 5$, so the corresponding polynomial systems have all degree $5! = 120$. On the other hand, the elements of (10) are three quadratic polynomials and a quintic one, so the corresponding system has degree 40. Some polynomials (in particular the quadratic polynomial P_1) appear in various sets, but in general, moving from (7) to (10), the polynomial sets become more and more simple, where simplicity is measured according to the number of terms of the support of each polynomial. In conclusion, the representation given by (10) seems to be the best one among the approaches we have considered here.

3   The Rational Normal Quartic in $\mathbb{P}^4$

In this section we use the computational method introduced in [15, 16] (and recalled in Sect. 2) to provide a representation of the rational normal quartic as a set-theoretic complete intersection of quadrics.

Let C^4 in $\mathbb{P}^4$ be the rational normal quartic, whose parametric representation is given by:

$$x_0 = a^4, \ x_1 = a^3 b, \ x_2 = a^2 b^2, \ x_3 = ab^3, \ x_4 = b^4$$

Let $I_4 = I(C^4)$ be the defining ideal of C^4 in $\overline{P} = \mathbb{K}[x_0, x_1, x_2, x_3, x_4]$. We recall that I_4 is generated by the 2×2 minors of the matrix $\left(\begin{smallmatrix} x_0 & x_1 & x_2 & x_3 \\ x_1 & x_2 & x_3 & x_4 \end{smallmatrix} \right)$, which are

$$\begin{array}{llll}
\varphi_{12} = x_0 x_2 - x_1^2 & \varphi_{13} = x_0 x_3 - x_1 x_2 & \varphi_{14} = x_0 x_4 - x_1 x_3 \\
\varphi_{23} = x_1 x_3 - x_2^2 & \varphi_{24} = x_1 x_4 - x_2 x_3 & \varphi_{34} = x_2 x_4 - x_3^2
\end{array} \tag{11}$$

We consider the affine chart defined by $x_0 \neq 0$; let $y_i = \frac{x_i}{x_0}$, for $i = 1, \ldots, 4$, and let Γ^4 be the affine curve in $\mathbb{A}^4$ defined parametrically by the equations:

$$y_1 = b, \ y_2 = b^2, \ y_3 = b^3, \ y_4 = b^4$$

Finally, let $I(\Gamma^4)$ be the prime ideal which defines Γ^4 in $P = \mathbb{K}[y_1, y_2, y_3, y_4]$. We observe that C^4 is the projective closure of Γ^4, which implies that $I_4 = I(C_4) = I(\Gamma_4)^{\mathrm{hom}}$. The following proposition yields a representation of Γ^4 as a set-theoretic complete intersection.

Proposition 1. *Let $P = \mathbb{K}[y_1, y_2, y_3, y_4]$, let Γ^4 in $\mathbb{A}^4$ be the affine curve defined by the parametric equations $y_i = b^i$, for $i = 1, \ldots, 4$, and let $I(\Gamma_4)$ be the prime ideal of P which defines Γ^4. Further, let $f_i = y_{i+1} - y_1 y_i$, for $i = 1, \ldots, 3$,*

$g = y_4 - 2y_1y_3 + y_2^2$, $h = y_2y_4 - y_3^2$, and let $J = (f_1, f_2^2, g)$. The following equalities hold true:

(a) $I(\Gamma_4) = \sqrt{J}$;
(b) $J = (f_1, h, g)$.

Proof. To prove (a) we observe that $I(\Gamma_4)$ is generated by (f_1, f_2, f_3), which is a regular sequence, implying that Γ_4 is a complete intersection. Further, we point out that the following equality $g = f_3 + y_2 f_1 - y_1 f_2$ holds true. It follows that $I(\Gamma_4) = (f_1, f_2, f_3) = (f_1, f_2, g)$, and so obviously $I(\Gamma_4) = \sqrt{J}$. To prove (b) it is sufficient to observe that $f_2^2 = -h + y_2 g - y_2^2 f_1$.

In the following theorem, we present an alternative constructive proof of the fact that C^4 is a set-theoretic complete intersection of quadrics.

Theorem 1. *Notation as in Proposition 1. Let $\overline{P} = \mathbb{K}[x_0, x_1, x_2, x_3, x_4]$, let C^4 be the rational normal quartic in $\mathbb{P}^4$ and let $I(\Gamma_4)$ be the ideal of $\overline{P}$ which defines C^4. Further, let $Q = x_0 x_4 - 2x_1 x_3 + x_2^2 \in \overline{P}$. Then:*

(a) $I(\Gamma_4)^{\mathrm{hom}} = \sqrt{(\varphi_{12}, Q, \varphi_{34})}$, *where φ_{12}, φ_{34} are given in (11);*
(b) *The curve C^4 is a set-theoretic complete intersection of the three quadrics of equations $\varphi_{12} = 0$, $Q = 0$, $\varphi_{34} = 0$.*

Proof. By using Proposition 1 and the commutative property of the operations of homogenization and radical over ideals, we have $I(\Gamma_4)^{\mathrm{hom}} = \sqrt{J^{\mathrm{hom}}} = \sqrt{(f_1, h, g)^{\mathrm{hom}}}$. In P we fix the term ordering $\sigma = \mathrm{DegRevLex}$ (see [7]) and get $\mathrm{LT}_\sigma(f_1) = y_1^2$, $\mathrm{LT}_\sigma(h) = y_3^2$, $\mathrm{LT}_\sigma(g) = y_2^2$, which are pairwise coprime. It follows that the polynomials f_1, h, g are the Gröbner basis (w.r.t. the term ordering σ) of J. Now, using Corollary 4.3.20 of [8] and homogenizing w.r.t. x_0 we get

$$\sqrt{(f_1, h, g)^{\mathrm{hom}}} = \sqrt{(f_1^{\mathrm{hom}}, h^{\mathrm{hom}}, g^{\mathrm{hom}})} = \sqrt{(\varphi_{12}, Q, \varphi_{34})}$$

which shows statement (a). Part (b) follows from statement (a) and the equality $I(C^4) = I(\Gamma^4)^{\mathrm{hom}}$.

Acknowledgements. I would like to thank Prof. D. Gallarati for bringing my attention to his paper [6], a joint work with Prof. A. Rollero, allowing me to write these notes. I deeply thank Prof. L. Robbiano and Prof. M. C. Beltrametti for numerous helpful discussions on the topic.

References

1. Badescu, L., Valla, G.: Grothendieck-Lefschetz theory, set-theoretic complete intersections and rational normal scrolls. J. Algebra **324**, 1636–1655 (2010)
2. Barile, M.: Certain minimal varieties are set-theoretic complete intersections. Comm. Algebra **35**(7), 2082–2095 (2007)
3. Barile, M.: On binomial set-theoretic complete intersections in characteristic p. Rev. Mat. Complut. **21**(1), 265–282 (2008)

4. Barile, M., Lyubeznik, G.: Set-theoretic complete intersections in characteristic p. Proc. Am. Math. Soc. **133**(11), 3199–3209 (2005)

5. Barile, M., Morales, M., Apostolos, A.: Set-theoretic complete intersections on binomials. Proc. Am. Math. Soc. **130**(7), 1893–1903 (2002)

6. Gallarati, D., Rollero, A.: Una osservazione sulle curve razionali normali. Atti dell'Accademia Ligure di Scienze e Lettere **XLV**, 131–132 (1988)

7. Kreuzer, M., Robbiano, L.: Computational Commutative Algebra, vol. 1. Springer, Heidelberg (2000)

8. Kreuzer, M., Robbiano, L.: Computational Commutative Algebra, vol. 2. Springer, Heidelberg (2005)

9. Moh, T.T.: A result on the set-theoretic complete intersection problem. Proc. Am. Math. Soc. **86**(1), 19–20 (1982)

10. Moh, T.T.: Set-theoretic complete intersections. Proc. Am. Math. Soc. **94**(2), 217–220 (1985)

11. Ohm, J.: Space curves as ideal-theoretic complete intersections. In: Seidenberg, A. (ed.) Studies in Mathematics, vol. 20, pp. 47–115. Math. Assoc. Amer., Washington, DC (1980)

12. Perron, O.: Über die Bedingungen, daß eine binäre Form n-ten Grades eine n-te Potenz ist, und über die rationale Kurve n-ter Ordnung im $\mathbb{R}_n$. Math. Ann. **118**, 305–309 (1941/1943)

13. Robbiano, L.: A problem of complete intersections. Nagoya Math. J. **52**, 129–132 (1973)

14. Robbiano, L.: Some properties of complete intersections in "good" projective varieties. Nagoya Math. J. **61**, 103–111 (1976)

15. Robbiano, L., Valla, G.: Some curves in $\mathbb{P}^3$ are set-theoretic complete intersections. In: Ciliberto, C., Ghione, E., Orecchia, F. (eds.) Algebraic Geometry — Open Problems. LNCS, vol. 997, pp. 391–399. Springer, Heidelberg (1983)

16. Robbiano, L., Valla, G.: On set-theoretic complete intersections in the projective space. Milan J. Math. **53**, 333–346 (1983)

17. Schenzel, P., Vogel, W.: On set-theoretic intersections. J. Algebra **48**(2), 401–408 (1977)

18. Schmitt, T., Vogel, W.: Note on set-theoretic intersections of subvarieties of projective space. Math. Ann. **245**(3), 247–253 (1979)

19. Verdi, L.: Le curve razionali normali come intersezioni complete insiemistiche. Bollettino U.M.I. **16–A**, 385–390 (1979)

Author Index

Printed in the United States
By Bookmasters